Computer Network Architectures and Protocols

SECOND EDITION

Applications of Communications Theory
Series Editor: R. W. Lucky, *AT&T Bell Laboratories*

Recent volumes in the series:

COMPUTER COMMUNICATIONS AND NETWORKS
John R. Freer

COMPUTER NETWORK ARCHITECTURES AND PROTOCOLS
Second Edition • Edited by Carl A. Sunshine

DATA TRANSPORTATION AND PROTECTION
John E. Hershey and R. K. Rao Yarlagadda

DEEP SPACE TELECOMMUNICATIONS SYSTEMS ENGINEERING
Edited by Joseph H. Yuen

DIGITAL PHASE MODULATION
John B. Anderson, Tor Aulin, and Carl-Erik Sundberg

DIGITAL PICTURES: Representation and Compression
Arun N. Netravali and Barry G. Haskell

ERROR-CORRECTION CODING FOR DIGITAL COMMUNICATIONS
George C. Clark, Jr., and J. Bibb Cain

FIBER OPTICS: Technology and Applications
Stewart D. Personick

FUNDAMENTALS OF DIGITAL SWITCHING
Second Edition • Edited by John C. McDonald

MODELING AND ANALYSIS OF COMPUTER
COMMUNICATIONS NETWORKS
Jeremiah F. Hayes

MODERN TELECOMMUNICATIONS
E. Bryan Carne

OPTICAL CHANNELS: Fibers, Clouds, Water, and the Atmosphere
Sherman Karp, Robert M. Gagliardi, Steven E. Moran, and Larry B. Stotts

OPTICAL FIBER TRANSMISSION SYSTEMS
Stewart D. Personick

PRACTICAL COMPUTER DATA COMMUNICATIONS
William J. Barksdale

Computer Network Architectures and Protocols

SECOND EDITION

Edited by

Carl A. Sunshine
Unisys West Coast Research Center
Santa Monica, California

PLENUM PRESS • NEW YORK AND LONDON

Library of Congress Cataloging in Publication Data

Computer network architectures and protocols / edited by Carl A. Sunshine. — 2nd ed.
   p.   cm. — (Applications of communications theory)
  Includes bibliographical references.
  ISBN 0-306-43189-0
  1. Computer network protocols. 2. Computer network architectures. I. Sunshine,
Carl A. II. Series.
TK5105.5.C638   1989   89-35230
004.6 — dc20   CIP

© 1989, 1982 Plenum Press, New York
A Division of Plenum Publishing Corporation
233 Spring Street, New York, N.Y. 10013

Printed in the United States of America

To
Vint Cerf
and
Harvey Gold

Contributors

Paul Bartoli
AT&T Bell Laboratories, Holmdel, New Jersey 07733

H. V. Bertine
AT&T Bell Laboratories, Holmdel, New Jersey 07733

Abhay K. Bhushan
Xerox Corporation, El Segundo, California 90245

Gregor V. Bochmann
Département d'IRO, University of Montreal, Montreal H3C 3J7, Canada

David E. Carlson
AT&T Bell Laboratories, Holmdel, New Jersey 07733

James W. Conard
Conard Associates, Costa Mesa, California 92626

John Day
Motorola, Inc., Schaumburg, Illinois 60196

Harold C. Folts
Omnicom, Inc., Vienna, Virginia 22180

Dennis G. Frahmann
Xerox Corporation, El Segundo, California 90245

Mario Gerla
Computer Science Department, UCLA, Los Angeles, California 90024

James P. Gray
IBM Corporation, Research Triangle Park, North Carolina 27709

Leonard Kleinrock
Computer Science Department, UCLA, Los Angeles, California 90024

Daniel A. Pitt
IBM Corporation, Research Triangle Park, North Carolina 27709

Diane P. Pozefsky
IBM Corporation, Research Triangle Park, North Carolina 27709

Antony Rybczynski
Data Networks Division, Northern Telecom, Ottawa, Ontario K2C 3T1, Canada

Mischa Schwartz
Center for Telecommunications Research, Columbia University, New York, New York 10027

Thomas E. Stern
Center for Telecommunications Research, Columbia University, New York, New York 10027

Carl A. Sunshine
Unisys, West Coast Research Center, Santa Monica, California 90406

Fouad A. Tobagi
Computer Systems Laboratory, Stanford University, Stanford, California 94305

Ronald P. Uhlig
Northern Telecom Inc., Richardson, Texas 75081

Charles E. Young
AT&T Bell Laboratories, Holmdel, New Jersey 07733

Preface

This is a book about the bricks and mortar from which are built those edifices that will permeate the emerging information society of the future—computer networks. For many years such computer networks have played an indirect role in our daily lives as the hidden servants of banks, airlines, and stores. Now they are becoming more visible as they enter our offices and homes and directly become part of our work, entertainment, and daily living.

The study of how computer networks function is a combined study of communication theory and computer science, two disciplines appearing to have very little in common. The modern communication scientist wishing to work in this area soon finds that solving the traditional problems of transmission, modulation, noise immunity, and error bounds in getting the signal from one point to another is just the beginning of the challenge. The communication must be in the right form to be routed properly, to be handled without congestion, and to be understood at various points in the network.

As for the computer scientist, he finds that his discipline has also changed. The fraction of computers that belong to networks is increasing all the time. And for a typical single computer, the fraction of its execution load, storage occupancy, and system management problems that are involved with being part of a network is also growing.

It is the objective of this book to provide a comprehensive text and reference volume that can be used in education, research, and development in this combined field of computer networks. The aim is to cover both theory and practice in a style that is instructive but highly readable. To this end, the majority of the volume is devoted to a presentation of structural principles and architectural concepts, with emphasis on the OSI international standards, and illustrated with brief examples from currently operating systems. This is followed by a detailed description of network systems developed by two leading vendors: Xerox and IBM.

An effort of this scope is beyond the capabilities of any single author. Building on the framework created in the First Edition by Paul Green, the "theory" chapters have been revised to reflect current applications. Chapters dealing with the more stable lower levels of the protocol architecture have been updated. Entirely new chapters were solicited covering the emerging higher-layer OSI standards and the IBM and Xerox network systems. The result represents a collection of tutorials by outstanding experts in each area, providing a more extensive coverage of the subject than any single author could.

Any summary of the present status of the computer network art must draw upon three main sources: research networks built by universities, often operating under government support; private networks provided by the computer manufacturers; and public network offerings provided by common carriers. In this volume all three sources of expertise have been tapped. Cooperation among these communities has led to the establishment of a high level of worldwide standardization on many aspects of network architectures and protocols, and these form the core of the new edition.

The book is organized along the lines of the now familiar "layered" view of networks, which is introduced in Part I. According to this scheme, the structure within any network "node" or "machine" may be broken into layers, with the raw transmission facilities of classical communications (for example, wires or satellite links) at the bottom, and the "users" (human or program) at the top. Part II discusses the lowest layer, the physical layer, by which data are transmitted between physically connected nodes. Part III presents the data link layer, which operates to deliver entire messages or "packets" from a particular node to one of its neighboring nodes. This includes the operation of "multiaccess links" and high-speed local area networks such as Ethernet, which are in such widespread use today.

In Part IV we see how packets make their way from the originating node to the destination node within the network layer, a process that can be a complex one when there are intermediate nodes and when there are many simultaneous users of network resources. Problems of routing, congestion control, and interconnecting multiple networks into "internets" are addressed. When we get to Part V, the fact that the path of the messages has been a sequence of nodes and links is no longer apparent. The higher layers deal with functions handled between end user nodes, such as error recovery, checkpointing, agreeing on a common format for exchanging data, and separating streams of data from different "conversations."

Part VI presents in detail the network systems developed by Xerox and IBM. These examples illustrate how two major vendors have chosen to apply the principles described earlier in the book. Finally, Part VII summarizes another increasingly important area: how to ensure the correctness of computer network protocols using formal specification and validation methods. This can help guarantee that the protocol designs are free of bugs and that specific products properly implement the protocols so that they will interoperate with other equipment.

Assembling this collection has taken long hours by many busy individuals. Some of the authors have borne the burden of revising their chapters several times to reflect new developments over the two-year period that the Second Edition has been underway. I wish to specially thank the Unisys Corporation and my family for supporting my own efforts on this edition. I am also grateful for the energy and patience of my colleagues, and of Sy Marchand at Plenum Press, in bearing with me through the lengthy gestation period of this volume, whose birth I hope will reflect the growing maturity and importance of computer network architectures and protocols.

Carl A. Sunshine

Santa Monica, California

Contents

PART I: INTRODUCTION

1. A Brief History of Computer Networking   3
   Carl A. Sunshine

2. The Reference Model for Open Systems Interconnection   7
   John Day

PART II: PHYSICAL LAYER

3. Physical Interfaces and Protocols 39
   H. V. Bertine

PART III: LINK CONTROL LAYER

4. Character-Oriented Link Control 83
   James W. Conard

5. Bit-Oriented Data Link Control 107
   David E. Carlson

6. Multiaccess Link Control 139
   Fouad A. Tobagi

PART IV: NETWORK LAYER

7. Circuit-Switched Network Layer 193
   Harold C. Folts

8. X.25 Packet-Switched Network Layer 211
   Antony Rybczynski

9. Routing Protocols 239
   Mischa Schwartz and Thomas E. Stern

10. Flow Control Protocols 273
    Mario Gerla and Leonard Kleinrock

11. Network Interconnection and Gateways 329
    Carl A. Sunshine

PART V: HIGHER-LAYER PROTOCOLS

12. OSI Transport and Session Layers 349
    Charles E. Young

13. OSI Presentation and Application Layers 377
    Paul D. Bartoli

14. Message Handling System Standards and Office Applications ... 399
    Ronald P. Uhlig

PART VI: NETWORK ARCHITECTURE EXAMPLES

15. Xerox Network Systems Architecture 417
    Abhay K. Bhushan and Dennis G. Frahmann

16. IBM's Systems Network Architecture 449
    Diane P. Pozefsky, Daniel A. Pitt, and James P. Gray

PART VII: FORMAL SPECIFICATIONS AND
THEIR MANIPULATION

17. Formal Methods for Protocol Specification and Validation 513
    Gregor V. Bochmann and Carl A. Sunshine

INDEX OF ACRONYMS 533

SUBJECT INDEX 537

I

Introduction

A Brief History
of Computer Networking

Carl A. Sunshine

Computer networking as we know it today may be said to have gotten its start with the ARPANET development in the late 1960s and early 1970s. Prior to that time there were computer vendor "networks" designed primarily to connect terminals and remote job entry stations to a mainframe. But the notion of networking between computers viewing each other as equal peers to achieve "resource sharing" was fundamental to the ARPANET design [1]. The other strong emphasis of the ARPANET work was its reliance on the then novel technique of packet switching to efficiently share communication resources among "bursty" users, instead of the more traditional message or circuit switching.

Although the term "network architecture" was not yet widely used, the initial ARPANET design did have a definite structure and introduced another key concept: protocol layering, or the idea that the total communications functions could be divided into several layers, each building upon the services of the one below. The original design had three major layers, a network layer, which included the network access and switch-to-switch (IMP-to-IMP) protocols, a host-to-host layer (the Network Control Protocol or NCP), and a "function-oriented protocol" layer, where specific applications such as file transfer, mail, speech, and remote terminal support were provided [2].

Similar ideas were being pursued in several other research projects around the world, including the Cyclades network in France [3], the

CARL A. SUNSHINE • Unisys, West Coast Research Center, Santa Monica, California 90406.

National Physical Laboratory Network in England [4], and the Ethernet system [5] at Xerox PARC in the USA. Some of these projects focused more heavily on the potential for high-speed local networks such as the early 3-Mbps Ethernet. Satellite and radio channels for mobile users were also a topic of growing interest.

By 1973 it was clear to the networking vanguard that another protocol layer needed to be inserted into the protocol hierarchy to accommodate the interconnection of diverse types of individual networks. Cerf and Kahn published their seminal paper describing such a scheme [6], and development of the new Internet Protocol (IP) and Transmission Control Protocol (TCP) to jointly replace the NCP began. Similar work was being pursued by other groups meeting in the newly formed IFIP WG 6.1, called the Internetwork Working Group [7].

The basis for the network interconnection approach developing in this community was to make use of a variety of individual networks each providing only a simple "best effort" or "datagram" transmission service. Reliable virtual circuit services would then be provided on an end-to-end basis with the TCP (or similar protocol) in the hosts. During the same time period, public data networks (PDNs) were emerging under the auspices of CCITT, aimed at providing more traditional virtual circuit types of network service via the newly defined X.25 protocol. The middle and late 1970s saw networking conferences dominated by heated debates over the relative merits of circuit versus packet switching and datagrams versus X.25 virtual circuits [8]. The computer vendors continued to offer their proprietary networks, gradually supporting the new X.25 service as links under their own protocols. Digital Equipment Corporation (DEC) was the notable exception, adopting the research community approach of peer-to-peer networking at an early date, and coming out with its own new suite of protocols (DECNET) [9].

By the late 1970s, a new major influence was emerging in the computer network community. The computer manufacturers realized that multivendor systems could no longer be avoided, and began to take action to satisfy the growing user demand for interoperability. Working through their traditional international body, the ISO, a new group (SC16) was created to develop standards in the networking area. Their initial charter was to define an explicit "architecture" for "Open Systems Interconnection" (OSI).

By the early 1980s there were three major players in the networking game: the ARPANET-style research community, the carriers with their PDNs in CCITT, and the manufacturers in ISO. The conference circuit became more acrimonious, with the research community lambasting the slow progress, ponderousness (7 layers!), lack of experimental support, and all-inclusiveness (five classes of transport protocol) of the ISO workers, while still taking occasional shots at the PDNs and X.25. The CCITT and

ISO had the big players and the dollars on their side, but TCP/IP protocols were included in the increasingly popular UNIX* operating system.

The outgrowth of the ARPANET, the DoD Internet, was beginning to face its own problems of success by the mid-1980s. With the hundreds of LANs and tens of thousands of workstation hosts, serious performance problems were emerging, and it was beginning to look like the critics of "stateless" datagram networking might have been right on some points. After the first ten years, the DoD had greatly reduced research funding for networks, and found that it had to hurriedly perform some engineering studies to maintain operations.

Now in the late 1980s, much of the battling seems over. CCITT and ISO have aligned their efforts, and the research community seems largely to have resigned itself to OSI. Bob Metcalfe, the creator of Ethernet and founder of 3Com, summed up the sentiment recently by stating that "it has not been worth the ten years wait to get from TCP to TP4, but OSI is now inevitable." OSI has created an internet sublayer within the network layer to accommodate the datagram internetting approach beside the CCITT X.25 approach. However, with the inevitable tendency of committee work toward all-inclusiveness, there remain some serious potential pitfalls in interoperability among those following different "profiles" of OSI protocols (e.g., how will TP2 users on X.25 networks talk to TP4 users on datagram networks?).

In hindsight, much of the networking debate has resulted from differences in how to prioritize the basic network design goals such as accountability, reliability, robustness, autonomy, efficiency, and cost effectiveness [10]. Higher priority on robustness and autonomy led to the DoD Internet design, while the PDNs have emphasized accountability and controllability.

It is ironic that while a consensus has developed that OSI is indeed inevitable, the TCP/IP protocol suite has achieved widespread deployment, and now serves as a de facto interoperability standard. Research is underway again, with new results on how to optimize TCP/IP performance over variable delay and/or very-high-speed networks. The TCP/IP Interoperability Conference (itself a new event started only in 1986) now features slick booths by the major computer and network vendors rather than the small niche vendors that once dominated the DoD networking marketplace. It appears that the vendors were unable to bring OSI products to market quickly enough to satisfy the demand for interoperable systems, and TCP/IP were there to fill the need.

The majority of this book deals with the OSI architecture and protocols because these will be dominant in the future. But efforts have been made wherever possible to show their roots in previous work, and the

---

*UNIX is a trademark of AT & T Bell Laboratories.

relation of OSI protocols to research and vendor efforts. The past decade has seen a tremendous rate of change in computing technology and in the networks for interconnecting computers. The architecture for the lower layers appears to be stabilizing, and the material in this book should carry us well into the next decade. The application layer is the new frontier, and will require another book to deal with in the future.

References

[1] L. Roberts and B. Wessler, "Computer network development to achieve resource sharing," *AFIPS Conf. Proc. (SJCC)*, vol. 36, 1970, pp. 543–549.

[2] S. Crocker *et al.*, "Function oriented protocols for the ARPA computer network," *AFIPS Conf. Proc.*, vol. 40, 1972, pp. 271–279.

[3] L. Pouzin, "Presentation and major design aspects of the Cyclades computer network," *Proc. 3rd Data Commun. Symp.*, St. Petersburg, Nov. 1973, pp. 80–85.

[4] R. Scantlebury and P. Wilkinson, "The National Physical Laboratory data communication network," *Proc. ICCC*, Stockholm, Aug. 1974.

[5] R. Metcalfe and D. Boggs, "ETHERNET: Distributed packet switching for local computer networks," *Commun. ACM*, vol. 19, July 1976, pp. 395–404.

[6] V. Cerf and R. Kahn, "A protocol for packet network intercommunication," *IEEE Trans. Commun.*, vol. COM-22, 1974, pp. 637–648.

[7] V. Cerf, A. McKenzie, R. Scantlebury, and H. Zimmermann, "Proposal for an international end-to-end protocol," *Computer Commun. Rev.*, vol. 6, ACM SIGCOMM, 1974, pp. 68–89.

[8] V. DiCiccio, C. Sunshine, J. Field, and E. Manning, "Alternatives for interconnection of public packet switching data networks," *Proc. 6th Data Commun. Symp.*, Nov. 1979, ACM/IEEE, pp. 120–125.

[9] S. Wecker, "DNA—The Digital Network Architecture," *Computer Network Architectures and Protocols*, P. Green, Jr., ed., New York: Plenum Press, 1982.

[10] D. Clark, "The design philosophy of the DARPA internet protocols," *Proc. SIGCOMM '88 Symp.*, Stanford, Aug. 1988, pp. 106–114.

The Reference Model for Open Systems Interconnection

John Day

I. Introduction

In the mid-1970s, the development and use of resource-sharing computer networks began to achieve considerable attention. The early successes of the ARPANET [1] and CYCLADES [2], the immediate commercial potential of packet switching, satellite, and local network technology, and the declining cost of hardware made it apparent that computer networking was quickly becoming an important area of innovation and commerce. It was also apparent that to utilize the full potential of such computer networks, international standards would be required to ensure that any system could communicate with any other system anywhere in the world.

In 1978, the International Organization for Standardization (ISO) Technical Committee 97 on Information Processing, recognizing the standards for networks of heterogeneous systems were urgently required, created a new subcommittee (SC16) for "open systems interconnection" (OSI). The term "open" was chosen to emphasize that by conforming to OSI standards, a system would be open to communication with any other system anywhere in the world obeying the same standards.

It was clear that the commercial endeavors to exploit the emerging communication technology would wait neither for SC16 to leisurely develop communication standards nor for the research community to answer many of the outstanding questions. If there was to be a consistent set of

---

JOHN DAY • Motorola, Inc., Schaumburg, Illinois 60196.

international standards, OSI would have to lead rather than follow commercial development and make use of the most recent research work when available. The size of the task would require the work to be divided among several working groups each developing standards while maintaining close overall coordination. It would also be critical to develop the standards in an order to meet the requirements of on-going implementation.

The first meeting of TC97/SC16 was held in March, 1978. Initial discussions revealed that a consensus could be reached rapidly on a basic layered architecture [3] that would satisfy most requirements of OSI and could be extended later to meet new requirements. SC16 decided to give the highest priority to the development of a standard model of architecture that would constitute the framework for the development of standard protocols.

After less than 18 months of discussions, this task was completed, and the Basic Reference Model of Open Systems Interconnection was transmitted by SC16 to TC97 along with recommendations to start a number of projects for developing an initial set of standard protocols for OSI. These recommendations were adopted by TC97 at the end of 1979 as the basis for development of standards for open systems interconnection within the ISO. The OSI reference model was also recognized by the Consultative Committee Internationale de Telephonie et Telegraphie (CCITT) Rapporteur's Group on Public Data Network Services. At this time, SC16 began development of standard OSI protocols for the upper four layers, while the development of the lower three layers was undertaken by SC6. These are discussed in more detail in subsequent chapters of this book.

In late 1980, SC16 recommended that the reference model be forwarded as a draft proposal (DP) for an international standard. After two rounds of comments, the reference model was progressed as a draft international standard (DIS) in the spring of 1982. Comments on this vote were processed late in 1982 and the basic reference model became an international standard (ISO 7498) [4] in the spring of 1983.

In most cases, the task of a standards committee is to take sets of commercial practices and the current research results and codify these procedures into a single standard that can be utilized by commercial products. However, given the speed of technological development in this area, SC16 was presented with a somewhat different problem: to develop a set of standards that emerging products could converge to before the commercial practices were in place and while many of the more fundamental research problems remained unsolved. It would be presumptuous to say that SC16 solved this problem. SC16 did, however, find a way to cope with the problem in such a way as to maximize flexibility and to minimize the impact of change brought on by new technologies or new techniques.

Central to this approach is the layered architecture, which is used to break up the problem into manageable pieces. The OSI reference model is a

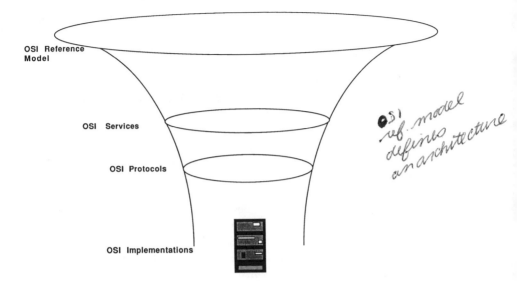

Fig. 1. The OSI reference model, services, and protocols define successively more detailed and therefore more constraining specifications.

framework for coordinating the development of OSI standards. In OSI, the problem is approached in a top-down fashion, starting with a description at a high level of abstraction, which imposes few constraints, and proceeding to more and more refined descriptions with tighter and tighter constraints. In OSI, three levels of abstraction are explicitly recognized: the architecture, the service specifications, and the protocol specifications (see Fig. 1).

The OSI architecture is the highest level of abstraction. The term "architecture" can be a very tricky term. It has been used to describe everything from a framework for development, to a particular form of organization, to hardware. A good way to think about the term is to consider the difference between an architecture and a building built to that architecture. For example, Victorian architecture is a set of rules and stylistic conventions that characterize a particular form. A Victorian building is a building built to those rules and conventions. You cannot walk into a Victorian architecture; you can walk into a Victorian building. In computer science we often mistakenly refer to a building as an architecture. More formally, this can also be considered as the distinction between the type of an object (architecture) and an instance of the object (building). The OSI reference model defines types of objects that are used to describe interprocess communications among open systems, the general relations among these types of objects, and the general constraints on these types of objects and relations. Specifications for the lower levels of abstractions may define other relations and tighter constraints for their purposes, but these must be consistent with those defined in the reference model.

The document [4] that describes the OSI architecture, ISO 7498, defines these objects, relations, and constraints and also defines a seven-layer model for interprocess communication constructed from these objects, relations, and constraints. These are used as a framework for coordinating the development of layer standards by OSI committees as well as the development of standards built on top of OSI.

The OSI service specifications represent a lower level of abstraction that defines in greater detail the service provided by each layer. The service specification will define tighter constraints than the reference model on the protocols and implementations that will satisfy the requirements of the layer. A service specification defines the facilities provided to the user of the service independent of the mechanisms used to accomplish the service. It also defines an abstract interface for the layer, in the sense that it defines the primitives that a user of the layer may request with no implication of how or if that interface is implemented.

The OSI protocol specifications represent the lowest level of abstraction in the OSI standard scheme. Each protocol specification defines precisely what control information is to be sent and what procedures are to be used to interpret this control information. The protocol specifications represent the tightest constraints placed on implementations built to conform to OSI standards.

As shown in Fig. 1, the three levels of abstraction used by OSI define successively tighter constraints on what will satisfy OSI. There are many services and protocols that satisfy the constraints required by the reference model. There are fewer protocols that satisfy both the reference model and the OSI service specifications. Finally, the protocol specifications constrain implementations sufficiently to allow open systems to communicate while still allowing differences in implementations.

Products may satisfy the much weaker constraints imposed by the reference model, but may not be able to communicate with open systems unless they also conform to the OSI services and protocols. The OSI reference model cannot be implemented, and it does not represent a preferred implementation approach. It is a model for describing the concepts for coordinating the parallel development of interprocess communication standards. One must remember that in the world of OSI, only OSI protocols can be implemented and products can conform only to OSI protocols. Thus the statement "this product conforms to or follows the OSI reference model" or the statement that "our products are built to an open architecture" does not imply the ability to interwork with other products that may make the same claim. Only the claim that particular products implement particular OSI protocols implies the ability to interwork.

The purpose of OSI is to allow any computer anywhere in the world to communicate with any other, as long as both obey OSI standards. OSI standards and the degree of compatibility required to meet this goal make formal description methods a necessity. Early in its work, SC16/WG1 on Architecture established a group to develop formal description methods for defining the protocols, so that they could be implemented unambiguously by people all over the world, without having to consult with a few experts on how to interpret the standard. Chapter 17 of this book discusses the formal description techniques developed for use with OSI.

In the remainder of this chapter, we describe the basic concepts used in the reference model, followed by a brief description of each of the seven layers, then a description of some of the major extensions to the model, and to identify a few of the outstanding architectural issues. Further details on the services and protocols for each layer may be fond in subsequent chapters of this book.

II. The Elements of the Architecture

ISO 7498, the document describing the basic OSI reference model, is divided into two major sections. The first of these describes the elements of the architecture. These constitute the building blocks that are used to construct the seven-layer model. The second describes the services and functions of the specific layered model constructed from the building blocks.

As mentioned above, the reference model is not intended to define a particular implementation. However, it is difficult to communicate the concepts and intent of OSI without discussing some sort of model that resembles an implementation. Thus, a strong distinction must be made between the description of the abstract and the real. The reference model represents an abstract description of communicating systems. In most cases, terms have been chosen for the reference model that are not used to describe real (existing) systems. The abstract model serves as a framework on which the constraints imposed by OSI standards can be hung. An implementation in the real world must only comply with the constraints. The reference model is only concerned with abstract objects. Other OSI standards, such as the internal organization of the network layer [9], are concerned with the mapping from the abstract to the real.

A. Systems, Layers, and Entities

The OSI reference model is an abstract description of interprocess communication. OSI is concerned with standards for communication among systems. In the OSI reference model, communication takes place among application processes running in distinct systems. A system is considered to be one or more autonomous computers and their associated software, peripherals, and users that are capable of information processing and/or transfer. Although OSI techniques could be used within a system (and it would be desirable for intra- and intersystem communication to appear as similar as possible to the user), it is not the intent of OSI to standardize the internal operation of a system. From the point of view of OSI, an open system may represent anything from a PC to a mainframe to a large network using nonstandard (OSI) "internal" protocols. The choice of where the boundary is drawn is left to the implementor.

Layering is used as a structuring technique to allow the problem of open systems interconnection to be logically decomposed into independent, smaller subsystems (see Fig. 2). Each individual system itself is viewed as being logically composed of a succession of subsystems, each subsystem corresponding to the intersection of the system with a layer. In other words, a layer is viewed as being composed of the subsystems of the same rank in

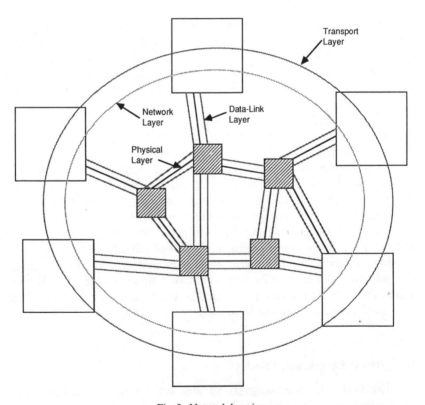

Fig. 2. Network layering.

all interconnected open systems. Each subsystem, in turn, is viewed as being made of one or several entities. A layer, therefore, comprises many entities distributed among interconnected open systems. Entities in the same layer are termed "peer entities." For OSI, the entity represents the active element within the layer which is operating in accordance with one or more OSI standards. The entity is the abstract representation of that part of a real system that implements OSI standards in that layer.

When describing general concepts that could be applied to any layer, the specific layer is referred to as the (N)-layer, while its next lower and next higher layers are referred to as the $(N-1)$-layer and the $(N+1)$-layer, respectively. The same notation is used to designate all concepts relating to layers, e.g., entities in the (N)-layer are termed (N)-entities; connections, (N)-connections; and so on (see Figs. 3 and 4).

The entities in a layer are further distinguished by applying the concepts of type and instance used in much of computer science. An (N)-entity type is defined as a particular class of all (N)-entities of an (N)-layer over all open systems. An (N)-entity represents the type of (N)-entity in a given (N)-subsystem. An (N)-entity-invocation represents a particular instantiation of an (N)-entity in an (N)-layer for a particular instance of communication.

Each (N)-layer treats the $(N-1)$-layer as a black box. The (N)-layer knows the service provided by the $(N-1)$-layer, but has no knowledge of the mechanisms used by the $(N-1)$-layer to provide those services. Each layer adds value to services provided by the set of lower layers in such a way that the highest layer is offered the full set of services needed to run distributed applications. Layering thus divides the total problem into smaller pieces.

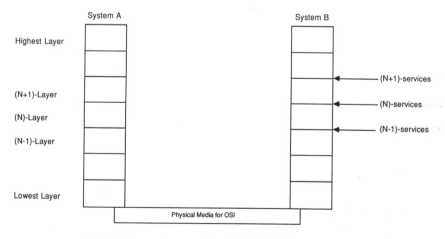

Fig. 3. Systems, layers, and services in the OSI environment.

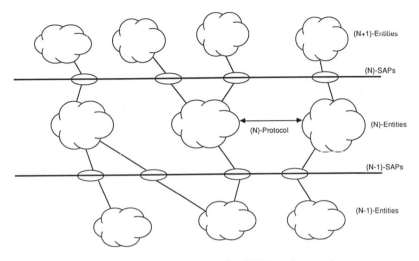

Fig. 4. Entities, service access points (SAPs), and protocols.

This approach ensures the independence of each layer by defining services provided by a layer to the next higher layer, independently of how these services are performed. This approach permits changes to be made in the way a layer or a set of layers operates, provided they still offer the same service to the next higher layer. This technique is similar to the layered designs of operating systems or to the concepts used in structured or object-oriented programming where only the functions performed by a module (and not its internal functioning) are known by its users.

B. Services and Service Access Points

Each layer provides services to the layer above (with the exception of the highest layer). A service is a capability of the (N)-layer that is provided to the $(N + 1)$-entities. But it is important to note that not all functions performed within the (N)-layer provide services. Only those capabilities that can be seen from the layer above are services.

(N)-entities distributed among the interconnected open systems work collectively to provide the (N)-service to $(N + 1)$-entities as illustrated in Fig. 4. In other words, the (N)-entities add value to the $(N - 1)$-layer and offer this enhanced service, i.e., the (N)-service to the $(N + 1)$-entities.

The (N)-services are offered to the $(N + 1)$-entities at the (N)-service access points, or (N)-SAPs for short, which represent the logical interfaces between the (N)-entities and the $(N + 1)$-entities. An $(N + 1)$-entity communicates with (N)-entities in the same system through (N)-SAPs. An (N)-SAP can be served by only one (N)-entity and used by only one

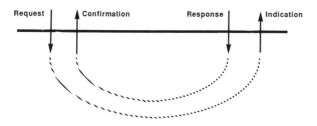

Fig. 5. Abstract service primitives.

$(N + 1)$-entity, but one (N)-entity can serve several (N)-SAPs and one $(N + 1)$-entity can use several (N)-SAPs. An (N)-SAP is located by its (N)-SAP-address (see Section II D).

OSI specifies the services of a layer in terms of abstract service primitives. These service primitives are defined using the form of a procedure call. The service defines an abstract interface. Unlike the definition of a real interface, OSI only defines primitives and parameters that are required for the operation of the protocol in the layer. For example, primitives and parameters required by the local system for local flow control between layers in the same system, querying the status of the layer (locally), and so on are not defined in an OSI service definition. There are four forms of service primitives defined for connection-oriented operation: request, indication, response, and conformation. Figure 5 illustrates these four types. The service primitives represent a somewhat different view of the data units described below. The user data in a service primitive corresponds to a (N)-service-data-unit. The other parameters in a service primitive used to control the protocol correspond to (N)-interface-control-information (see below). OSI has adopted a philosophy of not specifying the internal operation of systems. This approach allows the implementor the greatest flexibility to build systems in any way as long as their external behavior conforms to the standards.

C. Functions and Protocols

An (N)-function is part of the activity of an (N)-entity. (N)-functions support (N)-services. Not all (N)-functions directly support (N)-services. Only those (N)-functions whose behavior is visible from the layer above support (N)-services. Flow control, sequencing, data transformation are all examples of (N)-functions. Cooperation among (N)-entities is governed by (N)-protocols. An (N)-protocol is the set of rules and formats that govern the communication between (N)-entities performing the (N)-functions in different open systems. In particular, direct communication between the

(N)-entities in the same system, e.g., for sharing resources, is not visible from outside the system and thus is not covered by the OSI architecture.

D. Elementary Concepts of Naming

The purpose of OSI is to allow application processes to initiate, maintain, and terminate communication. To do this, (N)-entities in the initiating system must be able to refer to the names of the (N)-entities it needs to communicate with. (N)-entities use the (N)-SAP-address for naming these (N)-entities. The OSI architecture defines identifiers for entities, SAPs, and connections as well as relations between these identifiers.

Each (N)-entity is identified with a global title, which is unique and identifies the same (N)-entity anywhere in the network of open systems. Within more limited domains, an (N)-entity can be identified with a local title, which uniquely identifies the (N)-entity only in that domain. For instance, within the domain corresponding to the (N)-layer, (N)-entities are identified with (N)-global titles, which are unique within the (N)-layer.

Each (N)-SAP is identified by an (N)-SAP-address which uniquely locates the (N)-SAP at the boundary between the (N)-layer and the ($N + 1$)-layer. The concepts of titles and addresses are illustrated in Fig. 6.

Bindings between (N)-entities and the ($N - 1$)-SAPs they use (i.e., SAPs through which they can access each other and communicate) are defined in an (N)-directory, which indicates correspondence between global titles of (N)-entities and (N)-addresses through which they can be reached.

Correspondence between the (N)-addresses served by an (N)-entity and the ($N - 1$)-addresses is performed by an (N)-mapping function. In addition to the simplest case of one-to-one mapping, the mappings may, in

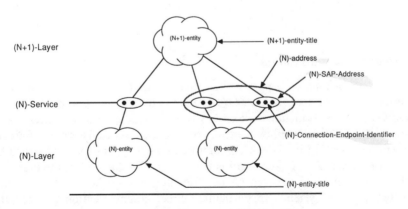

Fig. 6. Titles, addresses, and CEP identifiers.

particular, be hierarchical, with the (N)-address being made of an $(N - 1)$-address and an (N)-suffix.

E. Connections

A common service offered by all layers consists of providing connections among $(N + 1)$-entities. For many applications, the communications are best modeled as a logical pipe. Implementing this model requires an establishment phase to guarantee initial synchronization, after which information is exchanged in the data transfer phase, and when communication is complete a release procedure is invoked to ensure that the (N)-entities are left in a consistent state. The (N)-layer offers (N)-connections between (N)-SAPs as part of the (N)-services (see Fig. 7). The most common type of connection is the point-to-point connection, but there are also multiendpoint connections which correspond to multiple associations between entities (e.g., broadcast or multidrop communications). The end of an (N)-connection at an (N)-SAP is called an (N)-connection-endpoint or (N)-CEP for short. Several connections may coexist between the same pair of SAPs.

Each (N)-CEP is uniquely identified within its (N)-SAP by an (N)-CEP-identifier, which is used by the (N)-entity and the $(N + 1)$-entity on both sides of the (N)-SAP to identify the (N)-connection, as illustrated in Fig. 7. (N)-CEPI are local identifiers that are used by an (N)-entity to identify the end of a connection.

The basic reference model only describes communications between (N)-entities in a "connection mode." In this mode, the $(N - 1)$-service requires that an $(N - 1)$-connection be established between $(N - 1)$-SAPs before any communication between (N)-entities can take place. Conversely, when the (N)-entities no longer require communication, the $(N - 1)$-connection can be released. This connection-mode covers traditional

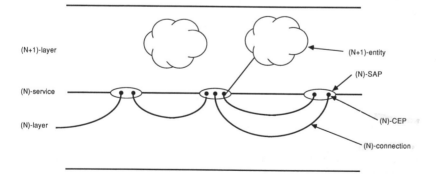

Fig. 7. Connections and connection endpoints (CEPS).

teleprocessing. For other applications, a "connectionless" mode extension (see Section IV) has been developed within ISO and a multipeer (broadcast) mode (see Section V) is currently being developed as a complement to the connection mode [6].

Establishment and Release of Connections. When an $(N + 1)$-entity requests the establishment of an (N)-connection from one of the (N)-SAPs it uses to another (N)-SAP, it must provide, at the local (N)-SAP, the (N)-address of the distant (N)-SAP. When the (N)-connection is established, both the $(N + 1)$-entity and the (N)-entity within one system will use the (N)-CEP-identifier to designate the (N)-connection.

(N)-connections may be established and released dynamically on top of $(N - 1)$-connections. Establishment of an (N)-connection implies the availability of an $(N - 1)$-connection between the two entities. If not available, the $(N - 1)$-connection must be established. This requires the availability of an $(N - 2)$-connection. The same consideration applies downwards until an available connection is encountered.

In some cases the (N)-connection may be established simultaneously with its supporting $(N - 1)$-connection provided the $(N - 1)$-connection establishment service permits (N)-entities to exchange information necessary to establish the (N)-connection.

Data Transfer on a Connection. Information is transferred in various types of data units between peer entities and between entities attached to a specific (N)-service-access-point. The data-units are defined below and the interrelationship among several of them is illustrated in Figs. 8 and 9.

(N)-service-data-unit, (N)-SDU, is the amount of (N)-interface-data whose identity is preserved from one end of an (N)-connection to the other. Data may be held within a connection until a complete service data unit is available.

(N)-interface-control-information, (N)-ICI, is information exchanged between an $(N + 1)$-entity and an (N)-entity to coordinate their joint operation.

(N)-interface-data is an amount of user data conveyed from the $(N + 1)$-layer to the (N)-layer in an (N)-IDU.

(N)-interface-data-unit, (N)-IDU, is the amount of information passed across the layer boundary as a single unit of interaction.

(N)-protocol-control-information, (N)-PCI, is information exchanged between two (N)-entities, using an $(N - 1)$-connection, to coordinate their joint operation.

(N)-user-data are the data transferred between two (N)-entities on behalf of the $(N + 1)$-entities for whom the (N)-entities are providing services.

An (N)-protocol-data-unit, (N)-PDU, is a unit of data that contains (N)-protocol-control-information and possibly (N)-user-data.

Thus, an (N)-SDU is conveyed from the $(N + 1)$-layer to the (N)-layer by making it (N)-interface-data along with any necessary (N)-ICI in one or

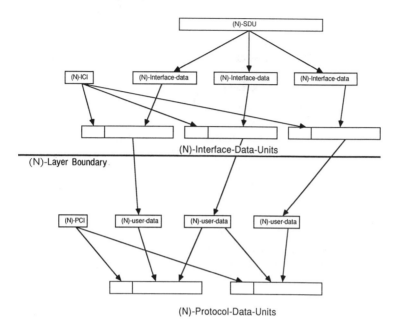

Fig. 8. Logical relationship among data units in adjacent layers.

more (N)-IDUs. The (N)-entity acts on the (N)-ICI and conveys the (N)-SDU to a peer (N)-entity by making the (N)-SDU, (N)-user-data in one or more (N)-PDUs. This is done in such a way that when the (N)-user-data is received by the peer, the (N)-SDU may be delivered to the peer $(N + 1)$-entity intact.

Expedited (N)-service-data-unit is a small (N)-service-data-unit whose transfer is expedited and not subject to the same flow control as normal

	Control	Data	Combined
(N)- (N) Peer Entities	(N)-Protocol- Control- Information	(N)-User- Data	(N)-Protocol- Data-Unit
(N)-(N-1) Adjacent Layer	(N-1)-Interface- Control- Information	(N-1)-Interface- Data	(N-1)-Interface- Data-Unit

Fig. 9. Relationships among data units.

data flow. The (N)-layer ensures that an expedited data unit will not be delivered after any subsequent service data unit or expedited data unit sent on that connection. In most cases, one would expect it to be delivered well before any subsequent (N)-SDUs. An expedited (N)-service-data-unit may also be referred to as an (N)-expedited-data-unit.

An (N)-protocol-data-unit may be mapped onto an ($N - 1$)-service-data-unit in a variety of ways. The (N)-PDU may be one-to-one, one-to-many (segmenting), and many-to-one (blocking). These operations generally require additional protocol control information to allow the inverse function to be performed by the receiving system (see Fig. 9).

Elements of Layer Operation. There are a number of functions that are part of layer operation in several layers. These include such things as multiplexing, flow control, and error control. As the reference model matures, other elements will be added.

Three particular types of construction of (N)-connections on top of ($N - 1$)-connections are distinguished:

1. One-to-one correspondence, where each (N)-connection is built on one ($N - 1$)-connection.
2. Multiplexing, where several (N)-connections are multiplexed on one single ($N - 1$)-connection.
3. Splitting, where one single (N)-connection is built on top of several ($N - 1$)-connections, the traffic on the (N)-connection being divided among the various ($N - 1$)-connections.

Two forms of flow control are recognized by the reference model: a peer flow control, which regulates the flow of (N)-protocol-data-units between entities within the same layer, and interface flow control, which regulates the flow of (N)-interface-data between an ($N + 1$)-entity and (N)-entity through an (N)-SAP. Since interface flow control occurs entirely within a single system, it is not the subject of any OSI standards.

A variety of error functions are recognized by the model, including acknowledgment, error detection, and error notification mechanisms. The model also describes a reset function to allow recovery from a loss of synchronization between communicating (N)-entities.

III. The Seven-Layer Model

In the previous section, the basic elements of the OSI reference model were developed. These serve as the building blocks for constructing the model of interprocess communication. In OSI, interprocess communication is subdivided into seven independent layers. Each (N)-layer uses the

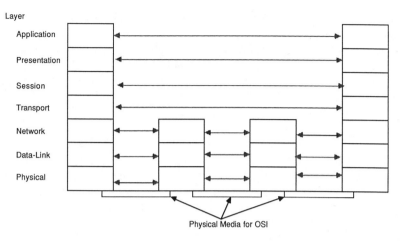

Fig. 10. The seven-layer OSI architecture.

services of the $(N - 1)$-layer to provide service to the $(N + 1)$-layer. Layers have been chosen to break up the problem into reasonably sized smaller problems that can be considered relatively independently. The seven layers are described briefly below (see Fig. 10).

The lower four layers are concerned with the reliable, transparent transmission of data. The upper three layers are concerned with providing general facilities in support of distributed applications. At this writing, standard protocols for all seven layers are international standards. Other chapters in this book describe these layers in more detail.

A. Application Layer

The Application Layer as the highest layer of OSI does not provide services to any other layer. The primary concern of the Application Layer is with the semantics of the application. From the point of view of OSI, only part of an application process is in the Application Layer. Those aspects of the application process concerned with interprocess communication (called the application-entity) are within the OSI environment. The rest of the application process is concerned with aspects other than OSI, such as database access or information processing. The application-entity is organized as sets of interacting service elements.

A large number of protocols have been and will be defined for the Application Layer. The Association-Control-Service-Elements (ACSE) and Commitment, Concurrency, and Recovery (CCR) are designed to be used as components of other applications. In addition, Application Layer protocols have been defined for file transfer (ISO 8571), virtual terminal (ISO 9040), job transfer (ISO 8831), OSI system management (ISO 9595), and directory services (ISO 9594), and others are being developed for specific industries, such as banking, business data interchange, and factory automation. The bulk of the application protocols will be defined by the users of OSI.

B. Presentation Layer

The primary purpose of the Presentation Layer is to provide application processes with independence from differences in syntax, i.e., data representation. The Presentation Layer protocol allows the user to select a "presentation context." The presentation context may be specific to an application such as a library protocol or virtual terminal, to a type of hardware such as a particular machine representation, or to some standard or canonical representation of information. This approach allows communication between similar systems to choose a syntax that is most efficient, while allowing unlike systems to negotiate a syntax that both understand. Thus, a user of OSI wanting to develop an OSI application protocol defines an application protocol using the relevant parts of the Association Control Service Elements and a presentation context that defines the representation of the data to be transferred. The OSI user may use an existing context or define his own and register it with ISO. In OSI, the Presentation Layer protocol is defined by ISO 8822 and ISO 8823. A common syntax for use with presentation has been defined in terms of an abstract syntax notation (ASN.1) described in ISO 8824 and a set of concrete encoding rules described in ISO 8825.

C. Session Layer

The primary purpose of the session layer is to provide the mechanisms for organizing and structuring the interactions between application processes. The mechanisms provided in the session layer allow for two-way simultaneous and two-way alternate operation, the establishment of major and minor synchronization points, and the definition of special tokens for structuring exchanges. In essence, the session layer provides the structure for controlling the communication. In OSI, the session layer is defined by ISO 8326 (service) and ISO 8327 (protocol).

D. Transport Layer

The purpose of the Transport Layer is to provide transparent transfer of data between end systems, thus relieving the upper layers from any concern with providing reliable and cost effective data transfer. In some cases, the transport/network layer boundary represents the traditional boundary between the carrier and the customer. From this point of view, the Transport Layer optimizes use of network services and provides any additional reliability over that supplied by the network service. In OSI, the transport service is defined by ISO 8072 and the transport protocol by ISO 8073 for connection-mode and ISO 8602 for connectionless.

E. Network Layer

The Network Layer provides independence from the data transfer technology and independence from relaying and routing considerations. The Network Layer masks from the Transport Layer all the peculiarities of the actual transfer medium. The Transport Layer needs to be concerned only with the quality of service and its costs, not with whether optical fiber, packet switching, satellites, or local area networks are being used. The Network Layer also handles relaying and routing data through as many concatenated networks as necessary while maintaining the quality of service parameters requested by the transport layer. The network layer functions can be categorized into four roles [9]:

1. The routing functions, concerned with routing and relaying among concatenated networks.
2. The subnetwork independent convergence (SNIC) functions, concerned with those functions that are not dependent on the particular subnetwork or data link technology necessary to enhance a particular subnetwork to allow data transfer across it to meet the requested quality of service parameters.
3. The subnetwork dependent convergence (SNDC) functions, concerned with those functions that are dependent on the particular subnetwork or data link technology necessary to enhance a particular subnetwork to allow data transfer across it to meet the requested quality of service parameters.
4. The subnetwork access (SNAC) functions, concerned with directly using the available data-link service to provide an abstract subnetwork.

This structure is required to make the transition from technology dependent to technology independent. The structure is sufficiently general to allow either the hop-by-hop harmonization or the internetworking approaches to achieve the goal of technology independence. In some cases, the service offered by an intervening subnetwork will not be equivalent to the two adjacent subnetworks. In this case, an SNDC protocol will be required to bring the intervening network up to the service of the two adjacent networks. In some cases, standard network access protocols provide access to a subnetwork, but the protocols within the subnetwork are proprietary. X.25 1980 is an example of a SNAC protocol that requires an SNDC in order to provide the OSI network service. More detail on the internal structure of the network layer can be found in Part IV of this book.

ISO 8348 defines the network service along with two addenda on connectionless and network layer addressing. ISO 8648 describes the internal organization of the network layer. X.25 (ISO 8208) and ISO 8473, the OSI Internetwork protocols are examples of protocols found in the network layer.

F. Data Link Layer

The purpose of the Data Link Layer is to provide the functional and procedural means to transfer data between network entities and to detect and possibly correct errors that may occur in the Physical Layer. Typical data link protocols are HDLC for point-to-point and multipoint connections and IEEE 802 (ISO 8802) for local area networks. Data link protocols and services are very sensitive to the physical transfer technology. While in the upper layers there is one protocol specified per layer, in the lower layers this will not be the case. In order to ensure efficient and effective use of the variety of transmission technologies, data link protocols are designed to meet the specific requirements of the technologies.

G. Physical Layer

The Physical Layer provides the mechanical, electrical, functional, and procedural standards to access the physical medium. Typical Physical Layer standards are RS-232C, V.21, and V.24.

IV. Extensions to the Basic Reference Model

When ISO 7498 was drafted, it was realized that the development of standard OSI protocols could not be delayed until a complete model of all aspects of interprocess communication had been described. The basic reference model describes the common connection-oriented mode of communication and the architectural concepts to support it. Once this was completed, work could start on the connection-oriented protocols, while in parallel the reference model was extended to cover other aspects of interprocess communication. Some of these extensions have reached maturity and are described in this section. Others are still being developed, and the next section provides some insight into the nature of the work in these areas.

A. Connectionless Transmission

The first extension to the basic reference model covers connectionless mode operation [5]. Examples of the connectionless mode are represented by datagram networks, by local area networks that access broadcast transmission media, and by some kinds of applications, such as telemetry.

This mode of operation differs from connection-mode in two characteristics: first, the communication between two (N)-entities does not require prior establishment of an $(N - 1)$-connection by the $(N - 1)$-services; and second, each SDU is independent of any other SDUs. Therefore, resources in the $(N - 1)$-layer and below need not be reserved in advance. They can be allocated dynamically for the transmission of a single $(N - 1)$-service-data-unit. The only protocol mechanisms that can be used do not require any state information to be maintained between the arrival of subsequent SDUs.

Connectionless is a mode of communication as is connection-oriented described above or multipeer described below. A mode of communication is not tied to either the upper or lower layers, nor does the use of one mode in the upper layers (or lower layers) imply that that mode will be used in the lower layers (or upper layers). An application can use one mode and operate over lower layers using any other mode. The mode of communication for the upper layers is chosen to best reflect the nature of the application dialogue. For example, an (N)-party distributed database update protocol (a multipeer protocol) could be operated over lower layers that were connection-oriented, connectionless, or multipeer. The broadcast requirements on the lower layers could be met by simulating a multipeer mode by $N - 1$ two-party connections over a connection-oriented technology or directly by a LAN or radio technology. The mode of communication in the lower layers generally reflects the physical nature of the communications media. For example, a broadcast LAN technology can be used by applications operating connectionless, connection-oriented, or broadcast protocols. This addendum covering connectionless transmission has been approved as ISO 7498-1/AD1.

B. Security

In many commercial uses of OSI, the ability to transfer data securely will be important. The reference model has also been extended to provide an architecture for security services within OSI. The security architecture provides a framework for analyzing threats and locating security services and protocols (including encryption) within OSI.

OSI communications are subject to the same four kinds of security threats as any other communications: eavesdropping, modification, denial of service, and traffic analysis. The OSI security architecture defines several security services that may be offered:

   a. Authentication service provides protection against masquerading or
      unauthorized replay.
   b. Access control service provides protection against unauthorized use
      of resources available via OSI.
   c. Data confidentiality service provides protection during the data
      transfer phrase from eavesdropping and traffic analysis.
   d. Data integrity service provides protection during the data transfer
      phase against modification, insertion, deletion, and replay.
   e. Nonrepudiation service provides protection against the sender falsely
      denying having sent the data.

The nature of security threats is such that not all services will be
provided at all layers of the model. The security architecture provides
direction on what services may be provided in which layers.
   (1) The only services that may be provided at the Physical Layer are
connection confidentiality (encryption) and traffic flow confidentiality
(traffic analysis). These services are restricted to passive threats and can be
applied to point-to-point or multipeer modes of transmission in the Physical Layer.
   (2) The services that may be provided at the Data Link Layer are
connection confidentiality and connectionless confidentiality. These services are performed by the normal layer functions for transmission and
after the normal layer functions for receipt, i.e., the security mechanisms
rely on the proper operation of the data link protocol.
   (3) The services that can be provided at the Network Layer by the
subnetwork access functions are as follows:

   a. Peer entity authentication
   b. Data origin authentication
   c. Access control service
   d. Connection confidentiality
   e. Connectionless confidentiality
   f. Traffic flow confidentiality
   g. Connection integrity without recovery
   h. Connectionless integrity

   (4) The same services provided at the Network Layer can also be
provided at the Transport Layer. The advantage of providing a service at
the Transport Layer rather than the Network Layer is to isolate one
transport connection from all other transport connections.
   (5) No security services are provided at the session layer.

(6) No specific services are provided at the Presentation Layer. However, the Presentation Layer contains a number of security mechanisms that support the provision of security services in the Application Layer. These security mechanisms are applied to the transfer syntax.

(7) The greatest number of security services, however, is provided at the Application Layer:

    a. Peer entity authentication
    b. Data origin authentication
    c. Access control service
    d. Connection confidentiality
    e. Connectionless confidentiality
    f. Selective field confidentiality
    g. Traffic flow confidentiality
    h. Connection integrity with recovery
    i. Connection integrity without recovery
    j. Selective field connection integrity
    k. Connectionless integrity
    l. Selective field connectionless integrity
    m. Nonrepudiation with proof of origin
    n. Nonrepudiation with proof of delivery

This extension was approved in 1988 as ISO 7498-2.

C. Naming and Addressing

The basic reference model defines the naming constructs required for relating entities between layers. This was sufficient for the early work in OSI, but an extension was required to provide a more complete framework for naming and addressing. This extension describes three broad areas: the role of naming domains, naming authorities, and directories; the details of the address mappings required to resolve addresses in a layer; and the application of the general framework to the specific layers of the reference model [7]. In discussing naming in OSI, one must keep in mind that we are concerned with describing an approach that will work for a global network with no central authority. This requires a somewhat different approach from what has been used in other networks. There is insufficient space here to treat this subject in detail, but we will give a broad overview.

There are two types of names: primitive names and descriptive names. Primitive names are uniquely assigned by naming authorities. Primitive names are characterized by the fact that the user of a primitive name need not know anything about how the name was derived. It is simply a label bound to an object. Descriptive names, on the other hand, are an unordered set of assertions about an object. Thus, a descriptive name is the intersection of the sets defined by each of the assertions. Descriptive names are characterized by the fact that using the name may denote one object, many objects, or no objects. The assertions that compose descriptive names use primitive names.

As just mentioned, primitive names are assigned by naming authorities. Naming authorities are defined as very simple mechanisms. A naming authority produces on request a unique name. The authority only guarantees that the name is not in use in that naming domain. The authority does not perform the binding of the name to the object or keep any records of what the name was used for. Naming authorities may be defined such that they partition their name domain and assign responsibility for handing out names for each partition to subauthorities. This may lead to names being constructed as a string of components assigned by each subauthority. However, the user of the name need have no knowledge of how the names are constructed.

An (N)-directory performs a mapping from an (N)-entity-title to an (N)-address. A directory allows one given the name of an object to retrieve information for locating it. Directories have no necessary relation to naming authorities. They are viewed as "local" caches of location information containing entries relevant to the users of the directory. It is often assumed that directories are closely associated with naming authorities. The assumption generally is that there is a master directory associated with each naming domain or subdomain that would contain all directory information for that domain. OSI does not assume that this is the case. There is nothing to preclude it. In fact, such a "backbone" of directories could be quite useful. But in a worldwide system, especially during the transition to it, one cannot predicate the operation of OSI on its existence.

(N)-Addresses are used in OSI to distinguish among sets of $(N + 1)$-entities supporting different $(N + 1)$-protocols or sets of $(N + 1)$-protocols; to accommodate security or management requirements; and to distinguish among $(N + 1)$-entities bound to different (N)-SAPs in order to allow reestablishment of communication. Addresses are not used to distinguish among aspects of protocols, such as classes, subsets, quality of service, or protocol versions; to distinguish between connection-mode and connection-less-mode operation; to derive routing information; or to distinguish hardware components.

The addressing structure allows a presentation-address to identify the location of an application-entity without constraining the binding of the (N)-entities in the various layers within the same open system. An (N)-address identifies a set of (N)-SAPs all located in a single (N)-subsystem. The exact membership of the set is an issue local to that (N)-subsystem. The set membership is not known to other open systems and may change over time. (N)-addresses are passed in service primitives to the (N)-entity.

The semantics of the (N)-address are encoded in the (N)-PCI, called (N)-protocol-addressing-information.

An (N)-selector is an element of addressing information defined by the local administration of an end-system which identifies the set of (N)-SAPs local to the (N)-subsystem. With the exception of the network and application layers, (N)-addresses are built from (N)-selectors and the network-address. (N)-selectors are unambiguous within the scope of an (N)-entity and are chosen by the local administration of the end system. When an (N)-selector identifies a set of (N)-SAPs, the resolution of the (N)-selector is the responsibility of the recipient (N)-subsystem. The application title directory service contains information about application-entities, including the protocol addressing information required to access the application through a PSAP as a tuple of the form

$(P$-selector, S-selector, T-selector, set of network addresses$)$

The application-entity-title and the network-address are the key to addressing in OSI. The directory provides the PSAP-address to which an application is bound, and the NSAP-address locates the end system. Since only the application-entity-title and the NSAP-address must be globally administered and the (N)-selectors are chosen locally and have limited scope, this provides an addressing structure that is both very rich and flexible as well as being efficient and easy to administer.

The addressing structure is constructed such that a system can be organized around the (N)-selectors to accommodate local requirements in as simple or complex a manner as the system designer desires. This structure allows the addresses to provide access to the wide variety of forms that an open system may take on, such as a single computer, a network of computers, or a computer with its upper and lower layers implemented as physically independent components. This extension is described in ISO 7498-3.

D. OSI Management

OSI management is divided into two broad categories: OSI systems management and OSI layer management [10]. The OSI management framework is described in ISO 7498-4.

OSI layer management is the general term given to any management functions that are done to manage the operation of a particular layer of the reference model. OSI systems management includes all of the functions required to manage a collection of OSI end and intermediate systems. OSI systems management includes most of the functions normally associated

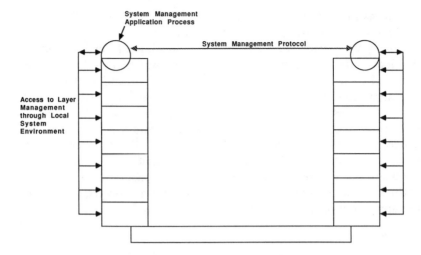

Fig. 11. OSI management architecture.

with network management, such as fault management, performance management, configuration management, error reporting, and accounting.

Systems management is performed entirely by application processes, called systems management application processes (SMAPs). Associated with each SMAP is a management information base (MIB). The MIB contains all management information for that system. Information in the MIB is used by SMAPs and layer management entities (MLEs). (See Fig. 11.) An LME is an (N)-entity found in each layer which is responsible for various forms of activation and error control. In some special circumstances, layer management protocols may be required to provide proper management of the services of the layer. SMAPs are also able to directly interact with LMEs through mechanisms provided by the local system (local IPC, systems calls, etc.). Precise specification of these mechanisms is outside the scope of OSI.

Systems management protocols are carried on between SMAPs using application layer protocols. In general, SMAPs interact among themselves to access necessary information either in remote MIBs or interact with LMEs through the local system. SMAPs do not communicate directly with LMEs, but communicate through the local system environment.

In some cases, interactions must take place among LMEs such that application protocols cannot be used. For example, interactions in which timing is critical, as in certain testing or loading functions; or interactions that cannot be subject to network congestion, such as routing update information. In these cases, special layer management protocols are used to provide these functions. However, the general rule is that layer management protocols are the exception, not the rule. Systems management protocols are the preferred mechanism for accomplishing all network management functions. Currently, two systems management protocols are being developed within OSI: a directory service protocol [DP 9594; X.500], and the common management information protocol (CMIP), DP 9595 and DP 9596.

A directory service protocol is used to access and update directory information. In general, a directory performs name-to-name or name-to-address mapping. The directory system being developed is a general template that can be used for different directory services, such as application-entity-title to PSAP-address, or user to application-entity-title, or application type to application-entity-title mapping. In addition, the directory service may provide additional information about the objects associated with the name, as well as "yellow pages" functions to allow information to be looked up based on the attributes of an object rather than its name.

The management information services provide a protocol that allows basic operations, such as get, set, action, and event to be performed among the SMAPs. These operations form the basis on which the management applications (fault, configuration, accounting, etc.) are built.

V. Future Directions for the OSI Architecture

In 1984, TC97 (Information Processing), the parent organization of SC16, recognizing the increased activity and importance of systems of standards that were emerging, undertook to reorganize TC97 to better address this new environment. Part of that reorganization was to increase the emphasis in OSI on distributed processing. Before the reorganization, other aspects of systems such as databases, graphics, and operating systems were in SC5 with the traditional programming languages. A new SC was formed, SC21 (Open Systems), which is responsible for the overall OSI architecture, the upper three layers of OSI, OSI Management, databases, and operating systems. The Transport Layer work was moved to SC6 (Data Communications), where development of the lower three layers based on the reference model was already underway. This reorganization focuses further OSI developments on supporting applications. This does not mean that OSI now includes databases and the like. OSI is still only concerned with interconnection. However, having all of this work in the same subcommittee greatly enhances the ability of OSI to develop the facilities necessary to support distributed applications. Some of these issues are touched on below.

Once the reference model was approved, SC16 began to get inquiries from other groups and organizations asking for clarification of various points in the model or questions about aspects not currently covered. SC16/WG1 instituted a procedure of establishing questions on the model to respond to these requests. Some of the questions led to major new extensions to the model, such as security and multipeer data transmission (broadcastlike transmissions). Others were answered with relatively short explanations that were not appropriate as extensions to the model [8]. However, SC16 (and now SC21) recognized that these interpretations would be useful to anyone trying to use the OSI reference model and other OSI standards. So the approved interpretations are collected into a single document. This document is updated after each meeting of the OSI Architecture Working Group (SC21/WG1), at which time any interpretations that have been approved since the last issue are included. This document is readily available from any ISO national member body, such as ANSI in the US, BSI in the UK, and AFNOR in France. Any group, corporation, or individual may submit a question on the OSI architecture; however, these should be directed to their national standards committee responsible for the OSI reference model for coordination, not to SC21 directly. Current outstanding issues can be grouped into five fairly broad categories:

1. Those of a general nature that affect the model as a whole
2. Lower layers
3. The upper layers
4. Security
5. Conformance testing

A brief word about some of these issues will give some idea of the topics being studied by SC21 and associated committees.

A. General Issues

1. Multipeer Data Transmission

There are a variety of communication technologies that are generally characterized by the terms "broadcast," "multicast," or "multidrop." The general property of these technologies is characterized by a single (N)-entity sending an (N)-protocol-data-unit that is received by several (N)-entities. A number of protocol disciplines have been developed over the years that are variations on this mode of transmission. These can be characterized by the following:

(a) One-way / two-way broadcast. Information is broadcast in one direction only (as in radio or TV) or in both directions.

(b) Centralized / decentralized. Centralized represents the classic master/slave relation where the master may or may not be fixed, or decentralized where all of the correspondents have equal status in the dialogue.

(c) Dynamic / static. The population of correspondents is either fixed for the span of the dialogue or may vary.

(d) Known / unknown. The population of correspondents may or may not be known to the senders. For example, a radio transmitter may not know all receivers.

(e) Confirmed / unconfirmed. Transmissions may be confirmed or unconfirmed.

(f) Reliable / unreliable. Transmissions may or may not be assumed to lose data.

(g) Pairwise / global sequencing. Transmissions may provide sequencing only among pairs of correspondents or ensure that all correspondents receive data in the same order.

This work is currently being progressed as an addendum to the basic reference model. It turns out that almost all combinations of these parameters occur for some communication technology or application. A principal effort in this work will be to develop a small number of protocol sets, especially in the upper layers, that will accommodate the major combinations of characteristics in an efficient and effective way. This work will be important not only for the lower layers, but also to serve as a basis for future work supporting distributed applications.

2. Voice and Image

There is a general trend to integrate transmission, storage, and processing of information in its various forms, i.e., not only traditional computer data but also voice and image data. This trend can be recognized in the evolution of telecommunications networks towards the integration of services characterized by integrated services digital networks (ISDN). The impact of this integration on the reference model must be studied further so that OSI standards can also be applied in this area.

3. Conformance Testing

Although not strictly part of the architecture, the architecture group undertakes development of general tools that are used by all groups to develop their standards. For example, formal description techniques (a subject of another chapter in this book) were developed by the architecture group. However, the formal descriptions of specific protocols are done by the groups developing the protocols. Similarly, the architecture group has undertaken to develop a general methodology for developing conformance test suites. This work is currently underway and will produce an independent set of documents that will be used by the protocol groups to develop specific test suites for their protocols.

B. Lower Layer Issues

The current architecture issues in the lower three layers are concerned with the following problems:

1. How local area networks fit into the reference model
2. The internal architecture of the network layer, especially how the network layer organizes and accommodates the wide variety of network services
3. The architecture for internetworking
4. Architectural issues for routing and relaying
5. ISDN and circuit switching

C. Upper Layer and Distributed Applications

The architectural problems in the upper layers have been the most difficult to solve because this is the area where traditional data communication begins to overlap with the problems of distributed processing among heterogeneous systems and the problems in databases, operating systems, and programming languages. While great strides have been made in the understanding of these layers (see other chapters in this book), there still remain many problems related to the kind of support OSI can provide to facilitate and structure distributed applications and information processing. This work falls into two areas:

Upper Layer Architecture. A further explication of the distinction between syntax and semantics in the presentation and application layers, and an explication of the distinction and boundary between the application entity (i.e., that part of the application process concerned with OSI) and the application process as a whole.

Open Distributed Processing. A major new area that has begun in SC21 is open distributed processing. Many diverse application areas will be developing whole systems of application protocols to support their distributed processing requirements. The development of the upper layers of OSI has led to the recognition that there are many aspects that these application protocols have in common. At this writing, this work has just begun, so it is impossible to say with any certainty all of the areas it will address, but the method getting the answer is fairly clear.

As with the OSI work, the ODP work has chosen to take an architected approach to developing a complex area of work, which will lead to specifications at several levels of abstraction. The first step then is to develop a reference model for open distributed processing. This reference model will facilitate the consistent definition of models for specific application areas and will help to avoid confusion in the areas where they interact. It should also aid in identifying those aspects that are common to many applications and can be standardized once, rather than independently and slightly differently for each application. This should greatly facilitate the ability of distributed applications to interact. This early in the development of distributed systems it would be pure folly to make decisions of the sort, "This distributed application will never have to talk to that distributed process"—regardless of what "this" and "that" are. Providing a common basis for diverse distributed processing applications, such as banking, process control, office systems, radio/TV/hi-fi, household management, image manipulation, and command/control can do much to reduce the problems of having these applications exchange information or cooperate, if and when it becomes necessary. The development of common components will also reduce the cost of such systems and simplify the task of developing new ones.

Undoubtedly, the ODP work will roughly parallel the development of OSI in that once there is sufficient understanding a basic reference model will be defined so that the development of specific protocols and mechanisms can begin as further extensions to the reference model are developed in parallel.

VI. Conclusions

The high cost of software production, the requirement that any system in the world potentially be able to communicate with any other, and the urgent need for open systems standards, have caused the OSI effort to make a major departure from the traditional mode of developing standards. The rate of technological change has made it necessary to define standards that new systems could converge to rather than waiting to standardize procedures after the fact. It has been clear from the beginning that many standards would be needed, that they would have to be developed in parallel, and that the development efforts would have to be closely coordinated to ensure that they worked well together. SC16, and now SC21, have used the reference model as the primary means to accomplish these goals. The reference model has proven to be extremely useful in coordinating different groups working on the different parts of the problem. It has made it possible to recognize at an early stage what needs to be standardized and what does not. In addition, the reference model has proven to be a very powerful pedagogical tool that has also allowed us to gain a much better understanding of the fundamentals of communications. It has made it much easier to separate those aspects of communications that can be considered first-order effectors and those that have a lesser effect.

The initial two years that SC16 spent developing the basic reference model have more than paid off in the long run. Never has any group been able to produce a coordinated set of 30 standards (of well over 100 pages each) in less than 7 years; and many more are being written. It is expected that other groups developing standards for other areas or large corporate systems will benefit from this work and will use the reference model for OSI and other reference models to organize and coordinate their work in communications, databases, and large application systems.

References

[1]   L. G. Roberts and B. D. Wessler, "Computer network development to achieve resource sharing," *Proc. SJCC*, pp. 543–549, 1970.
[2]   L. Pouzin, "Presentation and major design aspects of the CYCLADES computer network," in *Proc. 3rd ACM-IEEE Comp Symp.*, pp. 80–87, 1973.
[3]   ISO, "Provisional Model of Open Systems Architecture," ISO TC97/SC16/N34, 1978.
[4]   ISO, "Information Processing—Open Systems Interconnection—Basic Reference Model," ISO 7498-1, 1983.
[5]   ISO, Information Processing—Open Systems—Basic Reference Model, Addendum 1 (Connectionless Data Transmission), 1986.
[6]   ISO, Information Processing—Open Systems—Security Architecture, ISO 7498-2, 1988.
[7]   ISO, Information Processing—Open Systems—Naming and Addressing, ISO 7498-3, 1988.
[8]   ISO, Approved Commentaries of the OSI Reference Model, ISO TC97/SC21/N2740, 1988.
[9]   ISO, Internal Organization of the Network Layer, ISO 8648, 1985.
[10] ISO, Information Processing Systems—Open Systems—OSI Management Framework, ISO 7498-4, 1988.

II

Physical Layer

Physical Interfaces and Protocols

H. V. Bertine

The physical layer is the most basic protocol layer in the hierarchy of data communication protocols. This layer is concerned with the physical interface between devices and the rules by which bits are passed from one to another. These devices may be, for example, a data terminal equipment (DTE) and a data circuit-terminating equipment (DCE, e.g., a modem).

This chapter describes the physical layer of the open systems interconnection (OSI) basic reference model and the underlying physical media (see Fig. 1). Particular attention is given to the large number of domestic and international standards that apply to this layer. The presentation is organized into six major parts:

- The physical layer of OSI.
- Characteristics of the physical layer.
- Traditional DTE/DCE interfaces.
- Integrated services digital network (ISDN) interfaces.
- Local area network (LAN) interfaces.
- Recent activities and future directions.

I. The Physical Layer of OSI

The CCITT* Recommendation X.25 protocol for access to packet switched data networks is probably the first internationally recognized data

*International Telegraph and Telephone Consultative Committee. CCITT is a committee of the International Telecommunications Union (ITU), a specialized agency of the United Nations. The CCITT work on the physical layer for data communications is focused in three

H. V. BERTINE • AT&T Bell Laboratories, Holmdel, New Jersey 07733.

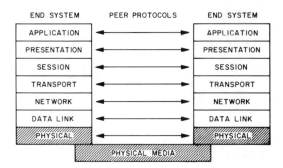

Fig. 1. OSI reference model.

communication protocol to use the concept of levels or layers. The 1976 version of Recommendation X.25 [1] defined the first layer, which it designated level 1, as "the physical, electrical, functional, and procedural characteristics to establish, maintain, and disconnect the physical link between the DTE and the DCE." ISO[†] and CCITT, in their joint work on open systems interconnection (OSI), have developed a seven-layer architectural model [2], [3]. This basic reference model, adopted in 1984, defines the physical layer as follows:

> The physical layer provides mechanical, electrical, functional, and procedural means to activate, maintain, and deactivate physical connections for bit transmission between data link entities. A physical connection may involve intermediate open systems, each relaying bit transmission within the physical layer. Physical layer entities are interconnected by means of a physical medium.
>
> A physical connection may allow duplex or half-duplex transmission of bit streams.

---

study groups. CCITT Study Group VII is responsible for data communications over data networks. Its work is contained in X-series recommendations. CCITT Study Group XVII is responsible for data communications over telephone facilities. Its work is contained in V-series recommendations. CCITT Study Group XVIII responsibilities include Integrated Services Digital Networks (ISDNs). Its work is contained in I-series recommendations.

[†] International Organization for Standardization. ISO is a voluntary nontreaty group made up of the principal standardization body of each represented nation. The U.S. member body is the American National Standards Institute (ANSI). ISO and the International Electrotechnical Commission (IEC) have recently combined their efforts on information technology standards in Joint Technical Committee 1 (JTC 1). The work on the physical layer for data communications is focused in JTC1/SC6, Telecommunications and Information Exchange Between Systems. JTC1/SC6 is responsible for standards pertaining to the lower four layers of OSI.

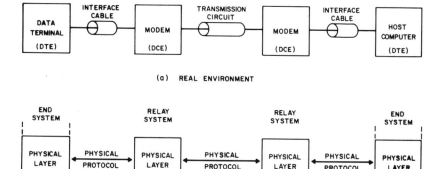

(a)   REAL ENVIRONMENT

(b)   OSI LOGICAL REPRESENTATION

Fig. 2. Relationship between real environment and OSI logical representation.

The OSI basic reference model also introduces the concept of a physical service data unit:

A physical service data unit consists of one bit in serial transmission and of "n" bits in parallel transmission.

The transmission of physical service data units (i.e., bits) may be synchronous or asynchronous.

The physical layer delivers bits in the same order in which they were submitted.

Figure 2 illustrates the relationship between real world equipment and the logical concepts of OSI. The text for the physical service definition [4], [5] further clarifies the distinction between the physical layer and the physical media by stating:

Actual data transmission takes place over the physical media. The mechanical, electromagnetic, and other media dependent characteristics of the physical media connection are defined at the boundary (interface) between the physical layer and the physical media.

The precise definition of the physical layer continues to be a subject of ongoing debate. The major issue is what functions belong in the data link layer and what functions belong in the physical layer. However, it is clear

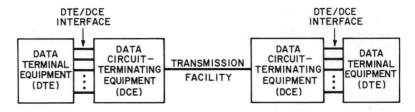

Fig. 3. DTE/DCE interface.

that the physical layer has four important characteristics, which are designated here as

- Mechanical.
- Electrical.
- Functional.
- Procedural.

Each of these characteristics is discussed in detail in Section II.

Prior to the creation of OSI, there was a very large and growing base of domestic and international standards pertaining to physical interfaces. A very important and widely standardized interface is the one between the DTE and the DCE as depicted in Fig. 3. The most familiar DTE/DCE interface in the USA is that described by EIA* RS-232-C [6]. There are other serial data DTE/DCE interfaces, such as EIA RS-449 [7], CCITT X.20 [8], and CCITT X.21 [8]. There are also what are known as parallel data DTE/DCE interfaces, such as CCITT V.19 [9] and V.20 [9]. Other types of data communication interfaces are important, such as EIA RS-366-A [10], which covers the DTE interface to automatic calling equipment (ACE). There is also a physical interface between the DCE (e.g., a modem) and the transmission facility (e.g., telephone line) as specified in CCITT V-series modem recommendations [9] and in EIA RS-496 [11]. Another

---

*Electronic Industries Association. EIA is a trade association that represents manufacturers in the U.S. electronics industry. The EIA work on data communications is carried out by Technical Committee TR30, Data Transmission Systems and Equipment. EIA standards on data communications were previously published in the RS-series but are now published in the EIA-series. In addition, EIA publishes supplementary material in Industrial Electronics Bulletins. Recently, a portion of the standards work of EIA, including TR30, was transferred to the Telecommunications Industry Association (TIA). TIA is a trade association that represents manufacturers in the U.S. telecommunications industry. Owing to their long-established usage in industry, the older EIA and RS-series designations will be used throughout this chapter.

important interface is the signaling interface between networks such as that specified by CCITT Recommendation X.75 [8].

Parameters associated with the physical layer also have been standardized. For example, ANSI* X3.1 [12], and CCITT V.5 [9], V.6 [9], X.1 [8], and X.10 [8] set forth the signaling rates (i.e., bits/s) for the physical layer of the DTE/DCE interface. The alignment of data and timing signals for synchronous operation are specified in EIA RS-334-A [13]. Signal quality for asynchronous operation is specified in EIA RS-363 [14], EIA RS-404 [15], and ISO 7480 [16]. Space does not permit further discussion of these standards.

Two specific interfaces, EIA RS-449 and CCITT X.21, are discussed in detail in Section III to provide a flavor for the considerations that go into the development of the physical layer. Two major new series of standard interfaces have been published in 1985—those for integrated services digital networks (ISDNs) and local area networks (LANs). These are covered in Sections IV and V, respectively. Work on standardizing the physical layer is continuing unabated; Section VI provides a look at this effort.

II. Characteristics of the Physical Layer

As mentioned previously, the four principal characteristics making up the physical layer are mechanical, electrical, functional, and procedural. Each is briefly described in this section along with examples from traditional DTE/DCE interfaces.

A. Mechanical Characteristics

The mechanical aspects pertain to the point of demarcation. Typically, this is a pluggable connector, but other arrangements, such as screw terminals, are sometimes used. Included are the specifics of the connector, the assignment of interchange circuits (see Section II C) to pins, the connector latching arrangements and so on. The location of the interface connector (e.g., close to or on the DCE) is often specified as well as the provision of cabling (e.g., interface cabling is generally considered part of the DTE).

---

*American National Standards Institute. ANSI is a nonprofit, nongovernmental organization. It serves as the national clearing house and coordinating activity for voluntary standards in the USA. The work on the physical layer for data communications is focused in Technical Committee X3S3 of the ANSI Accredited Standards Committee X3 (Information Processing Systems). Technical Committee X3S3, Data Communications, is responsible for standards pertaining to the lower four layers of OSI. American National Standards resulting from this work are contained in the X3.-series.

The following are the various mechanical characteristics that have been standardized by ISO for traditional interfaces:

- ISO 2110 [17]: 25-pin connector used for serial and parallel voice-band modems, public data network interfaces, telegraph (including Telex) interfaces, and automatic calling equipment. EIA RS-232-C and EIA RS-366-A are compatible with ISO 2110.
- ISO 2593 [18]: 34-pin connector used for the CCITT Recommendation V.35 [9] wide-band modem. Although there is no equivalent EIA standard, this interface is used within the USA for operation at 56 kbit/s.
- ISO 4902 [19]: 37-pin and 9-pin connectors used for serial voice-band and wide-band modems. EIA RS-449 is compatible with ISO 4902.
- ISO 4903 [20]: 15-pin connector used for public data network interfaces specified by CCITT Recommendations X.20, X.21, and X.22 [8]. There is no equivalent standard in the USA.

The various connectors and their relative sizes are illustrated in Fig. 4. All connectors, except for the 34-pin connector, belong to the same connector family.

The newer standards (ISO 4902 and ISO 4903) contain additional specifications to solve many of the mechanical interface problems experienced with implementations of the earlier standards. A key provision is the specification of an inexpensive DCE latching block (see Fig. 4), which enables latching and unlatching to be done either with or without a tool. This innovation should avoid the incompatibilities associated with the wide variety of latching devices used with the 25-pin connector. Another improvement is the placing of limitations on the size of the DTE connector including cover, cable clamp, and latching arrangement. This permits compact mounting arrangements involving multiple DCE connectors while assuring adequate clearances.

The EIA RS-449 and CCITT X.21 interfaces discussed in Section III use these connector specifications.

B. Electrical Characteristics

In the early standards (EIA RS-232-C, CCITT Recommendation V.28 [9]), the electrical characteristics were defined at the point of demarcation. More recent standards (EIA RS-422-A [21] and RS-423-A [22], CCITT Recommendations V.10 [9]/X.26 [8] and V.11 [9]/X.27 [8]) specify the electrical characteristics of the generators and receivers and give guidance with respect to the interconnecting cable. The latter situation, while simplifying the job of the integrated circuit manufacturer, has been criticized because there is no specification at the point of demarcation. The absence

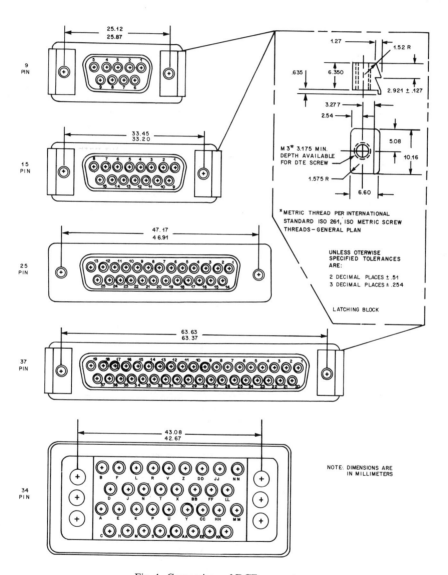

Fig. 4. Comparison of DCE connectors.

of this specification hampers conformance testing and sectionalization of trouble.

The following are the various electrical characteristics that have been standardized by CCITT for traditional interfaces:

- V.10/X.26: Unbalanced electrical characteristics. EIA RS-423-A, FED-STD 1030A [23], and MIL-STD 188-144 [24] are compatible with V.10/X.26.
- V.11/X.27: Balanced electrical characteristics. EIA RS-422-A, FED-STD 1020A [25], and MIL-STD 188-114 are compatible with V.11/X.27.
- V.28: Unbalanced electrical characteristics. EIA RS-232-C is compatible with V.28.
- V.31 [9]: Electrical characteristics for interchange circuits controlled by contact closure. Used in parallel modems (CCITT V.20 [8]). EIA RS-410 [26] is a similar standard.
- V.35 [9]: Balanced electrical characteristics used on the data and timing circuits of the CCITT V.35 modem and deployed for other applications using data rates above 20 kbit/s. Although there is no equivalent EIA standard, these electrical characteristics are used within the USA for operation at 56 kbit/s.

The use of the latter two electrical characteristics is limited and, therefore, they will not be discussed further. Figure 5 provides a comparison of V.28, V.10, and V.11. The key item to note is that V.10 provides a transitional mechanism since it is interoperable with both V.28 and V.11.

The V.10 and V.11 electrical characteristics were developed to provide improved performance in terms of supporting higher bit rates and longer cable distances compared with V.28 and RS-232-C. Integrated circuit manufacturers were active in the development of these new electrical characteristics to ensure their practical realization in state-of-the-art technology.

The electrical characteristics of V.28/RS-232-C specify a single-ended generator that produces a 5- to 15-V signal (negative for binary 1, positive for binary 0) with respect to signal ground (common return). A single common return lead is used for all interchange circuits. Generator rise time is relatively fast such that the time for the signal to pass through the ± 3-V transition region does not exceed 1 ms and, for data and timing interchange circuits, also does not exceed 3% (for V.28; 4% for RS-232-C) of the nominal signal element duration. A single-ended receiver is specified having a dc resistance between 3 and 7 kΩ. These electrical characteristics are generally limited to data signaling rates below 20 kbit/s and cable distances shorter than 15 m.

RS 232C   RS 449
unbal
20 Kbps
15 m
        RS 423   RS 422
        unbal   bal
        360 Kbps   10 Mbps
        1600 m   1000 m

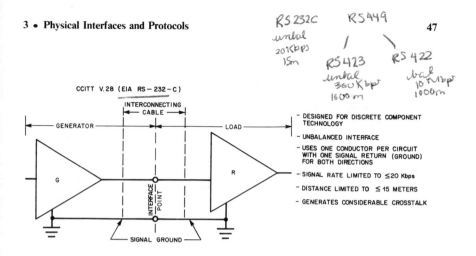

CCITT V.28 (EIA RS-232-C)

- DESIGNED FOR DISCRETE COMPONENT TECHNOLOGY
- UNBALANCED INTERFACE
- USES ONE CONDUCTOR PER CIRCUIT WITH ONE SIGNAL RETURN (GROUND) FOR BOTH DIRECTIONS
- SIGNAL RATE LIMITED TO ≤ 20 Kbps
- DISTANCE LIMITED TO ≤ 15 METERS
- GENERATES CONSIDERABLE CROSSTALK

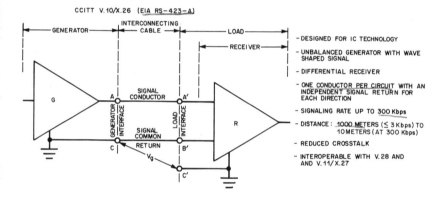

CCITT V.10/X.26 (EIA RS-423-A)

- DESIGNED FOR IC TECHNOLOGY
- UNBALANCED GENERATOR WITH WAVE SHAPED SIGNAL
- DIFFERENTIAL RECEIVER
- ONE CONDUCTOR PER CIRCUIT WITH AN INDEPENDENT SIGNAL RETURN FOR EACH DIRECTION
- SIGNALING RATE UP TO 300 Kbps
- DISTANCE: 1000 METERS (≤ 3 Kbps) TO 10 METERS (AT 300 Kbps)
- REDUCED CROSSTALK
- INTEROPERABLE WITH V.28 AND AND V.11/X.27

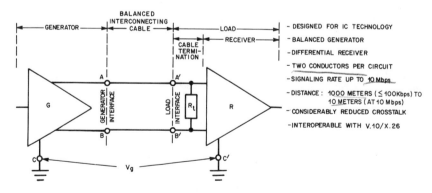

CCITT V.11/X.27 (EIA RS-422-A)

- DESIGNED FOR IC TECHNOLOGY
- BALANCED GENERATOR
- DIFFERENTIAL RECEIVER
- TWO CONDUCTORS PER CIRCUIT
- SIGNALING RATE UP TO 10 Mbps
- DISTANCE: 1000 METERS (≤ 100Kbps) TO 10 METERS (AT 10 Mbps)
- CONSIDERABLY REDUCED CROSSTALK
- INTEROPERABLE WITH V.10/X.26

Fig. 5. Comparison of electrical characteristics.

The V.10 unbalanced electrical characteristics specify a low-impedance ($\leq 50\ \Omega$) single-ended generator that produces a 4- to 6-V signal (negative for binary 1, positive for binary 0) with respect to the common return. A single common return lead for each direction of transmission can be used across the interface. Waveshaping of the generator output signal is used to control the level of near-end crosstalk to adjacent circuits in the interconnection. Data signaling rates up to 3 kbit/s can be used over cable distances up to 1000 m. For data signaling rates above 3 kbit/s, the cable distance decreases with increasing signaling rate to 10 m at 300 kbit/s.

The V.11 balanced electrical characteristics specify a low-impedance ($\leq 100\ \Omega$) balanced generator that produces a 2- to 6-V differential signal (A terminal negative with respect to the B terminal for binary 1, opposite polarity for binary 0). Each interchange circuit requires a pair of wires for balanced operation. Data signaling rates up to 100 kbit/s can be used over cable distances up to 1000 m. For data signaling rates above 100 kbit/s, the cable distance decreases with increasing signaling rate to 10 m at 10 Mbit/s.

The V.10 and V.11 electrical characteristics are identical for the receiver. They specify a differential receiver that has a high input impedance ($\geq 4\ k\Omega$) and a small transition region (± 0.2 V).

The correlation between the binary 1 and 0 states given above for each of the electrical characteristics and the states of the interchange circuits is shown in Fig. 6.

A key feature built into the V.10 and V.11 electrical characteristics is an evolution path from the existing V.28/RS-232-C electrical characteristics. The V.10/RS-423-A specifications were specifically designed to permit interoperation with both V.28/RS-232-C and V.11/RS-422-A. The EIA RS-449 and CCITT X.21 interfaces discussed in Section III make use of this capability.

C. Functional Characteristics

Interchange circuit functions are typically classified into the following broad categories: data, control, timing, and ground. Further classification

BINARY	1	0
DATA	MARK	SPACE
CONTROL	OFF	ON

Fig. 6. Signal state correlation table.

into primary and secondary channel functions is made for those DTE/DCE interfaces employing a secondary channel.

The following two CCITT recommendations define the functions of interchange circuits for traditional interfaces:

- V.24 [9]: DTE/DCE and DTE/ACE interchange circuits. Originally developed for use with modems and automatic calling equipment associated with modems, they may also be used with digital networks. EIA RS-232-C and RS-449 are compatible with V.24 for DTE/DCE interchange circuits and EIA RS-366-A is compatible with V.24 for DTE/ACE interchange circuits.
- X.24 [8]: DTE/DCE interchange circuits. Developed for use with public data networks (CCITT Recommendations X.20, X.21, and X.22). There is no equivalent standard in the USA.

The V.24 interchange circuits have been used for several decades. They employ the concept of one function per interchange circuit. Over the years, the list of interchange circuits has grown steadily. Recommendation V.24 presently defines 44 interchange circuits for use in various DTE/DCE interfaces and 12 interchange circuits for the DTE/ACE interface.

In the 1968–1972 CCITT study period, work started on interface standards (X.20 and X.21) specifically designed for the emerging duplex data networks. The technology to be employed in these networks favored a "compact" interface where the ACE functions, DCE control functions, and data were multiplexed over a single "data" interchange circuit in each direction. The result of this work was Recommendation X.24, which defines a small set of interchange circuits. This set includes a data and a control circuit in each direction plus a single bit timing circuit from the DCE. In addition, X.24 defines a framing circuit from the DCE for X.22, an optional byte timing circuit from the DCE for X.21, and an optional bit timing circuit from the DTE for use in direct connection of two X.21 DTEs (i.e., without intermediate DCEs).

The EIA RS-449 interface described in Section III A uses V.24 interchange circuits and the CCITT X.21 interface described in Section III B uses X.24 interchange circuits.

D. Procedural Characteristics

The final aspect of the physical layer is the set of procedures for using the interchange circuits. These procedures are the ones that need to be performed to enable the transmission of bits so that the higher-layer functions (described in subsequent chapters) can take place. The exact division between which procedures are part of the physical layer and which procedures are higher-layer procedures is an area of debate.

The following are the various CCITT recommendations that define procedures at the physical layer for traditional interfaces.

- V.24: Procedures for DTE/DCE interchange circuits. EIA RS-232-C and RS-449 contain equivalent procedures.
- V.25 [9]: Procedures for use with automatic calling and answering equipment when a separate DCE/ACE interface is provided. EIA RS-366-A contains equivalent procedures.
- V.25 bis [9]: Procedures for automatic calling and answering equipment when a separate DCE/ACE interface is not provided. There is no equivalent standard in the U.S.
- V.54 [9]: Procedures regarding maintenance test loops. EIA RS-449 contains equivalent procedures.
- V-series modems [9]: Modem-specific procedures for the use of interchange circuits. Several Federal Standards contain equivalent procedures.
- X.20: Procedures for asynchronous operation on a public data network. There is no equivalent standard in the USA.
- X.20 bis [8]: Procedures for asynchronous operation on a public data network for DTEs designed to interface with V-series asynchronous modems. EIA RS-232-C contains equivalent procedures.
- X.21: Procedures for synchronous operation on a public data network. There is no equivalent standard in the USA.
- X.21 bis [8]: Procedures for synchronous operation on a public data network for DTEs designed to interface with V-series synchronous modems. EIA RS-232-C and RS-449 contain equivalent procedures.
- X.22: Procedures for synchronous operation on a public data network whereby several circuits are time division multiplexed. There is no equivalent standard in the USA.
- X.150 [8]: Procedures regarding maintenance test loops for public data networks. There is no equivalent standard in the USA.
- ISO 9067 [27]: Procedures for automatic fault isolation using test loops. There is no equivalent standard in the USA.

Two examples of these procedures are given in Section III.

You may have wondered what happened to CCITT Recommendation X.25 [8], which was discussed at the beginning of this chapter. X.25, which specifies the packet mode interface to packet switched public data networks, does contain a section on the physical layer. This section, however, is quite brief, as it mainly references the appropriate sections of X.21, X.21 bis, V-series and I-series recommendations.

III. Traditional DTE/DCE Interfaces

In this section two examples are given of the physical layer for DTE/DCE interfaces. The first is EIA RS-449, which was developed to replace EIA RS-232-C. The second is CCITT Recommendation X.21, which was developed specifically as a synchronous interface to public data networks. In each example, the four characteristics of the physical layer—mechanical, electrical, functional, and procedural—are clearly evident.

Before taking up these interfaces, it is appropriate to briefly review EIA RS-232-C, the dominant DTE/DCE interface in use today. The first

	CIRCUIT	NAME	DIRECTION	DESCRIPTION
GROUND	AA	PROTECTIVE GROUND		ELECTRICALLY BONDS TOGETHER THE EQUIPMENT FRAMES
GROUND	AB	SIGNAL GROUND OR COMMON RETURN		ESTABLISHES THE COMMON GROUND REFERENCE POTENTIAL FOR ALL INTERCHANGE CIRCUITS
DATA	BA	TRANSMITTED DATA	TO DCE	CONVEYS DATA SIGNALS FOR TRANSMISSION TO THE COMMUNICATIONS CHANNEL
DATA	BB	RECEIVED DATA	TO DTE	CONVEYS DATA SIGNALS RECEIVED FROM THE COMMUNICATIONS CHANNEL
CONTROL	CA	REQUEST TO SEND	TO DCE	REQUESTS ABILITY TO TRANSMIT DATA TO THE COMMUNICATIONS CHANNEL
CONTROL	CB	CLEAR TO SEND	TO DTE	INDICATES WHETHER OR NOT THE DCE IS READY TO TRANSMIT DATA TO THE COMMUNICATIONS CHANNEL
CONTROL	CC	DATA SET READY	TO DTE	INDICATES WHETHER OR NOT THE DCE IS IN THE DATA MODE
CONTROL	CD	DATA TERMINAL READY	TO DCE	CONTROLS THE SWITCHING OF THE DCE TO AND FROM THE COMMUNICATIONS CHANNEL
CONTROL	CE	RING INDICATOR	TO DTE	INDICATES WHETHER OR NOT A "RINGING SIGNAL" IS BEING RECEIVED BY THE DCE
CONTROL	CF	RECEIVED LINE SIGNAL DETECTOR	TO DTE	INDICATES WHETHER OR NOT THE DCE IS RECEIVING A LINE SIGNAL FROM THE COMMUNICATIONS CHANNEL
CONTROL	CG	SIGNAL QUALITY DETECTOR	TO DTE	INDICATES WHETHER OR NOT THERE IS A HIGH PROBABILITY OF ERROR IN THE RECEIVED DATA
CONTROL	CH	DATA SIGNAL RATE SELECTOR (DTE SOURCE)	TO DCE	SELECTS BETWEEN TWO DATA SIGNALING RATES OR RANGES OF RATES
CONTROL	CI	DATA SIGNAL RATE SELECTOR (DCE SOURCE)	TO DTE	INDICATES ONE OF TWO DATA SIGNALING RATES OR RANGES OF RATES
TIMING	DA	TRANSMITTER SIGNAL ELE-MENT TIMING (DTE SOURCE)	TO DCE	PROVIDES TIMING SIGNALS FOR TRANSMITTED DATA
TIMING	DB	TRANSMITTER SIGNAL ELE-MENT TIMING (DCE SOURCE)	TO DTE	PROVIDES TIMING SIGNALS FOR TRANSMITTED DATA
TIMING	DD	RECEIVER SIGNAL ELEMENT TIMING (DCE SOURCE)	TO DTE	PROVIDES TIMING SIGNALS FOR RECEIVED DATA
SECONDARY	SBA	SECONDARY TRANSMITTED DATA	TO DCE	EQUIVALENT TO CIRCUIT BA EXCEPT IT APPLIES TO THE SECONDARY CHANNEL
SECONDARY	SBB	SECONDARY RECEIVED DATA	TO DTE	EQUIVALENT TO CIRCUIT BB EXCEPT IT APPLIES TO THE SECONDARY CHANNEL
SECONDARY	SCA	SECONDARY REQUEST TO SEND	TO DCE	EQUIVALENT TO CIRCUIT CA EXCEPT IT APPLIES TO THE SECONDARY CHANNEL
SECONDARY	SCB	SECONDARY CLEAR TO SEND	TO DTE	EQUIVALENT TO CIRCUIT CB EXCEPT IT APPLIES TO THE SECONDARY CHANNEL
SECONDARY	SCF	SECONDARY RECEIVED LINE SIGNAL DETECTOR	TO DTE	EQUIVALENT TO CIRCUIT CF EXCEPT IT APPLIES TO THE SECONDARY CHANNEL

Fig. 7. EIA RS-232-C interchange circuits.

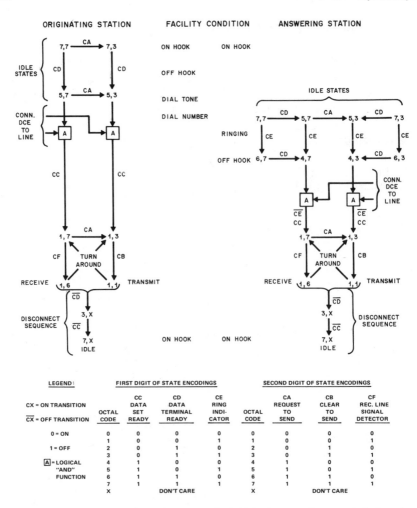

Fig. 8. EIA RS-232-C control lead sequences for half-duplex operation on switched service.

version of this standard, RS-232, was adopted in May, 1960. It was revised
three times—in October, 1963 as RS-232-A, in October, 1965 as RS-232-B,
and in August, 1969 as RS-232-C.*

RS-232-C defines 21 interchange circuits. Each circuit provides a single
function as summarized in Fig. 7. Not all circuits are needed in every
application. For example, the timing circuits are omitted for nonsyn-
chronous applications, certain control circuits are omitted for nonswitched

*Section VI discusses the recently approved fourth revision known as EIA-232-D.

applications, and the five secondary channel circuits are omitted when secondary channel operation is not employed.

The interchange circuit procedures contained in RS-232-C are more fully described in a separate Application Notes document [28]. Included is a series of charts giving control circuit state diagrams for a number of applications. An example illustration of these procedures is given in Fig. 8. This figure, covering half duplex operation over the switched network, shows the major states and transitions for the six principal RS-232-C control circuits.

RS-232-C includes the specification of electrical characteristics for the interchange circuits. These unbalanced characteristics were described above in Section II B. They apply at the point of demarcation between the DTE and DCE (i.e., at the 25-pin connector). Interface operation is generally limited to data signaling rates below 20 kbit/s and cable distances shorter than 15 m.

A. EIA RS-449

RS-232-C was recognized by EIA in 1973 to be a limiting factor in many user environments. The principal new capabilities and benefits desired were

- Improved performance, longer interface cable distances, and a significantly higher maximum data rate (to be achieved with the new electrical characteristics).
- Additional interface functions, such as loopback testing.
- Resolution of the mechanical interface problems which had led to a proliferation of designs, many of which were incompatible with one another.

The first approach examined was to update RS-232-C. Creating an RS-232-D would require a degree of compatibility with RS-232-C that would have severely compromised the desired new capabilities and benefits.* Therefore, the decision was made to develop a new interface. Two major approaches for this new interface were studied at the outset and were reviewed many times thereafter. One was to follow the basic concepts of RS-232-C. The other was to seek alignment with the developing CCITT work on Recommendation X.21 (see Section III B).

The principal advantage of the first approach is the ability to interoperate with RS-232-C. This would not be possible with the X.21 approach.

---

*The opposite was true for the automatic calling equipment interface, RS-366. It has been updated as RS-366-A.

The principal advantage of the X.21 approach is a lower cost interface achieved through a substantial reduction in the number of interchange circuits. However, there are significant technical and performance problems associated with the adoption of X.21 for the modem interface.* Therefore, the first approach was taken with two principal objectives:

- The ability to interoperate the new equipment with the presently existing RS-232-C equipment (no modification to RS-232-C equipment permitted).
- To obtain the new capabilities cited earlier when two new equipments are interfaced.

These objectives were satisfied (as described below) and RS-449 was published by EIA in November, 1977, after international agreement was reached in CCITT and ISO. To simplify the following discussion, the EIA terminology will be used. The listings given in Section II can be used for reference to the equivalent international standards.

1. Functional

One of the problems with RS-232-C was that many equipments included interface circuits in addition to those defined in RS-232-C. New Sync (now known as New Signal) is one example. More importantly, there

---

*One major problem is the significant reduction in throughput for half-duplex operation and for multipoint polling systems. This occurs since X.21 does not provide immediate recognition of a specific control signal. That is, X.21 requires the recognition of a bit pattern in contrast to the instant recognition of a signal level on an individual control lead.

A second problem with using X.21 is the loss of functionality, because there is no means to pass control information during data transfer. Two examples concerning signaling to the DTE while the receive direction is in the data transfer phase illustrate this problem. In this situation, X.21 circuit I is ON (indicating data transfer phase) and circuit R carries user data. Thus, there is no way to convey to the DTE information about the receive direction, such as Signal Quality (RS-232-C circuit CG). In addition, there is no way to convey to the DTE information about the transmit direction, such as Clear to Send (RS-232-C circuit CB). The impact of this latter problem is illustrated by a centralized multipoint system operating with the use of continuous carrier from the master station. After the remote DTE detects its poll, it responds by turning circuit C ON (a function equivalent to Request to Send in present day modems). However, as discussed above, there is no means to convey to the DTE when the DCE is prepared to accept data (i.e., the Clear to Send function). Since this time interval varies with modem type, this loss of capability is significant.

Other flexibilities of the EIA RS-232-C interface, such as separate send and receive timing circuits, would be lost if X.21 were used without change. Also, quite a few RS-232-C and V.24 interchange circuit functions that apply outside of the data transfer phase are not accommodated by X.21. Examples include data signaling rate selection, selection and indication of standby facilities, select frequency, and loopback testing. Either these functions would be lost or X.21 would require modification to accommodate them. In addition, a means would need to be provided for handling a secondary channel. A separate connector for the secondary channel would probably be required.

EIA RS-449		EIA RS-232-C		CCITT RECOMMENDATION V.24	
SG	SIGNAL GROUND	AB	SIGNAL GROUND OR COMMON RET.	102	SIGNAL GROUND OR COMMON RET.
SC	SEND COMMON			102a	DTE COMMON RETURN
RC	RECEIVE COMMON			102b	DCE COMMON RETURN
IS	TERMINAL IN SERVICE				
IC	INCOMING CALL	CE	RING INDICATOR	125	CALLING INDICATOR
TR	TERMINAL READY	CD	DATA TERMINAL READY	108/2	DATA TERMINAL READY
DM	DATA MODE	CC	DATA SET READY	107	DATA SET READY
SD	SEND DATA	BA	TRANSMITTED DATA	103	TRANSMITTED DATA
RD	RECEIVE DATA	BB	RECEIVED DATA	104	RECEIVED DATA
TT	TERMINAL TIMING	DA	TRANSMITTER SIGNAL ELEMENT TIMING (DTE SOURCE)	113	TRANSMITTER SIGNAL ELEMENT TIMING (DTE SOURCE)
ST	SEND TIMING	DB	TRANSMITTER SIGNAL ELEMENT TIMING (DCE SOURCE)	114	TRANSMITTER SIGNAL ELEMENT TIMING (DCE SOURCE)
RT	RECEIVE TIMING	DD	RECEIVER SIGNAL ELEMENT TIMING	115	RECEIVER SIGNAL ELEMENT TIMING (DCE SOURCE)
RS	REQUEST TO SEND	CA	REQUEST TO SEND	105	REQUEST TO SEND
CS	CLEAR TO SEND	CB	CLEAR TO SEND	106	READY FOR SENDING
RR	RECEIVER READY	CF	RECEIVED LINE SIGNAL DETECTOR	109	DATA CHANNEL RECEIVED LINE SIGNAL DETECTOR
SQ	SIGNAL QUALITY	CG	SIGNAL QUALITY DETECTOR	110	DATA SIGNAL QUALITY DETECTOR
NS	NEW SIGNAL			136	NEW SIGNAL
SF	SELECT FREQUENCY			126	SELECT TRANSMIT FREQUENCY
SR	SIGNALING RATE SELECTOR	CH	DATA SIGNAL RATE SELECTOR (DTE SOURCE)	111	DATA SIGNALING RATE SELECTOR (DTE SOURCE)
SI	SIGNALING RATE INDICATOR	CI	DATA SIGNAL RATE SELECTOR (DCE SOURCE)	112	DATA SIGNALING RATE SELECTOR (DCE SOURCE)
SSD	SECONDARY SEND DATA	SBA	SECONDARY TRANSMITTED DATA	118	TRANSMITTED BACKWARD CHANNEL DATA
SRD	SECONDARY RECEIVE DATA	SBB	SECONDARY RECEIVED DATA	119	RECEIVED BACKWARD CHANNEL DATA
SRS	SECONDARY REQUEST TO SEND	SCA	SECONDARY REQUEST TO SEND	120	TRANSMIT BACKWARD CHANNEL LINE SIGNAL
SCS	SECONDARY CLEAR TO SEND	SCB	SECONDARY CLEAR TO SEND	121	BACKWARD CHANNEL READY
SRR	SECONDARY RECEIVER READY	SCF	SECONDARY RECEIVED LINE SIGNAL DETECTOR	122	BACKWARD CHANNEL RECEIVED LINE SIGNAL DETECTOR
LL	LOCAL LOOPBACK			141	LOCAL LOOPBACK
RL	REMOTE LOOPBACK			140	LOOPBACK/MAINTENANCE TEST
TM	TEST MODE			142	TEST INDICATOR
SS	SELECT STANDBY			116	SELECT STANDBY*
SB	STANDBY INDICATOR			117	STANDBY INDICATOR

*In the 1988 version of Recommendation V.24, this function is provided by circuit 116/1, "Backup switching in direct mode."

Fig. 9. Equivalency of interchange circuits.

was a strong need to incorporate additional capabilities in the interface for loopback testing and other functions. These problems were solved in RS-449 by the addition of new interchange circuits following the philosophy used in RS-232-C of one function per interchange circuit. Figure 9 provides a complete listing of the 30 RS-449 interchange circuits and gives the equivalent interchange circuits in RS-232-C and CCITT Recommendation V.24.*

A new set of interface circuit names and mnemonics is used in RS-449. The names were chosen to more accurately describe the function performed and to eliminate the term "data set," which is no longer appropriate. The mnemonics were chosen to be easily related to the circuit names and to be unique from those used in RS-232-C to avoid confusion.

*CCITT Recommendation V.24 does not include one RS-449 interchange circuit (Terminal in Service).

Briefly, the new circuits are as follows:

- Send Common (SC)—provides a signal common return path for all unbalanced interchange circuits employing one wire used in the direction toward the DCE.
- Receive Common (RC)—provides a signal common return path for all unbalanced interchange circuits employing one wire used in the direction toward the DTE.
- Terminal in Service (IS)—indicates to the DCE whether or not the DTE is operational. A major use is to make an associated port on a line hunting group busy if the DTE is out of service.
- New Signal (NS)—indicates to the DCE when the DCE receiver should be prepared to acquire a new line signal. A major use is to improve the overall response time of multipoint polling systems.
- Select Frequency (SF)—controls the DCE transmit and receive operation with respect to two frequency bands. Its purpose is to allow selection of the frequency mode of the DCE in multipoint circuits where all stations have equal status.
- Local Loopback (LL)—requests the DCE to initiate a loopback of signals in the local DCE toward the local DTE. Its purpose is to allow checking of the functioning of the DTE and the local DCE.
- Remote Loopback (RL)—requests the DCE to initiate a loopback of signals in the remote DCE toward the local DTE. Its purpose is to allow checking of the functioning of the DTE, local DCE, transmission channel, and the remote DCE.
- Test Mode (TM)—indicates to the DTE when a test condition has been established involving the local DCE. Its purpose is to distinguish test conditions from other nondata mode conditions of the DCE.
- Select Standby (SS)—requests the DCE to replace regular facilities with predetermined standby facilities. Its purpose is to facilitate the rapid restoration of service when a failure has occurred.
- Standby Indicator (SB)—indicates to the DTE whether regular facilities or standby facilities are in use. This may be in response to activation by circuit SS or by other means.

2. Procedural

The text of RS-449 contains the procedures for using the interchange circuits. The basic RS-232-C procedures were carried over into RS-449. The state diagrams (e.g., see Fig. 8) prepared in an Application Note to RS-232-C [28] can be applied to RS-449.

The procedures for the new test and standby interchange circuits are based on action–reaction pairs. For example, the Local Loopback circuit is turned ON by the DTE (the action) to request a local loopback. The DTE now waits. When the DCE has established the loopback, it turns the Test Mode circuit ON (the reaction), indicating that the loop has been established and any data sent by the DTE on the Send Data circuit should be returned to the DTE on the Receive Data circuit. The DTE can now begin sending test data. A similar action–reaction sequence is followed when deactivating the loopback.

3. Electrical

As stated earlier, interoperability with EIA RS-232-C was a principal objective in the design of EIA RS-449. This is achieved by permitting the use of the unbalanced RS-423-A electrical characteristics on interchange circuits when the data rate is less than 20 kbit/s, the upper limit for RS-232-C. Unlike X.21 (see Section III B 1), this flexibility to use the unbalanced electrical characteristics is allowed for both the DTE and DCE. To provide good performance for data rates above 20 kbit/s (where interoperability with EIA RS-232-C does not apply), EIA RS-449 designates certain interchange circuits which must be operated with the balanced RS-422-A electrical characteristics. This enables EIA RS-449 to be used for data rates up to 2 Mbit/s.

The key to obtaining this flexibility is the use of two wires for each of the following interchange circuits (designated by RS-449 as Category I circuits):

SD—Send Data	RS—Request to Send
RD—Receive Data	CS—Clear to Send
TT—Terminal Timing	RR—Receiver Ready
ST—Send Timing	TR—Terminal Ready
RT—Receive Timing	DM—Data Mode

Either RS-422-A or RS-423-A generators can be used on these circuits for data rates below 20 kbit/s. For data rates above 20 kbit/s, these circuits are RS-422-A. All other interchange circuits (designated by RS-499 as Category II circuits) always use RS-423-A and thus have one wire per interchange circuit with a common signal return lead. Figure 10 summarizes this arrangement.

Two important benefits are achieved. For DTEs or DCEs designed for operation at speeds of 20 kbit/s or less, a manufacturer may choose to implement the unbalanced RS-423-A electrical characteristics on all interchange circuits. With this design, a single RS-449 implementation can

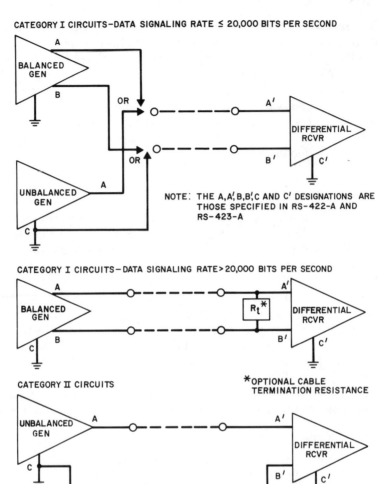

Fig. 10. EIA RS-449 interface connections of generators and receivers.

operate with another RS-449 device or interoperate with an RS-232-C device (see Section III A 5). Alternatively, a manufacturer may choose to implement the balanced RS-422-A electrical characteristics on the Category I interchange circuits. With this design, a single RS-449 implementation can operate with another RS-449 device at all bit rates up to 2 Mbit/s with maximum performance. This is contrasted with the variety of different interfaces (RS-232-C, V.35, etc.) required in the past, each applying to a narrow range of data rates.

4. Mechanical

The RS-449 connectors come from the same connector family as the familiar 25-pin connector used with RS-232-C. This selection was made because of the favorable experience associated with the use of the 25-pin connector. In order to satisfy the requirements of some foreign administrations, two connectors are used. A 37-pin connector is used for the basic interface. If secondary channel operation is used, these leads appear on a separate 9-pin connector. An important side benefit of the 9-pin and 37-pin connectors is that they are different from the present 25-pin and 34-pin connectors. This prevents the accidental interconnection of incompatible electrical characteristics, which may result in physical damage to interface generators and receivers.

The mechanical enhancements described in Section II A involving standardization of the DCE latching block and maximum DTE connector envelope size are incorporated in RS-449. The pin assignment plan was carefully chosen to minimize crosstalk in multipair cable (i.e., one Category I circuit or two Category II circuits in the same direction are assigned to a pair) and to facilitate the design of an adapter when interworking with RS-232-C is desired.

Finally, provision was made for the use of shielded interface cable. Pin 1 of the interface connector is used to ensure continuity of the shields between tandem connections of shielded interface cable.

5. Interoperability

Interoperability with RS-232-C, when desired, may be accomplished by means of a simple passive adapter and a few additional design criteria for the RS-449 interchange circuits. The adapter specification and the detailed design criteria are contained in [19] and [29]. No modifications are needed for the RS-232-C equipment. Performance for interoperability is that associated with RS-232-C interfaces.

B. CCITT Recommendation X.21

CCITT Recommendation X.21 will be used in this section as a second illustration of the four characteristics of the physical layer. Recommendation X.21 is one of the set of five DTE/DCE recommendations (see Fig. 11) developed by CCITT for access to circuit-switched public data networks.

CCITT Recommendation X.21 contains two distinct parts. One part specifies a "general purpose" DTE/DCE interface for synchronous operation on public data networks. This is the physical layer part of Recommendation X.21 and is applicable to both circuit-switched and packet-switched services. The second part of Recommendation X.21 specifies the call

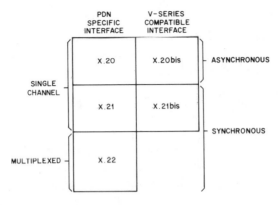

Fig. 11. DTE/DCE interface for circuit-switched public data networks.

control procedures for circuit-switched services. These call control elements of X.21 involve data link layer (e.g., for character alignment and parity) and network layer (e.g., for addressing and call progress signals) functions.

1. Electrical

One of the objectives for Recommendation X.21 was to permit interface operation over distances considerably greater than that available with Recommendation V.28. To achieve this objective at the synchronous data rates given in Recommendation X.1,* the new balanced electrical characteristics (Recommendation X.27) were specified for the DCE side of the interface. To allow flexibility in DTE design at the four lower data rates, the DTE is permitted to use either the new balanced or the new unbalanced (Recommendation X.26) electrical characteristics. For rates above 9600 bit/s, only the balanced electrical characteristics are permitted to ensure good performance.

2. Mechanical

The mechanical interface for X.21 is specified by ISO 4903. The mechanical enhancements described earlier for RS-449 involving standardization of the DCE latching block, maximum DTE connector envelope size, and the use of pin 1 for shield also apply to ISO 4903. Similarly, the 15-pin interface connector comes from the same family of connectors as the familiar 25-pin connector.

*CCITT Recommendation X.1 specifies data rates of 600, 2400, 4800, 9600, and 48,000 bit/s for Recommendation X.21. The 1988 version of X.1 adds 64,000 bit/s to this list.

Another major enhancement is the result of careful assignment of interchange circuits to connector pin numbers. The pin assignments provide for the connection of interchange circuits to multipaired interconnecting cable so that each interchange circuit operates over a pair. Of particular importance is the use of two wires for each interchange circuit even when interworking between a DTE using X.26 electrical characteristics and a DCE using X.27 electrical characteristics. This eliminates the need for either options inside the equipment or a special interface cord that connects certain pins together. Also, this provides a performance level when interworking that approximates the performance level when X.27 is used by both equipments.

3. Functional

Another objective in the design of Recommendation X.21 was to considerably reduce the number of interchange circuits while at the same time folding into the interface the automatic calling function. Thus, as illustrated pictorially in Fig. 12, X.21 contains five basic interchange circuits. A transmit (T) circuit and a receive (R) circuit are used to convey both user data and network control information depending on the state of the control (C) circuit and the indication (I) circuit. Bit timing is continuously provided by a signal element timing (S) circuit. A sixth interchange circuit, which provides byte timing· information, is optional. A signal ground circuit is also provided. Detailed definitions of these interchange circuits are contained in CCITT Recommendation X.24.

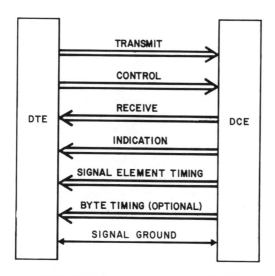

Fig. 12. CCITT Recommendation X.21 DTE/DCE interface.

4. Procedural

As mentioned earlier, some of the procedures in X.21 are above the physical layer. However, the procedures associated with the quiescent phase of X.21 are within the physical layer. A major feature incorporated into X.21 is the definition of three quiescent signals for each direction of the DTE/DCE interface. Two "not ready" signals are defined to distinguish between a nonoperational condition and a condition in which the DTE or DCE is operational but is temporarily out of service.

DCE ready indicates that the DCE (network) is ready to enter operational phases. *DCE ready* is signaled with continuous binary 1 on circuit *R* and the OFF condition on circuit *I*.

DCE not ready indicates that no service is available. It is signaled whenever possible during network fault conditions and when network test loops are activated. *DCE not ready* is signaled with continuous binary 0 on circuit *R* and the OFF condition on circuit *I*.

DCE controlled not ready indicates that, although the DCE is operational, it is temporarily unable to render service. *DCE controlled not ready* is signaled with a continuous bit stream of alternate binary 0 and binary 1 bits (i.e., 0101 ...) on circuit *R* and the OFF condition on circuit *I*.

DTE ready indicates that the DTE is ready to enter operational phases. *DTE ready* is signaled with continuous binary 1 on circuit *T* and the OFF condition on circuit *C*.

DTE uncontrolled not ready indicates that the DTE is unable to enter operational phases because of an abnormal condition. *DTE uncontrolled not ready* is signaled with continuous binary 0 on circuit *T* and the OFF condition on circuit *C*.

DTE controlled not ready indicates that, although the DTE is operational, the DTE is temporarily unable to enter operational phases. *DTE controlled not ready* is signaled with a continuous bit stream of alternate binary 0 and binary 1 bits (i.e., 0101 ...) on circuit *T* and the OFF condition on circuit *C*.

To ensure proper detection of these signals, X.21 requires that the DTE and DCE be prepared to send these signals for a period of at least 24 bit intervals. Detection of these signals for 16 contiguous bit intervals is required.

The various combinations of three DCE quiescent signals and the three DTE quiescent signals provide for the set of quiescent states of the X.21 interface as shown in Fig. 13. The implementations of X.21 by some networks do not allow all the possible transitions between these states. Therefore, Fig. 13 only shows those transitions that are valid for all networks.

X.21 also contains provisions to ensure proper interpretation of the interface under fault conditions (e.g., power off, disconnection of the

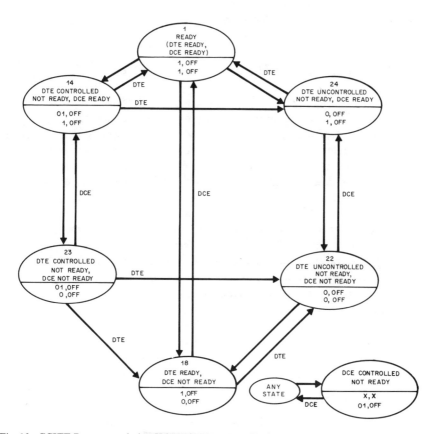

Fig. 13. CCITT Recommendation X.21 quiescent states. Each state is represented by an ellipse wherein the state name and number are indicated, together with the signals on the four interchange circuits that represent that state. Each state transition is represented by an arrow and the equipment responsible for the transition (DTE or DCE) is indicated beside that arrow. Here n is the state number, t is the signal on transmit interchange circuit, c is the signal on control interchange circuit, r is the signal on receive interchange circuit, i is the signal on indication interchange circuit; 0 means steady binary 0 condition, 1 means steady binary 1 condition, 01 means alternate binary 0 and binary 1 conditions, OFF means continuous OFF (binary 1) condition, ON means continuous ON (binary 0) condition, and X means any value.

interface cable, failure of an interchange circuit, and loss of incoming line signal to the DCE). Finally, X.21 defines the interface state for each of the various maintenance test loops.

IV. Integrated Services Digital Network Interfaces

An integrated services digital network (ISDN) is a network that provides end-to-end digital connectivity to support a wide range of services via a small set of multipurpose user-network interfaces. ISDN is an evolution of the telephony network that will include circuit-switched, packet-switched, and nonswitched connections and will support voice, data, and other services. Major elements of an ISDN include 64 kbit/s channels and common channel signaling.

The ISDN user-network reference points are depicted in Fig. 14 [31, 32]. The boxes represent sets of logical functions in ISDN user access arrangements. The R, S, T, and U reference points may correspond to physical interfaces between pieces of equipment. When the NT2 functional group is not present, the S and T reference points coincide. In this section, we will focus on the physical layer aspects of interfaces at reference points S and T.

Two user-network interfaces have been defined and they are applicable at both the S and T reference points. They are the basic interface and the primary rate interface.

A. Basic Interface

The basic interface [33] consists of two 64-kbit/s B channels (information channels) and a 16-kbit/s D channel for signaling and packet data. Both point-to-point (NT to one TE) and point-to-multipoint (NT to multiple TEs via a passive bus wiring arrangement) configurations are supported.

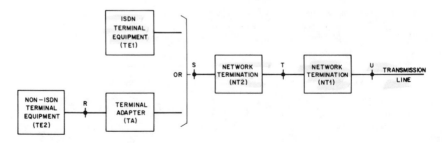

Fig. 14. ISDN user-network reference points. Note: NT1 provides the transmission line termination. Examples of NT2 are PBX's, and terminal controllers.

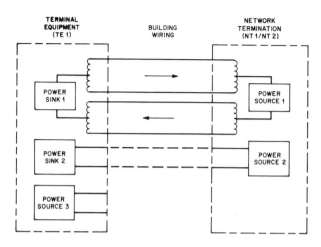

Fig. 15. ISDN basic interface.

1. Procedural

Three major procedures are defined for the basic interface. One procedure provides an orderly means for a number of terminals connected in the point-to-multipoint configuration to gain access to the D-channel. A second procedure, known as activation/deactivation, provides means for TE and NT equipment to be placed in a low power consumption mode when no calls are in progress. The third procedure provides for frame alignment.

2. Functional

The basic interface consists of four mandatory leads and four optional leads as depicted in Fig. 15. The four mandatory leads provide for two pairs, one for each direction of digital signal transmission. The four mandatory leads and the four optional leads may be used to convey power in various optional arrangements. Power source 1 provides for power transfer from NT to TE via the phantom circuit. Power source 2 (or a source located at an intermediate point, e.g., a wiring closet) can provide additional power to the TE. Power source 3 can provide for power transfer in TE to TE interconnections.

The transmit and receive digitally multiplexed signals operate at 192 kbit/s. These 192-kbit/s signals consist of 64 kbit/s for the first B-channel, 64 kbit/s for the second B-channel, 16 kbit/s for the D-channel, and 48 kbit/s for framing and other physical layer functions such as D channel contention resolution and interface activation/deactivation. The frame structure is different for the two directions of transmission.

3. Electrical

The physical medium is a metallic cable. In the point-to-multipoint case, it is considered to be one continuous cable run with jacks for the TEs attached directly to the cable or using stubs less than 1 m in length. A terminating resistor of 100 Ω is used at both ends of the cable on the digital transmit and receive pairs.

The 192-kbit/s bit stream is transmitted using Alternate Space Inversion coding. The nominal pulse amplitude (zero to peak) is 750 mV and the nominal pulse width is 5.21 μs.

4. Mechanical

The connector for the basic interface is specified in ISO 8877 [34]. It is an 8-pin plug and jack arrangement (see Fig. 16) that uses the same modular plug and jack technology that is widely deployed in the United States for telephony.

B. Primary Rate Interface

The primary rate interface consists of one or more information channels (B or H channels) and may include a 64-kbit/s D-channel for signaling or packet data. One D-channel can serve more than one primary rate interface.

Two primary rate interfaces [35] are defined—one is the North American/Japanese 1.544-Mbit/s interface and the other is the European 2.048-Mbit/s interface. Both interfaces provide time slots that operate at 64 kbit/s, making this one of the fundamental building blocks of ISDN.

The 1.544-Mbit/s interface provides 24 time slots of 64 kbit/s (numbered 1 to 24) plus 8 kbit/s for framing and other functions. If the signaling channel is present, it is assigned to time slot 24. The 2.048-Mbit/s interface provides 32 time slots of 64 kbit/s (numbered 0–31). Time slot 0 is used for framing and time slot 16 may be used only to provide the signaling channel.

Both primary rate interfaces can support multiple 64-kbit/s B channels and/or multiple 384-kbit/s H0 channels (6 time slots). In addition, the 1.544-Mbit/s interface can support one 1.536-Mbit/s H11 channel (24 time slots) and the 2.048-Mbit/s interface can support one 1.920-Mbit/s H12 channel (30 time slots) or one H11 channel and other channels.

The primary rate interface operates only in the point-to-point configuration. No activation/deactivation procedures apply since the interface is active at all times. Two pairs, one for each direction, are used for the transmission of digital signals. Work is currently underway in ISO to specify an interface connector.

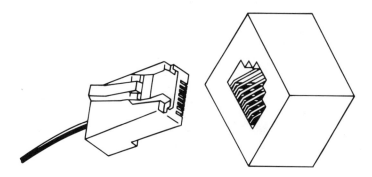

Fig. 16. Connector for ISDN basic interface.

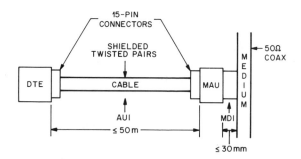

Fig. 17. Physical configuration for CSMA/CD media access method.

C. Support of Existing Interfaces

DTEs with existing interfaces are accommodated on ISDN by means of terminal adapters. CCITT has specified the conversions (including rate adaptation) necessary to support X.25 DTEs, X.21 DTEs, and DTEs designed to interface with V-series modems [36]–[41].

V. Local Area Network Interfaces

This section presents a brief overview of the physical layer and physical medium characteristics of the three types of local area networks (LANs) that have been standardized by the IEEE*:

- IEEE Standard 802.3-1985, carrier sense multiple access with collision detection (CSMA/CD) [42].
- IEEE Standard 802.4-1985, token-passing bus access method [43].
- IEEE Standard 802.5-1985, token ring access method [44].

Owing to space limitations, the following presentation is restricted to contrasting the different media access arrangements and their physical elements.

A. CSMA / CD Media Access Method

The CSMA/CD media access method uses a bus transmission medium and operates as follows. To transmit, a station (DTE) waits until there is a quiet period on the medium (i.e., no other station is transmitting) and then sends the intended message. If, after initiating a transmission, the message collides with that of another station, then each transmitting station intentionally sends a few additional bytes to ensure propagation of the collision. The station then remains silent for a random amount of time before attempting to transmit again.

The initial IEEE Standard 802.3 only covers a 10-Mbit/s baseband coaxial cable bus. The basic configuration is illustrated in Fig. 17. As shown in the figure, two interfaces are defined. One is the medium dependent interface (MDI) and the other is the attachment unit interface (AUI).

The coaxial cable is installed in a building such that it will be within 50 m of each station. Each station is attached to the coaxial cable by means of a medium attachment unit (MAU), an electronic coupling device. In

*Institute of Electrical and Electronics Engineers. IEEE is a professional society. Local area network standards are developed by the Technical Committee on Computer Communications of the IEEE Computer Society.

order not to disturb the transmission line characteristics of the cable significantly, the MAU is located as close to the coaxial cable as possible (≤ 30 mm). Normally, the MAU and the means of attachment to the coaxial cable are one assembly.

The coaxial cable has marks at 2.5-m spacings. A MAU may be connected only at a mark, and no more than 100 MAUs are permitted on a 500-m maximum length coaxial cable segment. Coax segments may be connected together with a repeater set, subject to a number of specified restrictions. The 50-Ω coaxial cable is terminated at each end in a 50-Ω impedance.

The station is connected to the MAU by means of a connectorized cable consisting of individually shielded twisted pairs of wires enclosed by an overall shield. The connectors are 15-pin and employ a slide latching arrangement. The five interchange circuits (each consisting of a twisted pair plus shield) are as follows:

- Data In—carries encoded data to DTE.
- Control In—carries encoded control information to DTE.
- Data Out—carries encoded data to MAU.
- Control Out (optional)—carries encoded control information to MAU.
- Voltage (optional)—provides for power transfer to MAU.

The data circuits are used exclusively for data transfer and the control circuits are used exclusively for control message transfer. The data and control circuits are independently self-clocked, thereby eliminating the need for separate timing circuits. Manchester encoding is used for the data circuits and a simpler encoding mechanism is used for the control circuits. The messages defined for each control circuit are as follows:

Control Out circuit:

- Normal—instructs MAU to be in normal mode.
- MAU Request (optional)—requests that MAU be made available.
- Isolate (optional)—instructs MAU to be in monitor mode.

Control In circuit:

- MAU Available—indicates that MAU is ready to output data.
- MAU Not Available (optional)—indicates that MAU is not ready to output data.
- Signal Quality Error—indicates that MAU has detected an error on input data.

B. Token-Passing Bus Media Access Method

The token bus media access method uses a bus transmission medium and operates as follows. A token controls the right of access to the medium; the station (DTE) that holds the token has momentary control over the medium. Sequential access to the medium is achieved by passing the token from station to station in a logically circular fashion. It is possible to have stations on the medium that can receive frames but cannot initiate a transmission if they will never be sent a token. Steady state operation consists of a data transfer phase and a token transfer phase.

The initial IEEE Standard 802.4 covers three different methods of physical layer signaling:

- Phase Continuous Frequency Shift Keying.
- Phase Coherent Frequency Shift Keying.
- Multilevel Duobinary Amplitude Modulation/Phase Shift Keying.

Three major distinguishing characteristics between these methods are summarized in Fig. 18. The medium in the first method consists of a long unbranched bidirectional trunk cable, which connects to stations by way of tee connectors and very short stubbed drop cables (see Fig. 19a). Extension of the topology to a branched trunk is usually accomplished by way of active regenerative repeaters which are connected to span the branches.

The medium in the second method consists of a CATV-like trunk and drop structure with branching possible. The trunk cable is connected to stations by way of nondirectional passive impedance-matching networks (taps) and drop cables (see Fig. 19b). Extension of the topology or size is accomplished by way of active regenerative repeaters, which are connected in series in the trunk calling.

The third method is designed to operate as conventional bidirectional (by frequency) CATV-like broadband coaxial cable systems (see Fig. 19c). Communication between stations occurs through the transmission of a frequency F1 in the low-band toward the head-end, reception at the

CHARACTERISTIC	PHASE CONTINUOUS FSK	PHASE COHERENT FSK	MULTILEVEL DUOBINARY AM/PSK
TOPOLOGY	OMNIDIRECTIONAL BUS		DUAL CHANNEL (DIRECTIONAL) BUS WITH ACTIVE HEAD-END REPEATER
CABLE CONFIGURATION	LONG UNBRANCHED TRUNK CABLE WITH VERY SHORT STUB DROP CABLES	CATV – LIKE SEMIRIGID TRUNK CABLE AND FLEXIBLE DROP CABLES	
DATA RATE	1 Mbit/s	5 Mbit/s 10 Mbit/s	1 Mbit/s IN 1.5 MHz BANDWIDTH 5 Mbit/s IN 6 MHz BANDWIDTH 10 Mbit/s IN 12 MHz BANDWIDTH

Fig. 18. Characteristics of token-passing bus media access methods.

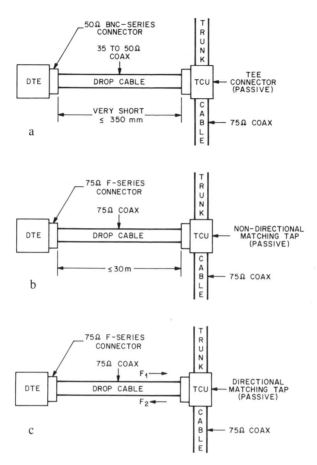

Fig. 19. Physical configurations for token-passing bus media access methods. (a) Single-channel phase-continuous FSK bus. (b) Single-channel phase coherent FSK bus. (c) Broadband bus (multilevel duobinary AM/PSK) [note: either a single (bidirectional cable) or a dual (two unidirectional cables) system may be used].

head-end and remodulation at a high-band frequency F2, and reception by stations. Thus a network really uses a pair of directional channels, the low-frequency (reverse) channel having many transmitters and one receiver, and the high-frequency (forward) channel having one transmitter and many receivers. The head-end remodulator, located at the "root" of the treelike topology structure, serves as a relay station.

C. Token Ring Media Access Method

The token ring media access method has stations (DTEs) serially connected by a transmission medium and operates as follows. A station gains the right to transmit its information onto the medium by capturing the token and transferring information onto the ring. The information circulates from one station to the next with each station regenerating and repeating each bit. The addressed destination station(s) copies the information as it passes. Finally, the station that transmitted the information effectively removes the information from the ring. At the completion of information transfer, the station initiates a new token, which provides other stations the opportunity to gain access to the ring. A token holding timer controls the maximum period a station can use the medium before passing the token.

The initial IEEE Standard 802.5 covers operation at data signaling rates of 1 Mbit/s and 4 Mbit/s. Differential Manchester coding is employed. The particular trunk cable medium (e.g., twisted pair, coaxial cable, optical fiber) is left for further consideration. However, performance specifications are defined at the medium interface connector (MIC, see Fig. 20).

The MIC has four signal contacts plus a ground (shield) contact. It is hermaphroditic in design so that two identical units will mate when oriented 180° with respect to each other (see Fig. 21). When disconnected, each connector provides an automatic looping capability by shorting pins together (i.e., transmit pair to receive pair).

A mechanism to insert or bypass a station is contained in the trunk coupling unit (TCU). A station controls the mechanism via a dc phantom voltage on the signal pairs. The presence of the dc voltage effects a switching action in the TCU which inserts the station in the ring. Absence of the dc voltage will cause the TCU to bypass the station and loop the signal from the station back to the station.

It should be noted that IEEE Standard 802.5 specifies that each octet of the information field is transmitted most significant bit (MSB) first. This convention is reversed from that used in IEEE Standard 802.3 and IEEE Standard 802.4, which specify that octets are transmitted least significant bit (LSB) first. The interconnection of an LSB network and an MSB

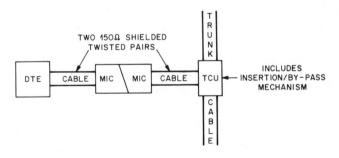

Fig. 20. Physical configuration for token ring media access method.

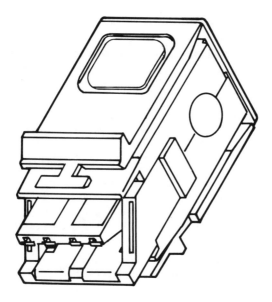

Fig. 21. Connector for token ring media access method.

network requires bit reordering to be performed in the gateway between the two networks.

VI. Recent Activities and Future Directions

The preceding sections have reviewed the basic characteristics of the physical layer and have described a number of the standardized interfaces. There is also a large body of recent and ongoing standardization activity that applies to the physical layer. Some highlights of this effort are summarized in this section.

A. Traditional Interfaces

The RS-449 and X.21 interfaces described in Section III have not been widely deployed in the USA. Consequently, X.21 has not been adopted as a U.S. standard and RS-449 is now being recommended only for rates above 20 kbit/s.* The major problem with RS-449 is that the larger 37-pin connector is contrary to the increased miniaturization of DTEs and DCEs. Since RS-449 is no longer "the standard toward which industry should

---

*CCITT deleted the RS-449 equivalent option from their 1984 voiceband modem recommendations [9] while retaining it in their V.36 [9] and V.37 [9] wideband modem recommendations. ISO is making a similar modification to connector standard ISO 4902 [19].

evolve," TIA Technical Committee TR 30 is pursuing work on a number of projects. A sampling is given below.

TR 30 has revised RS-232-C as EIA-232-D [45]. One objective of this revision is to provide increased consistency with the equivalent set of international standards. The major changes from RS-232-C are as follows:

- Specification of the 25-pin interface connector.
- Inclusion of the local loopback, remote loopback, and test mode interchange circuits.
- Removal of the protective ground interchange circuit.
- Provision for shielded interface cable.
- A few terminology changes.

TR 30 has also developed a new high-speed interface (20 kbit/s to 2 Mbit/s) that is based on the 25-pin connector of RS-232-C but uses the RS-422-A and RS-423-A electrical characteristics employed by RS-449. It is EIA-530 [46]. It provides many of the advantages of RS-449 but without the larger connector. This is accomplished by using all 10 Category I interchange circuits but only a few of the 20 Category II interchange circuits in RS-449 (see Fig. 22). One major consequence is that the interface is restricted in its range of applications.

TR 30 is developing a small connector RS-232-C-like interface, which will be known as EIA-561. This asynchronous serial interface would use only a very few of the RS-232-C interchange circuits with the 8-pin

CIRCUIT TYPE	INTERCHANGE CIRCUIT	EQUIVALENT RS-449 CIRCUIT	ELECTRICAL CHARACTERISTICS
GROUND	AB SIGNAL GROUND	SG	–
DATA	BA TRANSMITTED DATA	SD	
	BB RECEIVED DATA	RD	
TIMING	DA TRANSMIT SIGNAL ELEMENT TIMING (DTE SOURCE)	TT	
	DB TRANSMIT SIGNAL ELEMENT TIMING (DCE SOURCE)	ST	
	DD RECEIVER SIGNAL ELEMENT TIMING (DCE SOURCE)	RT	RS-422-A
CONTROL	CA REQUEST TO SEND	RS	
	CB CLEAR TO SEND	CS	
	CC DCE READY	DM	
	CD DTE READY	TR	
	CF RECEIVED LINE SIGNAL DETECTOR	RR	
	LL LOCAL LOOPBACK	LL	
	RL REMOTE LOOPBACK	RL	RS-423-A
	TM TEST MODE	TM	

Fig. 22. EIA-530 interchange circuits.

connector that has been standardized for the ISDN basic interface (see Fig. 16). A new 5-V version of the EIA-232-D (CCITT V.28) electrical characteristics is being developed for this interface and will be known as EIA-562.

It is well recognized that significant benefits can be achieved through the serial transfer of automatic calling information across the transmitted data and received data interchange circuits of RS-232-C rather than using a separate automatic calling unit interface (RS-366-A). CCITT Recommendation V.25 bis [9] and several "de facto standards" employ this concept. TR 30 is presently developing an asynchronous standard and a synchronous standard to cover this arrangement. Consideration is also being given to extension of this concept to include additional functions that can be conveyed prior to data transfer (e.g., setting of options).

TR 30 is also investigating development of a modem interface standard based on the physical layer of the ISDN basic interface. The objective is to specify a DTE/DCE interface so that a DTE can be connected either to an ISDN basic interface or to a modem without requiring an adapter.

B. Integrated Services Digital Network Interfaces

A major ISDN activity that has recently been completed in Committee T1* is an ANSI standard for layer 1 of the basic interface at the U reference point [47]. This is a two-wire digital transmission interface that employs echo canceling to achieve duplex transmission over considerable distances (e.g., customer premises to a telephone company central office).

Committee T1 has also completed an ANSI standard for layer 1 of the basic interface at reference points S and T [48]. Work is underway in Committee T1 on an ANSI standard for layer 1 of the primary rate interface at reference points S, T, and U.

C. Local Area Network Interfaces

The local area network activity within the IEEE 802 Committee has developed additional physical arrangements for inclusion in the existing set of standards.

For example, five additions have been approved for CSMA/CD operation [49], [50]. They are as follows:

- 802.3a, Medium attachment unit and baseband medium specifications, type 10BASE2 (new Section 10).
- 802.3b, Broadband medium attachment unit and broadband medium specifications, type 10BROAD36 (new Section 11).

---

*ANSI Accredited Standards Committee T1 (Telecommunications). Within T1, Technical Subcommittee T1E1 is responsible for work on the physical layer of ISDN. American National Standards resulting from the work of Committee T1 are contained in the T1.-series.

- 802.3c, Repeater unit for 10 Mb/s baseband networks (revised Section 9).
- 802.3d, Medium attachment unit and baseband medium specification for a vendor independent fiber optic inter-repeater link (new section 9.9).
- 802.3e, Physical signaling, medium attachment, and baseband medium specifications, type 1BASE5 (new Section 12).

The 1BASE5 addition, also known as StarLAN, is quite different from the other LANs in that it is a low-cost 1-Mbit/s network that uses twisted pair wiring. Stations (DTEs) are connected to shared hubs and hubs are connected to other hubs (up to five levels of hubs can be cascaded) by point-to-point wiring resulting in a star topology network. Figure 23 illustrates a simple two-level configuration. Signals from a station are regenerated/retimed by each hub (or replaced by a hub generated unique collision presence signal if signals are concurrently present from two or more lower level inputs) and propagated upward to the single header hub. The signal is then broadcast downward to all stations with each intermediate hub regenerating/retiming the signal. Physical connections are made using the same 8-pin plugs and jacks that have been standardized for the ISDN basic interface (see Fig. 16).

Ongoing work in IEEE 802.3 includes a task force which is preparing a 10-Mbit/s version of StarLAN operating over twisted pair wiring (currently known as 10BASE-T [51]) and a study group (known as FOSTAR) exploring an all fiber optic LAN.

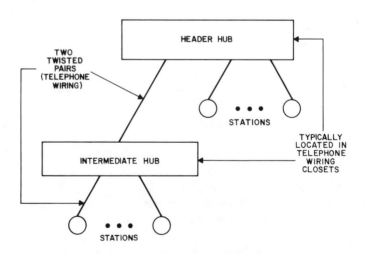

Fig. 23. Example "StarLAN" configuration.

Another area of standardization is the integrated voice/data LAN activity in IEEE 802.9 [52]. This work is based on the synthesis of two technologies—LANs and ISDNs. The current assumption is that each integrated voice/data workstation will be connected to an access unit via a separate IEEE 802.9 physical interface over unshielded twisted pair wiring. The access unit, in turn, will interconnect to a backbone LAN (e.g., IEEE 802.3, IEEE 802.4, IEEE 802.5, IEEE 802.6, FDDI I, or FDDI II) and/or to an ISDN [e.g., private branch exchange (PBX) or central office (CO)].

D. Higher Bandwidth Interfaces

Higher bandwidth data interface standards are also being developed. This includes the fiber distributed data interface (FDDI), metropolitan area networks (MANs), and broadband ISDN (B-ISDN). All of these activities are based on the use of fiber optic cable. FDDI [53]–[56] provides 100 Mbit/s general purpose interconnection among computers and peripheral equipment over 62.5 μm multimode fiber up to 200 Km total path length. FDDI uses the concepts employed in IEEE 802.5 token ring operation, adjusted to accommodate the higher speed. A second version, known as FDDI II, is also under development. It will accommodate TDM voice transmission as well as data.

The work on metropolitan area networks (MANs) in IEEE 802.6 [57] is now focused on standardizing a distributed queue dual bus (DQDB) arrangement operating at approximately 150 Mbit/s over optical fiber. Both buses are operated synchronously with fixed length slots dedicated to either isochronous or nonisochronous data. The physical layer specification will allow the MAN to incorporate public network high-capacity transport facilities such as the synchronous optical network (SONET) and broadband ISDN.

Broadband ISDN [58] is being designed to support a wide range of voice, video, and data applications. Both switched and nonswitched connections are supported as well as circuit-mode and packet-mode services. B-ISDN user-network interfaces will be standardized at two bit rates: approximately 150 Mbit/s and approximately 600 Mbit/s. In addition to supporting B, H0, H11, and H12 channels, B-ISDN will support an H21 broadband channel of 32.768 Mbit/s, an H22 broadband channel in the range of 43–45 Mbit/s, and an H4 broadband channel in the range of 132–138.240 Mbit/s.

References

[1] CCITT Recommendation X.25, "Interface between data terminal equipment (DTE) and data circuit-terminating equipment (DCE) for terminals operating in the packet mode on public data networks," fascicle VIII.2 of *CCITT Orange Book*, 1976.

[2]   CCITT Recommendation X.200, "Reference model of open systems interconnection for CCITT applications," fascicle VIII.5 of *CCITT Red Book*, 1984 (updated version to be published in *CCITT Blue Book*).

[3]   ISO International Standard 7498, "Information processing systems—Open systems interconnection—Basic reference model," 1984.

[4]   CCITT Recommendation X.211, "Physical service definition of open systems interconnection for CCITT applications" (to be published in *CCITT Blue Book*).

[5]   ISO International Standard 10022, "Information processing systems—Open systems interconnections—Physical service definition" (awaiting publication).

[6]   EIA Standard RS-232-C, "Interface between data terminal equipment and data communication equipment employing serial binary data interchange," August, 1969.

[7]   EIA Standard RS-449, "General purpose 37-position and 9-position interface for data terminal equipment and data circuit-terminating equipment employing serial binary data interchange," November, 1977, and Addendum 1 to RS-449, February, 1980.

[8]   CCITT X-Series Recommendations, fascicles VIII.2-VIII.7 of *CCITT Red Book*, 1984 (updated versions to be published in *CCITT Blue Book*).

[9]   CCITT V-Series Recommendations, fascicle VIII.1 of *CCITT Red Book*, 1984 (updated versions to be published in *CCITT Blue Book*).

[10] EIA Standard RS-366-A, "Interface between data terminal equipment and automatic calling equipment for data communication," March, 1979.

[11] EIA Standard RS-496, "Interface between data circuit-terminating equipment (DCE) and the public switched telephone network (PSTN)," May, 1984 (revision currently out for ballot).

[12] ANSI X3.1—1987, "Data transmission—Synchronous signaling rates."

[13] EIA Standard RS-334-A, "Signal quality at interface between data terminal equipment and synchronous data circuit-terminating equipment for serial data transmission," August 1981, and Addendum 1 to RS-334-A, May, 1983.

[14] EIA Standard RS-363, "Standard for specifying signal quality for transmitting and receiving data processing terminal equipments using serial data transmission at the interface with non-synchronous data communication equipment," May, 1969.

[15] EIA Standard RS-404-A, "Standard for start–stop signal quality for non-synchronous data terminal equipment," January, 1986.

[16] ISO International Standard 7480, "Information processing—Start–stop transmission signal quality at DTE/DCE interfaces," 1984.

[17] ISO International Standard 2110, "Data communication—25-pin DTE/DCE interface connector and pin assignments," 1980 (revision awaiting publication).

[18] ISO International Standard 2593, "Data communication—34-pin DTE/DCE interface connector and pin assignments," 1984 (revision awaiting publication).

[19] ISO International Standard 4902, "Data communication—37-pin and 9-pin DTE/DCE interface connectors and pin assignments," 1980 (revision awaiting publication).

[20] ISO International Standard 4903, "Data communication—15-pin DTE/DCE interface connector and pin assignments," 1980 (revision awaiting publication).

[21] EIA Standard RS-422-A, "Electrical characteristics of balanced voltage digital interface circuits," December, 1978.

[22] EIA Standard RS-423-A, "Electrical characteristics of unbalanced voltage digital interface circuits," December, 1978.

[23] FED-STD 1030A, "Electrical characteristics of unbalanced voltage digital interface circuits," January, 1980.

[24] MIL-STD 188-144, "Electrical characteristics of digital interface circuits," March 1976.

[25] FED-STD 1020A, "Electrical characteristics of balanced voltage interface circuits," January, 1980.

[26] EIA Standard RS-410, "Standard for the electrical characteristics of Class A closure interchange circuits," April, 1974.

[27] ISO International Standard 9067, "Information processing systems—Data communication—Automatic fault isolation procedures using test loops," 1987.

[28] EIA Industrial Electronics Bulletin No. 9, "Application notes for EIA Standard RS-232-C," May, 1971.

[29] EIA Industrial Electronics Bulletin No. 12, "Application notes on interconnection between interface circuits using RS-449 and RS-232-C," November, 1977.

[30] ISO Technical Report 7477, "Data communication—Arrangements for DTE to DTE physical connection using V.24 and X.24 interchange circuits," 1985.

[31] CCITT Recommendation I.411, "ISDN user-network interfaces—Reference configurations," fascicle III.5 of *CCITT Red Book*, 1984 (to be republished in *CCITT Blue Book*).

[32] CCITT Recommendation I.412, "ISDN user-network interfaces—Interface structures and access capabilities," fascicle III.5 of *CCITT Red Book*, 1984 (updated version to be published in *CCITT Blue Book*).

[33] CCITT Recommendation I.430, "Basic user–network interface—Layer 1 specification," fascicle III.5 of *CCITT Red Book*, 1984 (updated version to be published in *CCITT Blue Book*).

[34] ISO International Standard 8877, "Information processing systems—Interface connector and contact assignments for ISDN basic access interface located at reference points S and T," 1987.

[35] CCITT Recommendation I.431, "Primary rate user-network interface—Layer 1 specification," facicle III.5 of *CCITT Red Book*, 1984 (updated version to be published in *CCITT Blue Book*).

[36] CCITT Recommendation I.460, "Multiplexing, rate adaptation and support of existing interfaces," fascicle III.5 of *CCITT Red Book*, 1984 (updated version to be published in *CCITT Blue Book*).

[37] CCITT Recommendation I.461 (X.30), "Support of X.21 and X.21 bis based data terminal equipments (DTEs) by an Integrated Services Digital Network (ISDN)," fascicle III.5 of *CCITT Red Book*, 1984 (updated version to be published in *CCITT Blue Book*).

[38] CCITT Recommendation I.462 (X.31), "Support of packet mode terminal equipment by an ISDN," fascicle III.5 of *CCITT Red Book*, 1984 (updated version to be published in *CCITT Blue Book*).

[39] CCITT Recommendation I.463 (V.110), "Support of data terminal equipments (DTEs) with V-series type interfaces by an Integrated Services Digital Network (ISDN)," fascicle III.5 of *CCITT Red Book*, 1984 (updated version to be published in *CCITT Blue Book*).

[40] CCITT Recommendation I.464, "Multiplexing, rate adaptation and support of existing interfaces for restricted 64 kbit/s transfer capability," fascicle III.5 of *CCITT Red Book*, 1984 (updated version to be published in *CCITT Blue Book*).

[41] CCITT Recommendation I.465 (V.120), "Support by an ISDN of data terminal equipment with V-series interfaces with provision for statistical multiplexing" (to be published in *CCITT Blue Book*).

[42] ANSI/IEEE Std. 802.3—1985, "Carrier sense multiple access with collision detection (CSMA/CD) access method and physical layer specifications" (see also ISO 8802-3).

[43] ANSI/IEEE Std. 802.4—1985, "Token-passing bus access method and physical layer specifications" (see also ISO 8802-4).

[44] ANSI/IEEE Std. 802.5—1985, "Token ring access method and physical layer specification" (see also ISO 8802-5).

[45] EIA-232-D, "Interface between data terminal equipment and data circuit-terminating equipment employing serial binary data interchange," January, 1987.

[46] EIA-530, "High speed 25-position interface for data terminal equipment and data circuit-terminating equipment," March, 1987.

[47] ANSI T1.601—1988, "Integrated services digital network (ISDN)—Basic access interface for use on metallic loops for application on the network side of the NT (Layer 1 specification)."

[48] ANSI T1.605—1989, "Integrated services digital network (ISDN)—Basic access interface for S and T reference points (Layer 1 specification)."

[49] ANSI/IEEE Std. 802.3a, b, c, and e—1988, "Supplements to carrier sense multiple access with collision detection (CSMA/CD) access method and physical layer specifications."

[50] ANSI/IEEE Std. 802.3d, "Supplement to carrier sense multiple access with collision detection (CSMA/CD) access method and physical layer specifications" (awaiting publication).

[51] Twisted pair medium attachment unit and baseband medium, type 10BASE-T, Draft E rev., 1989.

[52] Draft of proposed IEEE Standard 802.9, "Integrated voice/data LAN interface specifications," Draft 1.3, March, 1989.

[53] ANSI X3.148, "Fiber Distributed Data Interface (FDDI)—Physical Layer Protocol" (awaiting publication).

[54] ISO International Standard 9314-1, "Information processing systems—Fibre Distributed Data Interface (FDDI)—Part 1: Physical Layer Protocol (PHY)" (awaiting publication).

[55] ANSI X3.166, "Fiber Distributed Data Interface (FDDI), Physical Layer Medium Dependent (PMD)" (under development).

[56] ISO Draft International Standard 9314-3, "Information processing systems—Fibre Distributed Data Interface (FDDI)—Part 3: Physical Layer Medium Dependent (PMD)," October, 1988.

[57] IEEE standard 802.6, "Metropolitan Area Network (MAN) distributed queue dual bus media access control," Draft D7, May 7, 1989.

[58] CCITT Recommendation I.121, "Broadband aspects of ISDN" (to be published in *CCITT Blue Book*).

III

Link Control Layer

4

Character-Oriented Link Control

James W. Conard

1. Overview

A data link control protocol is a set of very specific rules governing the interchange of data over an interconnecting communication link between business machines.

The business machines may be computers, terminals, message or packet switches, concentrators, or any of a broad range of data terminal equipment in any mix. The interconnecting communication links may be assembled in any of several arrangements and may be comprised of private or common carrier multipoint, point-to-point, switched (dial), or non-switched (dedicated) facilities using cable, land line, microwave, or satellite channels. The data being interchanged can be represented in many forms and can serve batched, conversational, processor to processor, inquiry/response, or other typical applications.

The link control protocol rules typically define initialization of an already established physical link, control of normal data interchange, termination of the link at the end of the transaction, and, perhaps most important from the point of view of the user, techniques to control recovery from abnormal conditions such as invalid or no response, loss of synchronization, and faults resulting from anomalies in the communication link.

Link control protocols have traditionally been character-oriented. They utilize, either singularly or in sequence, defined character structures from a given code set to convey the information necessary to frame the data and supervise its interchange. Protocols which use defined character structures for supervisory control are also known as byte-oriented protocols.

JAMES W. CONARD • Conard Associates, Costa Mesa, California 92626.

Many variations of the basic character-oriented protocol are possible and form the subject of this chapter. A major subset, perhaps a separate class, uses combinations of characters and byte-length fields to supervise the link. These are known as byte-count or count-framed protocols. A totally different class of link control protocols uses positionally located control fields rather than code set combinations for supervisory control. These are known as bit-oriented link control protocols and are dealt with in Chapter 5.

Strictly speaking, the term "link control" excludes other levels within the commonly recognized standard layer model of communication control. Ideally, the link control level should be independent of the other levels, should be distinct as to functions performed and services offered, and should have clearly delineated interfaces with the physical level protocol below it and the network level protocol above it. Character-oriented protocols, having evolved with rapidly changing communication requirements, have certain characteristics such as intermixed message, device, and link control, which tend to blur the interface between logically independent layers. These characteristics led to the development of the many variations of character-oriented protocols and ultimately to the now emerging bit-oriented protocols.

It must be remembered that character-oriented protocols as control mechanisms are concerned solely with the transfer of data over an established communications link. They are not concerned with the physical processes necessary to establish a link at Level 1. Nor are they network protocols. They do not control the flow of information between end points of a multinodal network. They can, however, be applied between nodes or between a node and an end point user.

Character-oriented protocols are suitable for two-way alternate and, less often, for two-way simultaneous operation using a variety of data link configurations, including full and half-duplex, multipoint, switched, and dedicated. The two facility configurations most commonly encountered in association with character-oriented link control are illustrated in Fig. 1. A point-to-point facility is one which interconnects two and only two stations. Point-to-point facilities may be either nonswitched, sometimes referred to as private line or dedicated, or they may be switched. The difference between switched and nonswitched is one of facility acquisition. In the switched case the facility must be acquired by the lower physical level protocol prior to the transfer of data and released at the end of the transfer. Nonswitched facilities are dedicated and usable on demand.

A multipoint facility very common for these applications consists of a single master and two or more remote stations. Transmissions from the master are received by all remotes. Transmissions from the remotes are received only by the master. This multipoint arrangement normally requires four-wire channels.

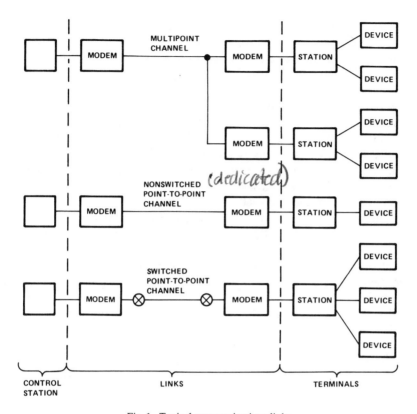

Fig. 1. Typical communications links.

Many special and hybrid combinations of interconnect arrangements are possible and often encountered. The system designer must be aware of the type and characteristics of the interconnecting link since these characteristics often directly influence the choice of protocol and its operational procedures.

II. Protocol Perspective

Data link control protocols are as old as data communications. Over the years these protocols have been evolving typically to fulfill the requirements of a particular application. Early systems, using Baudot code, had no inherent link control capability. They relied totally on sequences of data characters to implement supervisory functions. The advent of other character sets led to protocols using controls derived from these sets. Each manufacturer developed protocols reflecting the needs of its product line and usually optimized for a specific implementation. Many users groups also developed protocols to meet their unique requirements. All of these various protocols were character-oriented in approach and generally incompatible with each other.

Standards organizations in the United States and abroad recognized the problem and struggled to resolve the incompatibilities. The American National Standards Institute (ANSI) and the International Standards Organization (ISO) were especially active in this effort. For lack of standardization, the protocols developed by the larger dominant manufacturers tended to fill the vacuum by becoming, in effect, de facto standards. This has certainly been the case with IBM's BSC (Binary Synchronous Communication) developed in the late 1960s.

The standards organizations finally reached agreement with the publication in 1971 of ANSI's X3.28 on the use of ASCII control characters for link control and of ISO's IS1745. These activities, among standards bodies, manufacturers, and users groups, continue to produce revised, updated, and even new character-oriented protocols to keep pace with evolving technology and requirements. At the present time the most widely used and familiar link protocols are those briefly described in the following paragraphs.

ANSI X3.28 [1]: This protocol standard carries the rather lengthy title of "Procedures for Use of the Communication Control Characters of American National Standard Code for Information Interchange in Specified Data Communication Links." It was first promulgated in 1971 and updated in 1976. This standard specifies a group of control protocols, called categories, each designed to meet the requirements of a specific combination of link configuration and message transfer application. These procedures are based on the use of ten communication control characters provided in the ASCII code set.

ISO IS1745 [2]: This internationally accepted protocol is titled "Basic Mode Control Procedures for Data Communication Systems." It, too, is based on the use of ten communications control characters to supervise a data link. Link control is organized into phases of connection, link establishment, information transfer, termination, and disconnect. The standard defines formats of messages and supervisory sequences for each of these phases. It is designed for two-way alternate operation. The current version of this standard was released in 1975.

ECMA-16 [3]: The European Computer Manufacturers Association (ECMA) also standardized a character-oriented protocol: "Basic Mode Control Procedures for Data Communication Systems Using the ECMA 7-bit Code." This standard closely resembles IS1745 in definition of formats, supervisory sequences, and phases.

IBM BSC [4]: The Binary Synchronous Communication (BSC) protocol is the most widely implemented of the protocols developed by the various manufacturers. It is character-oriented and designed for two-way alternate operation over point-to-point or multipoint links. It can be implemented using control characters from any of three code sets: EBCDIC, ASCII, or Six-bit Transcode. BSC utilizes nine of the "standard" communication control characters and supplements these with six additional two-character sequences to provide additional link control functions.

IATA SLC [5]: Less well known but very widespread in use is the Synchronous Link Control standardized by the International Air Transport Association. This protocol supervises two-way simultaneous data transfer over full duplex links. It utilizes a combination of communication control characters and character length fields to form control blocks for link supervision. The control blocks permit identification of message blocks, sequence numbers, priorities, and other parameters.

DEC DDCMP [6]: Digital Equipment Corporation's "Digital Data Communications Message Protocol" (DDCMP) is, perhaps, the best known of the byte-count-oriented protocols. It combines communication control characters and control fields to frame information blocks with supervisory controls. Complete transparency to the information is achieved through the use of a byte-count mechanism rather than the more common escape sequences. DDCMP constitutes the data link control level of Digital Equipment Corporation's DNA architecture.

The protocols listed above represent only a very small sample of the character-oriented protocols in use today. Many variations have been developed and implemented. Often a protocol will be developed and optimized for a very specific parameter such as response time, efficiency over a specific facility, or throughput. While satisfying a particular requirement, such specialization usually limits widespread application.

III. Protocol Characteristics

Character-oriented protocols, despite the wide variety of application and implementation parameters, generally share a common set of characteristics. They are much alike in basic structure, functions performed, phases of operation, code set utilization, and control character definition. The various protocols differ in how these fundamental characteristics are applied to a particular situation. Before examining the details of protocol operation it is apropos to review these basic characteristics.

A. Functions

The fundamental task of any link control protocol involves the interchange of information between senders and receivers over a given intercon-

necting link. The integrity of the information being transferred is a
paramount consideration. Garbled or lost information is of no value to the
user and can often be disastrous. If the data interchange always took place
between two stations over an error-free point-to-point facility only a
rudimentary data link control procedure would be necessary. The data
communications environment is, however, far from ideal and exhibits
characteristics which must be accommodated by the protocol.

In addition to the requirements imposed by the link itself, the protocol
must contend with requirements which derive from the application, the
nature of the information, i.e., conversational, batch, inquiry/response, the
need for transparent operation, recovery techniques, flow control, and
others. To accomplish this task a basic set of link protocol functions have
evolved. These are described next.

Frame Control delimits the beginning and end of transmission blocks
by the use of delimiting characters and a character count in byte-count-oriented protocols. This is necessary since extremely long blocks of information are unlikely to survive transmission through the electrically noisy
medium without error. The block mechanism provides a method of implementing and controlling a block length chosen as most likely to survive and
thus keep retransmission to a minimum. The mechanism also provides the
ability to identify when information should, but may not, be present, and
finally, provides a convenient method of signaling when the checking
mechanism is to be active. Frame control characters are also commonly
used to acquire, maintain, and if necessary, reestablish synchronization
between sender and receiver. This is absolutely essential if the receiver is to
decode the information correctly. Note that bit synchronization, which is
no less critical, is a physical level function.

Error Control provides for the detection of errors, the acknowledgment
of correctly received blocks and messages, and the requests for retransmission of incorrectly received messages. The most commonly used error
detection techniques are vertical and longitudinal parity checks, and cyclic
redundancy checks. These are described later in this chapter. Another
method of error control is sequence control. Sequence control mechanizes a
method of numbering blocks and messages to facilitate proper retransmission and to eliminate or at least identify lost or duplicate messages.

Initialization Control governs the establishment of an active data link
over a communication facility that has been idle. It usually involves an
exchange of sequences identifying a particular sender or receiver among the
many present on a multipoint facility or among the almost infinite number
connectable through a switched facility. Polling and calling are typical of
initialization control.

Flow Control sequences regulate the flow of information across the
data link. They permit a receiver to exercise some control of the amount
and rate of information flowing into his system so as to avoid overwhelm-

ing his capacity to accept and process the incoming data. At link level, flow control is limited to the ability to accept or not accept information transfers.

A further discussion of flow control at the data link level and other flow control mechanisms at higher levels is given in Chapter 10.

Link Management sequences are used to supervise the links, by controlling transmission direction, establishing and terminating logical connections, and identifying which station is going to send and which is going to receive. Link management responsibility usually resides in a master or control station.

Transparency is a characteristic of some, but not all, character-oriented protocols. It allows the link control to be totally independent of the pattern or code structure of the information being transmitted. A transparent link control is able to transfer machine language data streams without the information interfering with link control functions. Character-oriented link protocols require escape or count mechanisms to implement transparent operation.

Abnormal Recovery controls supervise action to be taken to recover from abnormal occurrences such as illegal sequence, cessation of block flow, loss of responses, and other protocol defined exception conditions. Timeouts are a common method of detecting such conditions.

How the functions are implemented is the basis for the more detailed review of protocol operation in a later section.

B. Code Sets and Control Characters

Character-oriented protocols make use of defined characters from a given code set to execute communications supervisory functions. The most common code set in use is the American National Standard Code for Information Interchange (ASCII) which is defined in ANSI X3.4-1976 and reproduced here as Fig. 2. This code set is basically identical to the CCITT Alphabet 5 and the ISO Standard 646.

Of interest to the communicator are the ten characters of these code sets which are designated as communication control characters. The primary functions of these characters will be defined next, but first it will be used to clarify the terms "message" and "block." A message is an ordered sequence of characters arranged to convey information from originator to user. A message may be contained in one or more blocks. A block is a group of characters arranged for technical or logical reasons to be transmitted as a unit. A block may contain an entire message or part of a message.

SOH (Start of Heading). A control character which identifies the beginning of a sequence of characters which constitutes the heading of a message. The sequence usually contains addressing and routing information.

b7 b6 b5 →				0 0 0	0 0 1	0 1 0	0 1 1	1 0 0	1 0 1	1 1 0	1 1 1
b4	b3	b2	b1 COLUMN / ROW	0	1	2	3	4	5	6	7
0	0	0	0	0	NUL	DLE	SP	0	@	P	`	p
0	0	0	1	1	SOH	DC1	!	1	A	Q	a	q
0	0	1	0	2	STX	DC2	"	2	B	R	b	r
0	0	1	1	3	ETX	DC3	#	3	C	S	c	s
0	1	0	0	4	EOT	DC4	$	4	D	T	d	t
0	1	0	1	5	ENQ	NAK	%	5	E	U	e	u
0	1	1	0	6	ACK	SYN	&	6	F	V	f	v
0	1	1	1	7	BEL	ETB	'	7	G	W	g	w
1	0	0	0	8	BS	CAN	(	8	H	X	h	x
1	0	0	1	9	HT	EM	)	9	I	Y	i	y
1	0	1	0	10	LF	SUB	*	:	J	Z	j	z
1	0	1	1	11	VT	ESC	+	;	K	[	k	{
1	1	0	0	12	FF	FS	,	<	L	\	l	|
1	1	0	1	13	CR	GS	−	=	M	]	m	}
1	1	1	0	14	SO	RS	.	>	N	^	n	~
1	1	1	1	15	SI	US	/	?	O	___	o	DEL

Fig. 2. ASCII Code Set: Communications control characters are outlined.

STX (Start of Text). A character delimiting that part of a message which constitutes the text. An STX is often used to terminate the header which began with SOH.

ETX (End of Text). A control character used to delimit the end of a series of characters constituting the text of a message.

EOT (End of Transmission). This control character signifies the end of a transmission which may have contained one or more messages. It usually implies relinquishment of the data link.

ENQ (Enquiry). A communications control character used to solicit a response from another station. It may be used as a status request or as a request for identification, or both.

ACK (Acknowledgment). A control character which represents an affirmative response to a sender. It acknowledges error-free reception of a block or segment of a message.

DLE (Data Link Escape). A control character which changes the meaning of a limited set of contiguous following characters. It is used to provide supplementary control the most common of which is transparent operation.

NAK (Negative Acknowledgment). This character represents a negative response from the receiver to the sender. It indicates that a block of information has been received with errors and must be retransmitted.

SYN (Synchronous Idle). A communications control character used to establish and to maintain character synchronization between the sender and the receiver. It is also often used as a transmission idle in the absence of any data.

ETB (End of Transmission Block). This character signifies the end of block of data for communication control purposes.

These ten control characters are, in some protocols, combined with other characters to form a sequence for additional control purposes. For example DLE is combined with STX to indicate the beginning of a transparent data sequence which would end with the sequence DLE ETX.

Another code set commonly encountered is the Extended Binary Coded Decimal Interchange Code (EBCDIC). This code set also contains the communication control characters defined above. See Fig. 3.

		Most Significant Bits (Bit 8 Transmitted Last)															
		bit Positions 8,7,6,5)															
Bit Positions 4,3,2,1	CR	0000	0001	0010	0011	0100	0101	0110	0111	1000	1001	1010	1011	1100	1101	1110	1111
		0	1	2	3	4	5	6	7	8	9	10	11	12	13	14	15
0000	0	NUL	DLE	DS		SP	&	−						{	}	\	0
0001	1	SOH	DC1	SOS						a	j	~		A	J		1
0010	2	STX	DC2	FS	SYN					b	k	s		B	K	S	2
0011	3	ETX	DC3							c	l	t		C	L	T	3
0100	4	PF	RES	BYP	PN					d	m	u		D	M	U	4
0101	5	HT	NL	LF	RS					e	n	v		E	N	V	5
0110	6	LC	BS	ETB	UC					f	o	w		F	O	W	6
0111	7	DEL	IL	ESC	EOT					g	p	x		G	P	X	7
1000	8		CAN							h	q	y		H	Q	Y	8
1001	9	RLF	EM						\	i	r	z		I	R	Z	9
1010	10	SMM	CC	SM		¢	!	¦	:								
1011	11	VT				.	$	,	#								
1100	12	FF	IFS		DC4	<	*	%	@								
1101	13	CR	IGS	ENQ	NAK	(	)	_	'								
1110	14	SO	IRS	ACK		+	;	>	=								
1111	15	SI	IUS	BEL	SUB	|	¬	?	"								

Least Significant Bits (Bit 1 Transmitted First)

Fig. 3. EBCDIC code set.

C. Transmission Error Control

The three most common error detection methods used by character-oriented protocols are the Vertical Redundancy Check (VRC), the Longitudinal Redundancy Check (LRC), and the Cyclic Redundancy Check (CRC). VRC and LRC are often combined.

VRC is simply a parity scheme in which a bit is appended to the bits which comprise the character. The value of the appended bit is calculated to provide either an "odd" number of 1 bits in the character or an "even" number of 1 bits in the character. The transmitter calculates and appends the parity bit. The receiver also calculates the correct parity bit and compares it with the bit appended to the incoming character. Failure to compare indicates that a transmission error has occurred. VRC schemes can only detect an odd number of bit errors in a character.

LRC is identical in implementation to VRC except that it is computed on a sequence of successive characters. The result is transmitted as an extra check character called an LRC character. LRC schemes are vulnerable to double-bit errors in the row of characters.

Even when VRC and LRC techniques are combined there are many possibilities for errors to occur in such a way as to be undetected by the error detection technique. This shortcoming led to the development of the much more powerful detection technique known as Cyclic Redundance Check (CRC).

A complete treatment of the theory of cyclic codes is beyond the scope of this chapter. Very simply stated, however, a cyclic redundancy code is one which makes use of the mathematical properties of the block of data being transmitted. Any sequence of bits represents the coefficients of a polynomial. If the polynomial representing the message is divided by another polynomial which represents the cyclic generator polynomial the result will be a remainder which can be appended to the message as check bits and transmitted as part of the block. An identical process performed by the receiver on the incoming block should result in the same remainder. If it does not, a transmission error has occurred. Generator polynomials are very carefully chosen to be immune to the particular error properties of the transmission medium. The most commonly used CRCs in character-oriented protocols are the CRC-16 and the CCITT V.41 polynomial. CRC-16 is represented by the algebraic expression $X^{16} + X^{15} + X^2 + 1$. It will detect all errors occurring in bursts up to 16 bits in length and over 99 of bursts longer than 16 bits. The V.41 polynomial is represented as $X^{16} + X^{12} + X^5 + 1$ and has performance capability similar to that of CRC-16. Both have 16 check bits which are appended to the block as two 8-bit characters.

D. Phases of Link Control

Character-oriented protocols are generally structured as a series of well-ordered logical processes as illustrated in Fig. 4. ANSI calls these

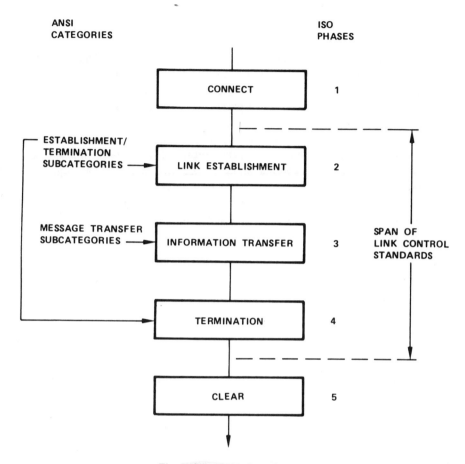

Fig. 4. Communications phases.

processes subcategories while ISO calls them phases. In X3.28, in fact, ANSI creates protocols by combining an information transfer category with an establishment/termination category. Note that of the five phases illustrated, two, the connect and clear phases, are associated with the level 1 protocol. Let us look at each of these phases:

Connect. The connect phase consists of those processes associated with establishing a connection over a switched facility such as provided by a common carrier. The process normally includes off-hook signaling, switching, and an exchange of identification. These functions, as stated, are normally provided by level 1 protocol and are not required on dedicated facilities.

Link Establishment. Link Establishment is the first phase in the span of link control protocols. This phase includes processes required to initialize data transfer over an already established physical link. Polling (inviting to send) and selecting (calling) are typical processes.

Information Transfer. This phase includes processes associated with the objective of link control; the transfer of data. It begins following link establishment and terminates with the end of the message or data transfer. It includes the actual transfer between connected station comprising the message and also includes the acknowledgment process.

Termination. The termination process consists of those functions associated with relinquishing control of the link following transmission of the message or messages. Control is normally returned to the master or control station in a multipoint link who can then return to link establishment phase to initialize transfer to another station. The termination phase can also initiate physical disconnect on a switched connection.

Clear. The clear phase functions to release the facility by signaling on-hook. These functions, like the connect phase, are normally part of level 1 protocol.

IV. Protocol Operation

Having reviewed the characteristics which form the common framework for the various character-oriented protocols we can now examine in more detail the "how" of these protocols.

A specific category of ANSI X3.28 has been chosen as being typical of the most common character-oriented protocols. This protocol will be described in terms of the establishment, information transfer, and termination phases described earlier.

The objective of this protocol is to supervise the transfer of messages over a two-way alternate dedicated multipoint link. One station is designated as the control or master station. The blocks of data being transferred may constitute an entire message or a part of a message. VRC/LRC is used as an error detection technique.

A. Establishing the Data Link

Since a dedicated communications facility has been assigned to the link, the control station may initiate the link without the need to acquire a facility as would be necessary on a nondedicated switched network. The control station may establish the link at any time by either polling or calling. The control procedure is illustrated in Fig. 5. Polling and calling are distinguished by the use of unique addresses in the prefix preceding ENQ. The control station [Fig. 5(1)] wishing to solicit input messages from one of the tributary stations sends a polling sequence [Fig. 5(2)] consisting of the control character ENQ and the address of the selected station.

The addressed tributary responds in one of two ways. If it has no message to send it responds with the terminating control character EOT

[Fig. 5(5), (14)], thus returning control to the control station, which may then poll or call another station.

If the addressed tributary has a message to send it enters the information transfer phase [Fig. 5(4)] described later.

A third possibility is that the control station receives either no reply or an invalid reply to its poll [Fig. 5(6)]. In this event, usually detected by a timeout, the control station terminates with an EOT [Fig. 5(14)] before resuming polling or calling.

The control station having messages to deliver to one of the tributaries selects or calls that tributary by sending a calling sequence [Fig. 5(7)] consisting of the control character ENQ preceded by SYN characters and the tributary's call address.

The addressed tributary, recognizing its call address, has one of two choices. If it is ready to receive messages it sends the acknowledgment sequence ACK 0 [Fig. 5(8)], which is represented by the sequence DLE 0. The control station, upon detecting this reply, enters the information transfer phase.

The two-character acknowledgment sequence is a technique used to provide an additional check on transmission integrity by the use of alternating acknowledgments. The first and all odd-numbered blocks of a received sequence are acknowledged with the sequence ACK 1 transmitted as DLE 1. The second and all even-numbered blocks of the sequence are acknowledged by the sequence ACK 0. The receipt of two successive even or odd acknowledgments, e.g., ACK 0, ACK 0, indicates the loss of a transmission block. To return to our called tributary, if it is not ready to receive traffic it responds with the control character NAK [Fig. 5(9)]. The control station has now the option of calling the same tributary again [Fig. 5(11), (7)] or terminating the exchange with an EOT [Fig. 5(12), (14)].

Again the possibility exists that the control station will receive an invalid reply or no reply [Fig. 5(10)] to its calling sequence. In this event the control station may either terminate with an EOT [Fig. 5(14)], recall the same tributary [Fig. 5(11), (7)], or exit to a recovery procedure [Fig. 5(13)]. Usually a number of retries are made before initiating recovery.

B. Information Transfer

Information transfer begins under control of the master station following successful establishment of the data link. The master station can be the control station which has messages to send and which has successfully selected a tributary station. The master station could also be one of the tributary stations which has been successfully polled and assigned master status when it responded with an indication that it had traffic to send.

The information transfer phase is illustrated in Fig. 6. The station begins transmission of the first message block with either SOH [Fig. 6(2)] if

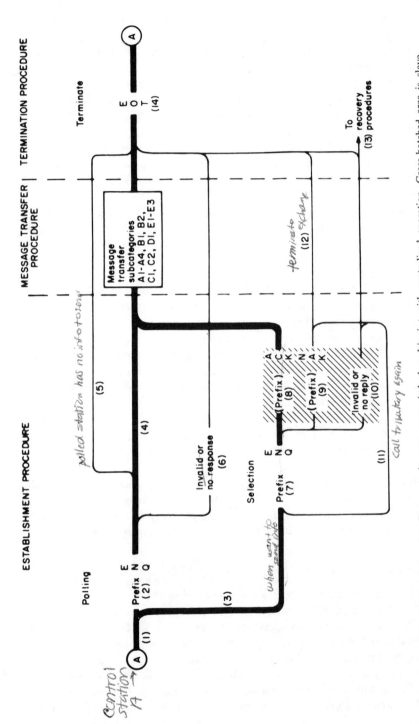

Fig. 5. Subcategory 2.4: Two-way alternate, nonswitched multipoint with centralized operation. Cross-hatched area is slave response.

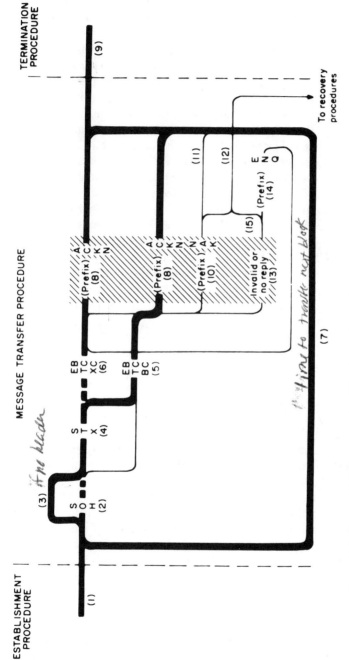

Fig. 6. Subcategory B2: Message-associated blocking, with longitudinal checking and alternating acknowledgments.

the message has a heading, or STX if it does not [Fig. 6(3), (4)]. Following, i.e., intermediate blocks, which begin or continue the text of a message are started with STX.

Transmission of the block continues with data characters until the system-defined block length is reached. At this point the station appends the end-of-block sequence ETB BCC [Fig. 6(5)] or the end-of-message sequence ETX BCC [Fig. 6(6)] if the block is the last block of the message. The BCC is the block check character, in this case a single LRC character. At the receiver this incoming stream of characters is examined and appropriate action taken. The SOH character is recognized as both the start of block and the start of message character. As the start of block character it initiates the parity-checking mechanism. The STX character was recognized as the start of block character and initiated parity accumulation. ETB and ETX initiate parity check comparison and line turnaround for a reply.

If the first block is received correctly and the station is ready to receive another block it transmits the positive acknowledgment sequence ACK 1 [Fig. 6(8)]. ACK 0 would be used as the positive reply to the second block using the alternate acknowledgment method discussed earlier. The master station receiving the acknowledgment may transmit the next block [Fig. 6(7)], or, if the acknowledged block was the last of the message, initiate termination [Fig. 6(9)].

Were the block received incorrectly and not accepted, the receiving station would respond with NAK [Fig. 6(10)], indicating to the master station that retransmission is required. Usually a particular block will be retransmitted [Fig. 6(11), (7)] several times before a recovery procedure [Fig. 6(12)] is invoked. Note that the NAK does not alter the sequence of the alternating acknowledgments. The receiver acknowledges the retransmitted block with the same ACK 1 or ACK 0 it would have used for a successful first transmission. Were the master to receive the wrong ACK, e.g., ACK 1 when ACK 0 was expected, it would retransmit that block as though a NAK had been received.

If the master station fails to receive a reply or receives an invalid reply [Fig. 6(13)] to a transmitted block it will usually send an ENQ [Fig. 6(14)] after an appropriate timeout. The receiving station will then repeat its last reply. Several unsuccessful attempts at sending an ENQ to solicit a reply will result in initiation of a recovery procedure [Fig. 6(15)].

C. Termination of the Data Link

The data link is terminated by the transmission of the control character EOT. Termination may be initiated by the master following reception of a positive acknowledgment to the last block of a message. Termination is also initiated by a control station which has received an invalid or no reply to a poll or a call, or a NAK to a call. Termination can be initiated by a polled station which has no traffic to send.

D. Abnormal Conditions and Fault Recovery

Any of several causes such as line noise, line breaks, operator errors, or equipment malfunction may cause operation to violate the established protocol procedures. This usually results in an abnormal or fault condition. Recovery procedures are intended to aid in reestablishing normal operation. Recovery procedures can often only detect the condition and sometimes require normal intervention to correct the problem. More sophisticated systems may permit a degree of automatic recovery.

Timers play a very significant role in the detection of and recovery from fault conditions. In the protocol just described the prudent system designer would have implemented at least two timers. These timers and their role in fault recovery are as follows:

1. Response Timer. This timer protects a sending station against an invalid or missing response. It would be started with the transmission of any control character requiring a response such as ETX, ETB, or ENQ. It would be stopped when the valid reply was received. If it expires appropriate action such as retransmission *N* times is initiated. If link recovery does not occur, higher-level intervention by an operator or recovery program is required. Typical values for a response timer are in the range of 2–3 s. Timer values are, of course, highly dependent on the properties and parameters of the specific system.

2. Receive Timer. A receive timer will protect against failure to receive or recognize an end-of-block character ETB or ETX. Such a timer is started on detection of a start of block character SOH or STX, restarted with continuing data input, and stopped upon recognition of an end-of-block character. Expiration of this timer, usually in the range of 500 ms, initiates a search for character synchronization, discarding of the faulted block, and notification to higher level.

Other possible timer uses are to detect no activity on a line and to detect a missed DLE EOT where this sequence is used to initiate a switched circuit disconnect. Failure to clear a switched circuit connection can be very expensive!

E. Variations

Many variations in procedure and many options to the standard procedure are available. Most are designed for particular applications and situations but those described here have come into fairly widespread use. Most are subject to bilateral agreement, which is to say that both stations involved in the interchange must be aware of their presence.

1. Block Abort. A block abort occurs when the sending station in the process of sending a block decides for whatever reason to end transmission and terminate prior to the normal end of block. An abort is accomplished by transmitting the control character ENQ and then halting. The receiving station detects the ENQ (instead of ETB or ETX), discards the partial block, sends a NAK, and waits for the sender to resume.

2. Send Abort. Should a sending station decide to terminate a transmission after receiving an acknowledgment for a block but before sending the next block it may do so by sending EOT. This will cause the receiving terminal to discard the incomplete incoming message and wait for the control station to reestablish the link.

3. Receive Abort. A receiving station can abort a sequence of incoming blocks by sending EOT in lieu of its normal acknowledgment response. It may do this, for example, if it has become unable to receive owing to a fault or lack of storage. The sending station would normally try to recall the station and resend the message.

4. Temporary Delay. This variation allows a receiver to temporarily delay blocks coming from a sender. It is signaled by sending DLE; in place of the usual acknowledgment. This sequence, known as WACK, indicates to the sender that the block has been received correctly but that a temporary delay is required. The sender will send an ENQ to which the receiver will respond with another DLE. This will continue until the receiver responds to an ENQ with the appropriate ACK or until the sender loses patience and initiates recovery procedures.

5. Reverse Interrupt. Another variation in procedure allows a receiving station to ask a sending station to terminate the transmission so that the receiving station may gain access to the line. This would be useful for high-priority traffic as an example. The receiver asks for a reverse interrupt by transmitting the sequence DLE < in lieu of its normal acknowledgment sequence. The sender treats the DLE < as a positive acknowledgment and releases the line by transmitting EOT at the earliest opportunity.

6. Synchronization Sequence. If the receiver is to accurately locate and decode data characters from an incoming bit stream it must acquire and maintain character synchronization with the transmitter. In synchronous systems this is accomplished by the transmission of a synchronizing character called SYN at the beginning of a data stream, or following any period when no characters have been transmitted. Since there is a high probability that the first character of a stream of data will be distorted or errored, most protocols require that a minimum of two SYN characters precede the control or data characters. Many implementations require three or even four SYN characters. This allows the receiver to search for two successive SYN characters before declaring that synchronization has been acquired.

For the sake of simplicity, the SYN characters have been omitted from control sequences described in this chapter. It must be remembered that sequences such as ENQ, SOH, STX, EOT, and ACK are actually transmitted with two or more preceding SYNs, as for example SYN SYN ENQ or SYN SYN EOT.

7. Pad Characters. Pad characters are used by some protocols to assure that the first and last characters of a transmission are fully and properly transmitted by the associated data set. At the start of a data stream the pad assures that the receiver is prepared to receive and searching for synchronization. At the end of the data stream, the pad guarantees that the last significant character has been transmitted by the data set before it turns off for line turnaround. Most protocols define the leading pad as a SYN character and the trailing pad as an all-1's character.

8. Prefixes. In many protocols the control sequences consist of a prefix followed by the communication control character. The use of a prefix must be by bilateral agreement. When used they usually consist of up to 15 characters, which must be other than communication control characters. Prefixes are often used to convey status information or station identity especially when operating on switched network facilities where it is important to verify proper connection. The control character EOT is never preceded by a prefix. DLE EOT, however, is often used in switched network operation to initiate a disconnect of the switched circuit.

F. Transparency

One of the major and most often encountered variations to the basic data link control protocol is transparent operation. As defined earlier, transparent operation provides the link control with the ability to treat all transmitted and received characters, including normally restricted control characters, as data.

In character-oriented protocols transparent operation is implemented by the use of the Data Link Escape (DLE) character to form what are known as code extension sequences. Transparent operation is illustrated in Fig. 7 and operates as described next.

In the transparent mode each control character is preceded by a DLE character to form two-character sequences, examples of which are:

DLE STX	Initiates the transparent mode for the following data block.
DLE ETB	Terminates a block of transparent data.
DLE ETX	Terminates the last block of transparent data.
DLE SYN	Character synchronization sequence inserted into transparent transmitted blocks at approximately 1-s intervals; not accumulated in the BCC.
DLE ENQ	Aborts the transmission of a block of transparent data; used by the transmitting terminal.
DLE DLE	Permits the transmission of a DLE as data within a transparent block.

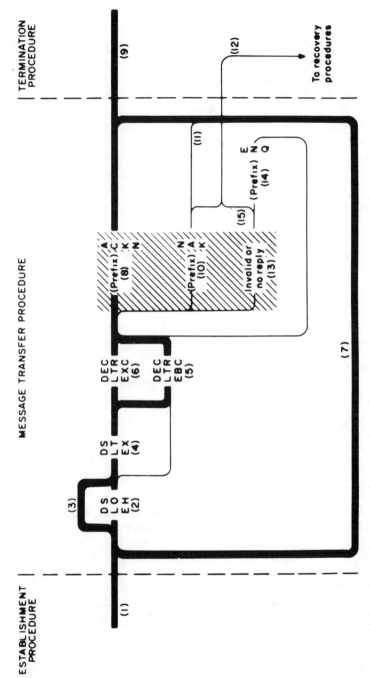

Fig. 7. Subcategory D1: Message-independent blocking, with cyclic checking, alternating acknowledgments, and transparent heading and text.

The DLE character effectively instructs the receiver to recognize the next character as a control character and to ignore control characters (which may appear in transparent text) not preceded by DLE. Since the character DLE itself may appear in transparent text, the receiver would decode the following character as a control character. This is prevented by having the transmitter insert an extra DLE following one which appears in text, creating the two-character sequence DLE DLE. The receiver recognizing this sequence deletes one DLE, restoring the original data stream, and treats the other as valid data.

G. Byte Count Protocols

The use of escape sequences, based on DLE, introduces considerable complexity to the character-oriented protocols. Great care is required in either hardware or software implementation to avoid misinterpretation of control sequences.

Protocols, such as DDCMP, described in Chapter 10, have been devised which solve the transparency problem without the use of escape characters. These protocols are known as byte-count-oriented protocols since they achieve transparency by keeping track of character count and transmitting this information with each block.

The character count is normally transmitted as a positionally located field usually immediately following the SYN characters. The field length is in character increments and indicates the number of characters comprising the block. The receiver then counts characters instead of searching for a control character to determine the location of the check characters and the end of the block.

H. Acknowledgment

Variations in the technique used to acknowledge blocks are quite common. The majority of these fall into one of the following categories:

No Acknowledgment. Protocols are in operation which use no acknowledgment method at all. Usually these would be found in situations where only one-way facilities may be used and where the data itself is highly redundant so that the receiver could discard an obviously errored block.

Single Acknowledgment. In this method the transmitter sends a block of information and waits for a reply. The reply is a single character providing a positive (ACK) or negative (NAK) reply. If positive the transmitter sends the next block. If negative the transmitter retransmits the last block.

Alternating Acknowledgments. This technique is the same as the single acknowledgment method with the exception that positive acknowledgments alternate odd and even blocks with a two-character sequences such as ACK 1 and ACK 0. This provides an additional protection against lost blocks.

Block Numbering. A much more sophisticated acknowledgment technique involves the numbering and sequence of each transmitted block. Using this method, sometimes referred to as "pipelining," the transmitter may send blocks continuously, each block having been assigned and identified with a block number. The receiver, using a reverse channel, sends acknowledgments containing the block numbers correctly received. A NAK from the receiver causes the transmitter to retransmit all blocks sent after the last correctly received block. Many variations of this technique have been implemented. They all require a backward channel, usually a full duplex channel. This technique is of course much more efficient than block-by-block acknowledgment.

V. Implementation Considerations

Despite the existence of standards, the link control implementor must be aware of the many characteristics of the communications environment that are either subject to bilateral agreement or influence the behavior of the control protocol. Among these are the following:

- The physical facility and its interface requirements.
- The precise characteristics of the stations on the link in terms of the protocol options and variations implemented.
- The actual code set utilized.
- The formats of the messages being interchanged.
- Additional or optional link controls being implemented.
- The recovery procedures being used by all stations.
- The requirement for synchronization.
- The maximum block length accommodated and provisions for short blocks.

One way to assure that all parameters, characteristics, and functional behavior are well understood is to create a protocol specification for the specific application. This specification can be based on the "standard" protocol but is expanded to include all options, timers, bilateral agreements, and unique characteristics if any. The document can then be reviewed and revised until all parties sharing the communication link agree on its content.

VI. Limitations

Even though character-oriented protocols represent the vast majority of link control protocols in use today, it has long been recognized that they suffer from many deficiencies. Among these are the following:

1. The necessity to distinguish between data and control characters within a code set places a burden on hardware and software implementation.
2. The assignment of characters for link control subtracts from the combinations otherwise available for information transfer.
3. The character orientation means that they are not naturally transparent to the structure or encoding of the text.
4. Transparency can only be achieved by invoking complicated escape techniques and at the expense of incompatibility with nontransparent protocols.
5. The mixture of message control, device control, and link control forces a significant amount of processing at a low functional level, and blurs the interface between these logically independent functions.
6. Error checking is usually done only on the text, thus exposing supervisory sequences to undetected errors which complicate error recovery.
7. The inherent two-way alternate nature of these protocols does not economically utilize full-duplex facilities.
8. The rigid structure of character-oriented protocols lacks flexibility and expandability.

Many deviations and variations have been devised in attempts to improve the character-oriented protocols. To a great extent these have been unsuccessful. The explosion in information technology combined with the rapid evolution in hardware have begun to overwhelm the ability of character-oriented protocols to keep pace.

This inability to overcome the inherent deficiencies of the character-oriented protocols was the impetus for the development of a whole new family of link control protocols now known as bit-oriented protocols.

Despite their deficiencies and despite the rapid emergence of the bit-oriented protocols, the character-oriented protocols can be expected to be with us well into the future, primarily because of widescale implementation. They have served our industry well.

References

[1] ANSI X3.28-1976: "American National Standard procedure for the use of the communication control characters of American National Standard code for information interchange in specified data communication links." American National Standards Institute, Inc., New York.

[2] IS 1745-1975, "Basic mode control procedures for data communication systems," International Standards Organization, Geneva, Switzerland.

[3] ECMA-16, "Basic mode control procedures for data communication using the ECMA 7 bit code," European Computer Manufacturing Association.

[4] IBM BSC, "General information—Binary synchronous communications," Publication GA27-3004-1, IBM System Reference Library.

[5] IATA SLC, "Synchronous Link control procedures," ATA/IATA Interline Communications Manual, International Air Transport Association, Montreal, Canada.

[6] DEC DDCMP, "Digital Data Communication message protocol," Digital Equipment Corporation.

[7] ANSI X3.4-1977, "American National Standard code for information interchange," American National Standards Institute, New York.

Bit-Oriented Data Link Control

David E. Carlson

I. Introduction

A new breed of data link control has achieved widespread acceptance. Known under a variety of names and mnemonics—ADCCP, HDLC, LAPB, LAPD, BDLC, SDLC, UDLC, LLC, etc.—it is based on a bit-oriented, rather than character-oriented, organization and format. It offers a high level of flexibility, enhanceability, adaptability, reliability, and efficiency of operation for today's as well as for tomorrow's synchronous data communications needs.

A. Historical Background

Bit-oriented data link control procedures had their beginnings approximately a decade and a half ago. It was then that it became evident that the various existing character-oriented data link control procedures (Chapter 4) that had served so well in so many applications (and still do in some) were not well suited for many of the newer interactive applications being pursued. Technology had provided more reliable transmission facilities, more intelligent and cost-effective computers and terminals, and new frontiers for their use in almost every segment of the business, industry, government, and academic environments. Extending or modifying the existing protocols to satisfy these needs was found to be generally inadequate. The character-oriented procedures were, from a control standpoint, basically two-way alternate ("half-duplex") in nature and batch-oriented in operation. They were inherently tied to the transmission code being used,

DAVID E. CARLSON • AT&T Bell Laboratories, Holmdel, New Jersey 07733.

and generally utilized unprotected control codes and sequences to perform link control management functions. Generally, only a single data link function was performed with each transmission unit sent [e.g., transfer data, acknowledge data, solicit (poll) data] and so large numbers of logical link turnarounds were often required. This, in turn, would lead to an unsatisfactory ratio of data transfer exchange to control exchange capability in many cases. Often, a character-oriented procedure would vary so much in format and function from one type of application or use to another, that for all intent and purposes they were no longer the same data link control procedure.

All in all, it was time for a new approach to data link control, an approach that would correct and improve the identified shortcomings present in the existing protocols, provide the features and services that this new environment demanded, and offer the ability for extension and enhancement in order to provide for the future. The bit-oriented data link control described in this chapter seems to provide a satisfactory solution to these problems for many synchronous data communications needs.

B. General Requirements and Capabilities

The principal requirement for this new data link control was that it support the emerging interactive operations. To this end, the following capabilities were identified as being essential:

1. Code-independent operation (transparency).
2. Adaptability to various applications, configurations, and uses in a consistent manner.
3. Both two-way alternate and two-way simultaneous ("full duplex") data transfer.
4. High efficiency (throughput).
5. High reliability.

Code independence means that the user should be able to choose the code set or bit patterns to be used for data transfer without concern for the data link control procedure being used. There should be no need to dedicate certain does or bit patterns from the user's set for data link control purposes (as had been the case with character-oriented protocols). The user's choice should be predicted solely on satisfying user-identified objectives.

Adaptability to various applications, configurations, and uses means that the composition of the procedures should be such that they are readily applicable to two-wire or four-wire equivalent physical circuits, in point-to-point or multipoint configurations, on switched or nonswitched circuits.

Adaptability and code independence also mean that there should be a sense of station independence as well, in that stations of different degrees of sophistication can coexist on the same link, so long as the controlling station is aware of the capabilities and limitations of each individual station and the station operations themselves do not interfere with one another. This should allow the combining of stations on a link on the basis of geographic location and traffic requirements, without concern for the type of stations involved.

The inclusion of *two-way simultaneous data transfer* capability means that more efficient operation should be possible, resulting in increased throughput and probably lower cost where two-way traffic flow requirements exist. It also means that fewer overall transmission paths should be needed in the resultant system configuration. Two-way simultaneous capability can be vitally important when operating in a long propagation delay situation, such as over satellite connections or very high speed links.

High efficiency means that the ratio of data transfer exchange to control exchange per unit of time should be high. The organization of the data link controls should allow multiple functions to be conveyed in each transmission, for example, transfer of data, acknowledgments for data received in earlier transmissions, plus in the case of a controlling station, a solicitation (poll) for a return transmission. High efficiency should also be realized by the use of a data link control organization that holds down the number of logical turnarounds required in the operation of the data link.

High reliability means that all transmissions, data and control, should be protected from transmission errors by a powerful error detection and correction mechanism. Recovery from transmission errors should be an automatic aspect of the procedures (for example, the execution of up to some design number N retransmission attempts before alarming). High reliability also means that data transfer sequence integrity should be maintained with respect to the order of the data that is passed to the higher level at the receiving station. Also, no data should be lost or duplicated without appropriate notification to the higher level.

C. Organization of Chapter

The balance of this chapter is presented in terms of the American National Standards Institute (ANSI) bit-oriented data link control procedure standard—ADCCP, the Advanced Data Communication Control Procedures (X3.66-1979) [1]. ADCCP is compatible with the High-Level Data Link Control (HDLC) standard that was developed by the International Organization for Standardization (ISO).

ADCCP is used as the baseline because it covers a wide scope of possible bit-oriented data link control procedure applications and has had the benefit of a broad base of input and comment from a large cross section of providers, users, and general interest organizations and individuals. This chapter describes the various link configurations, modes of operation, and station types that are covered by ADCCP. The composition of this bit-oriented procedure is described, including the frame structure, the repertoire of commands and responses, and the classes of procedure defined to date. To illustrate some of the principles of bit-oriented procedure operation, a few typical examples of operation are examined. The status of similar bit-oriented data link control procedure activity by other standards bodies (e.g., ISO, CCITT) is reviewed. The subject of compatibility with proprietary bit-oriented protocols (BDLC, SDLC, etc.) is touched on briefly. Finally, there is a crystal-ball view of possible future development and standardization in the subject area.

Throughout all sections of this chapter certain liberties have been taken in the level and completeness of description of the general operation so as not to cloud the overall picture with details of operation. The goal is to provide an overview of bit-oriented data link control operation in general, not an in-depth presentation of the ADCCP standard in particular. For details of specific ADCCP operation under various operating conditions, [1] should be consulted.

II. Configurations, Modes, and Station Types

During the development of the bit-oriented data link control procedure approach, every attempt was made to identify the needs and requirements of a general data link control procedure that would have widespread applicability in today's and tomorrow's marketplace. Taken into consideration were point-to-point and multipoint configurations, using two-way alternate and two-way simultaneous operation over switched and non-switched transmission lines. Both terrestrial and satellite connections were recognized as being part of the problem. Also included was communication between logical equals and communication between logical unequals.

To satisfy the above needs, three different data transfer modes of operation evolved and three different types of stations were identified. The three modes are:

1. The normal response mode (NRM) for use in point-to-point or multipoint configurations.
2. The asynchronous response made (ARM) for use in point-to-point or multipoint configurations.
3. The asynchronous balanced mode (ABM) for use in point-to-point configurations.

The three types of stations are:

1. The primary station (one per NRM or ARM operation).
2. The secondary station (one or more per NRM or ARM operation).
3. The combined station (two per ABM operation).

The normal and asynchronous response modes (NRM and ARM) provide an unbalanced type of data transfer capability between logically unequal stations (a single primary station and one or more secondary stations) operating in a centralized control environment. In both NRM and ARM, the role of the primary station is to control the overall data link operation. The primary station is responsible for initializing the link [activating the secondary station(s)], controlling the flow of data to and from the secondary station(s), recovering from system errors not recoverable by retransmission of the same data, and logically disconnecting the secondary station(s) when required. The secondary stations are subservient to the primary station at the data link level. Their role is generally passive and they have little or no capability for recovery from system errors. As a rule (and in many cases as an objective), the extent of their logical complexity is such that they can be significantly simpler and less costly than their primary station counterpart.

A primary station issues commands and receives expected responses. A secondary station receives commands and issues responses in accordance with the nature of the command received and the mode of operation used. In NRM, a secondary station initiates transmission *only* as a result of receiving explicit permission to do so from the primary station. Once permission is received, a secondary station response transmission must be initiated, with the end of the transmission being explicitly identified. The transmission may or may not include the transfer of data from the secondary station to the primary station, depending on the availability of data to transmit and the form of the explicit permission to send. In ARM, a secondary station is not required to receive explicit permission from the primary station in order to initiate transmissions (responses) of its own. ARM operation, therefore, is more freewheeling and less disciplined than NRM operation.

The normal response mode (NRM) is ideally suited for polled multipoint operation where ordered interaction between a central location and a number of outlying stations is required, or any situation where it is desirable for one station to be able to control the transmittability of other related station(s). Similarly, the asynchronous response mode (ARM) seems ideally suited for situations where a single primary station and a single activated secondary station wish to transmit freely to one another without the overhead of a polling control discipline.

Because of the asynchronous nature of secondary station transmissions when ARM is utilized in a multipoint environment, only one secondary station can be activated (on-line) at a time. Other secondary stations on the multipoint link must be kept in a quiescent disconnected mode (off-line) so as not to interfere with any transmission in progress.

The asynchronous balanced mode (ABM) provides a balanced type of data transfer capability between two logically equal stations (two combined stations) in a balanced control environment. Each combined station is capable of initializing the link, activating the other combined station, and logically disconnecting the link (deactivating the other combined station) when required, and is responsible for controlling its own data flow and recovering from its own system errors. A combined station can both issue commands and responses, and receive commands and responses. The asynchronous nature of the balanced mode of operation means that there is no operational overhead required to control transmission (start and stop data transfer) from the other combined station.

For a point-to-point configuration, the asynchronous modes (ARM and ABM) are usually more efficient than the normal response mode (NRM) because there is no polling overhead required. The choice of which asynchronous mode to specify is dependent on the relative level of data link control capability that is provided in each station: ABM operation for logical equals, ARM operation for logical unequals.

In a great many instances of NRM or ARM applications, the primary station will be a host computer. The secondary stations will be operator-controlled terminals, simple data collection or data display devices, or the like, depending on the needs of the data system. In many ABM applications, each combined station will be a host computer, an intelligent network node (e.g., a packet-switching node), or at least a highly intelligent terminal that has the capability to control the data link itself.

In addition to the data transfer modes cited above, there are also non-data-transfer modes that have been defined to complete the complement of bit-oriented data link control procedures. They include two disconnected modes and an optional initialization mode.

In the optional initialization mode, a primary/combined station may initialize or regenerate the link control of a secondary/combined station. Details regarding the nature of such initialization activities have been deemed to be system dependent and, therefore, are not structured or specified in a standard manner at this time.

Both disconnected modes have the stations logically disconnected from the link. The normal disconnected mode (NDM) applies to primary and secondary stations only. In the normal disconnected mode, the secondary station may not initiate any form of transmission until explicitly requested to respond by the primary station. When so requested, the response can only be one of a limited set of responses that either accepts the command, refuses the command, or requests some alternative action on the part of the primary station. The asynchronous disconnected mode (ADM) applies to combined stations as well as to primary and secondary stations. In the asynchronous disconnected mode, the secondary/combined station may generate a particular response on an asynchronous basis as a request for a mode setting command in order to establish a data transfer mode.

The choice and evolution of the data transfer modes provided by these bit-oriented data link control procedures was not driven totally by technical considerations. Other factors played a role as well. For instance, political considerations provided some of the motivation for inclusion of the asynchronous balanced mode (ABM). The NRM and ARM modes had been defined first, and were pretty well fixed in place, and generally accepted internationally, when it was observed that they did not quite satisfy all of the "requirements." True, they supported point-to-point and multipoint configurations, two-way alternate and two-way simultaneous operation over switched and nonswitched facilities. They also provided transparent, efficient, and reliable data transfer. However, each was built on a primary station/secondary station relationship that had the negative aspect associated with it that in two-station configurations one of the stations was operationally "secondary" to the other. The overall control of data flow and responsibility for system recovery resided in only one of the stations—the designated primary station. For many applications, this was considered to be unacceptable. For example, when interconnecting governments, corporations, independent systems, etc., the thought of being the "secondary" to another, dependent upon another for one's operation and livelihood, was generally unacceptable. Hence the asynchronous balanced mode (ABM) was defined to support fully balanced, independent data transfer between logical equals. As noted later, ABM has become a vital part of the bit-oriented data link control procedure solution.

III. Composition of the Bit-Oriented Procedures

This section provides a brief sketch of the major elements that make up the composition of bit-oriented data link control procedures. Included are the frame structure and transmission formats; the commands, responses, and parameters; and the resultant classes of procedures.

A. Frame Structure

The basic transmission unit is called a frame. All transmissions (data, control, or both) are in frames, and each frame conforms to one of the two following formats:

1. If there is an information field to transport,

F, A, C, Info, FCS, F

2. If there are only data link control sequences to transport,

F, A, C, FCS, F

where

F = flag sequence
A = address field
C = control field
Info = information field
FCS = frame check sequence

The *flag sequence* (F) is a unique eight-bit pattern (a 0 bit followed by six 1 bits ending with a 0 bit) used to synchronize the receiver with the incoming frame. It delimits the start and close of each transmitted frame and is also used by the sender to fill time between frames during a transmission of multiple frames.

To achieve transparency, the unique flag sequence is prohibited from occurring anywhere in the address, control, information, and FCS fields by having the transmitter and receiver perform the following action after sending and receiving, respectively, the opening flag sequence:

- Transmitter: insert a 0 bit following five continuous 1 bits anywhere before sending the closing flag sequence ("bit stuffing").
- Receiver: delete the 0 bit following five continuous 1 bits following a 0 bit anywhere before receiving the closing flag sequence ("destuffing").

The closing flag for one frame may also serve as the opening flag for the next frame.

The *address field* (A) identifies the station (secondary or combined) on that link that is to receive (or is sending) the frame. Command frames are always sent with the receiving station's address. Response frames are always sent with the sending station's address. Hence, the address field identifies a secondary station in the normal and asynchronous response modes and the response-generating portion of a combined station in the asynchronous balanced mode.

Two mutually exclusive address field options are defined—single octet and multiple octet addressing. Single octet addressing provides for up to 256 different addresses. Multiple octet addressing provides for greater than

CONTROL FIELD BITS

CONTROL FIELD FOR	1	2	3	4	5	6	7	8
INFORMATION TRANSFER COMMAND / RESPONSE (I FRAME)	0	N(S)			P/F	N(R)		
SUPERVISORY COMMANDS / RESPONSES (S FRAME)	1	0	S	S	P/F	N(R)		
UNNUMBERED COMMANDS / RESPONSES (U FRAME)	1	1	M	M	P/F	M	M	M

Fig. 1. Control field formats.

256 addresses and also allows for character-oriented encoding of the address field where such may be desirable. In the case of multiple octet addressing, the address field is recursively extendable with the first bit of each octet used to indicate which is the final octet of the address field. The final octet will have its first bit set to "1," and each preceding octet will have its first bit set to "0."

The all-ones address is specified as a global (broadcast) address that all stations will be responsive to. The all-zeros address is specified as a null address that no station will be responsive to. Group addressing is possible on a system-by-system basis.

The *control field* (C) identifies the function and purpose of the frame. Three different control field formats are defined: information transfer, supervisory, and unnumbered. Figure 1 depicts the general organization of the control field formats.

Bit 1 set to "0" identifies the I frame format. Bit 1 set to "1" with bit 2 set to "0" or "1" identifies the S frame format or the U frame format, respectively. Only the I frame format has a send sequence number $N(S)$ to uniquely identify the frame and to allow it to be kept in sequential order during the data transfer operation. Both the I and S frame formats have a receive sequence number $N(R)$, so both formats are usable to acknowledge I frames received. The two S bits in the S frame format provide for the specification of four supervisory functions. The five M bits in the U frame format provide for the specification of up to 32 commands and 32 responses to cover the remaining control functions required. The P/F bit in each format provides for a checkpointing mechanism that allows a response frame to be logically associated with the appropriate initiating command frame. (The bit is considered to be the P bit if the frame is a command and the F bit if it is a response.) Checkpointing is accomplished by setting the F bit equal to "1" in the response frame that is to be treated as the logical counterpart of the command frame sent with the P bit set to "1."

In the normal response mode, the P bit set to "1" serves to poll the secondary station(s) to which it is addressed. Similarly, in NRM, the F bit set to "1" serves to identify the final frame in a series of frames sent by the secondary in response to a received P bit set to "1". In the asynchronous response mode and asynchronous balanced mode, the receipt of a frame with the P bit set to "1" will cause the secondary station (ARM) or combined station (ABM) to set the F bit equal to "1" in the next appropriate frame transmitted.

The *information field* (Info) contains the data that are to be transferred across the link. The data may be of any length and may consist of any code or grouping of bits. The bit stuffing protocol ensures "transparency," i.e., it allows the information field to include any patterns, even patterns that look to the end users like flags (01111110), without producing a spurious action at the receiving end.

All frames include a 16-bit *frame check sequence* (FCS) field prior to the closing flag sequence to assist in the detection of transmission errors. The FCS (described in the preceding chapter) is performed on the contents of the address, control, and information fields of the frame, using the well-known CCITT V.41 generator polynomial: $x^{16} + x^{12} + x^5 + 1$. Prior to initiating the FCS check at the transmitter (and the receiver), the FCS register (or equivalent) is preset to all ones. In the absence of transmission errors, a unique, nonzero 16-bit pattern is then detected at the receiver. Frames failing to pass the FCS check are discarded and ignored.

An optional 32-bit FCS is now available for those applications requiring a higher degree of detection of transmission errors. The 32-bit generator polynomial is

$$x^{32} + x^{26} + x^{23} + x^{22} + x^{16} + x^{12} + x^{11} +$$
$$x^{10} + x^8 + x^7 + x^5 + x^4 + x^2 + x + 1$$

Should the transmitter of a frame determine during the course of the transmission that the frame should be discarded and ignored by the receiver, it may accomplish this in either of two ways. One method involves premature termination of the frame in the normal manner with a flag sequence, but purposely causing an incorrect FCS field to be included in the frame. The other method involves aborting the frame in progress by transmitting a continuous ones state (with no inserted zeros) that persists for at least seven bit intervals in length. Aborted frames are ignored by the receiver.

If the ones state on the link persists for 15 bit times or more, an idle link state is defined. The idle link state indicates that the sending station has relinquished the right to continue transmission. It is often used in conjunction with two-way alternate operation on half-duplex transmission facilities. The reappearance of a flag sequence defines reentry into the active link state, wherein an operational mode may be established and information transferred between stations.

B. Elements of Procedure

Each frame contains a command or a response that is either an information transfer frame, a supervisory frame, or a miscellaneous unnumbered control frame. Although the list of commands and responses seems long, in most cases only a few are needed and used to any extent in the provision of normal operation. The majority of them are of the unnumbered format variety and are either associated with various link establishment/disconnect procedures or the provision of optional features and capabilities to satisfy special needs. A brief description of the various commands and responses that are defined in the ADCCP standard are listed below. (The specific control field bit encodings are given in Table I.)

Information (I) Command / Response (I Format): I frames are used to transfer sequentially numbered information fields across a data link, and to acknowledge I frames already received from the other station.

Table I. Specified ADCCP I, S, and U Format Control Field Bit Encodings[a]

Frame format	Command	Control field bits								Response
		1	2	3	4	5	6	7	8	
I	I	0	$N(S)$			P/F	$N(R)$			I
S	RR	1	0	0	0	P/F	$N(R)$			RR
	REJ	1	0	0	1	P/F	$N(R)$			REJ
	RNR	1	0	1	0	P/F	$N(R)$			RNR
	SREJ	1	0	1	1	P/F	$N(R)$			SREJ
U	UI	1	1	0	0	P/F	0	0	0	UI
	SNRM	1	1	0	0	P	0	0	1	
	DISC	1	1	0	0	P/F	0	1	0	RD
	UP	1	1	0	0	P	1	0	0	
		1	1	0	0	F	1	1	0	UA
	Nonreserved	1	1	0	1	P/F	0	0	0	Nonreserved
	Nonreserved	1	1	0	1	P/F	0	0	1	Nonreserved
	Nonreserved	1	1	0	1	P/F	0	1	0	Nonreserved
	Nonreserved	1	1	0	1	P/F	0	1	1	Nonreserved
	SIM	1	1	1	0	P/F	0	0	0	RIM
		1	1	1	1	F	0	0	1	FRMR
	SARM	1	1	1	1	P/F	0	0	0	DM
	RSET	1	1	1	1	P	0	0	1	
	SARME	1	1	1	1	P	0	1	0	
	SNRME	1	1	1	1	P	0	1	1	
	SABM	1	1	1	1	P	1	0	0	
	XID	1	1	1	1	P/F	1	0	1	XID
	SABME	1	1	1	1	P	1	1	0	

[a] Note: All unassigned unnumbered control field bit encodings are reserved for possible future standardization.

Receive Ready (RR) Command / Response (S Format): RR frames are used to indicate readiness to receive I frames and to acknowledge I frames already received from the other station.

Receive Not Ready (RNR) Command / Response (S Format): RNR frames are used to indicate a temporary busy condition and to acknowledge I frames already received from the other station.

Reject (REJ) Command / Response (S Format): REJ frames are used to request retransmission of all I frames starting from a designated point in the numbering cycle and to acknowledge I frames already received from the other station.

Selective Reject (SREJ) Command / Response (S Format): SREJ frames are used to request retransmission of a single designated I frame previously transmitted and to acknowledge I frames already received from the other station.

Unextended Numbering Set Mode (SXXM) Commands (U Format): Unextended numbering set mode commands are used to establish the particular Modulo 8 sequence numbering mode of operation to be used. Upon accepting and acknowledging a set mode command, the receiving station's send and receive state variables are set to zero. Three unextended numbering set mode commands are defined:

SNRM—Set Normal Response Mode

SARM—Set Asynchronous Response Mode

SABM—Set Asynchronous Balanced Mode

Extended Numbering Set Mode (SXXME) Commands (U Format): Extended numbering set mode commands are used to establish the particular Modulo 128 sequence numbering mode of operation to be used. Upon accepting and acknowledging a set mode command, the receiving station's send and receive state variables are set to zero. Three extended numbering set mode commands are defined.

SNRME—Set Normal Response Mode Extended

SARME—Set Asynchronous Response Mode Extended

SABME—Set Asynchronous Balanced Mode Extended

Set Initialization Mode (SIM) Command (U Format): The SIM command is used to establish the initialization mode of operation, during which the link control may be initialized or regenerated, or operational parameters exchanged.

Disconnect (DISC) Command (U Format): The DISC command is used to logically terminate a previously established operational mode, and to cause the stations involved to assume the system predetermined disconnected mode.

Reset (RSET) Command (U Format): The RSET command is used to reset the send state variable at the transmitting station and the receive state variable at the receiving station to zero. The values of the state variables associated with the other direction of transmission remain unaffected.

Unnumbered Poll (UP) Command (U Format): The UP command is used to solicit response frames from one or more stations by establishing a special logical operational condition that exists at each addressed station for one respond opportunity. (Loop operation [2] is typical of the type of application that could utilize the UP command.) The UP command does not acknowledge any I frames (or UI frames) received from the other station.

Unnumbered Information (UI) Command / Response (U Format): UI frames are used to transfer information fields across a data link without impacting the send or receive state variables at any station. There is no specific link level acknowledgment provided for UI frames.

Exchange Identification (XID) Command / Response (U Format): XID frames are used to request and/or report a station's identity and, optionally, to convey the parameters, operational capabilities, and characteristics of the transmitting station.

Request Initialization Mode (RIM) Response (U Format): The RIM response is used to request that the initialization mode be established.

Request Disconnect (RD) Response (U Format): The RD response is used to request that the link be put in a disconnected mode.

Unnumbered Acknowledgment (UA) Response (U Format): The UA response is used to acknowledge receipt and execution of a mode setting, initializing, resetting, or disconnecting command.

Disconnected Mode (DM) Response (U Format): The DM response is used to indicate a request for a set mode command, or, if used in reply to a set mode command, as an indication that the set mode cannot be acted on at this time.

Frame Reject (FRMR) Response (U Format): The FRMR response is used to indicate that a frame received (command or response) was in error in a manner not recoverable by retransmission of the identical frame, such as

1. Receipt of a control field that is invalid or not implemented, or
2. Receipt of an information bearing frame with an information field that exceeds the maximum established length, or
3. Receipt of an $N(R)$ which either points to an I frame which has been transmitted and acknowledged or to an I frame which has not been transmitted and is not the next sequential I frame awaiting transmission.

In addition, the ADCCP standard sets aside four unnumbered command/response code points as *Nonreserved Commands* and *Nonreserved Responses*. This provides implementors of the standard with a set of unnumbered code points that can be used to define special system-dependent data link functions, that may be necessary for a particular application but that do not have general applicability and hence were not standardized, with the assurance that said code points will *not* at some future time be assigned to a standard function or feature by ANSI.

C. Classes of Procedure

The three modes of operation, NRM, ARM, and ABM, provide the framework for the definition of three corresponding classes of procedure—the Unbalanced Normal Class (UNC), the Unbalanced Asynchronous Class (UAC), and the Balanced Asynchronous Class (BAC), respectively. Classes of procedure are defined in order to provide organization and direction for the application of the bit-oriented data link control procedures. Certain commands and responses are identified as belonging to the basic repertoire of theses classes. Other commands and/or responses are viewed as being optional, either adding capability to the basic set or restricting utility of a general function for specific applications. Figure 2 depicts the three classes of procedures, the basic command and response repertoire, plus the optional functions, that are presently defined in the ADCCP standard.

In addition to certain of the commands and responses being optional functions, there is also an optional function concerning addressing (option 7) and two optional functions that deal with restrictions on the use of the I frame as a command only or a response only (options 8 and 9).

The following notation is used to identify a class of procedures and the optional functions that are supported:

- UNC,3,4—depicts the unbalanced, normal response mode class of procedures with the selective reject (SREJ) feature plus the ability to send nonsequenced unnumbered information (UI) frames.
- BAC,2,8—depicts the balanced, asynchronous balanced mode class of procedures with the reject (REJ) feature plus a restriction on the use of *I* frames as commands only.

Class UNC,3,4 might be typical of a multipoint configuration where the two-way alternate data transfer requirements are such that each secondary station transmission consists of multiple I frames, where the transmission error statistics indicate that the occurrence of transmission errors is low and that when errors do occur they generally only affect one frame out of a multiple frame transmission, and where there is occasionally a need to send some useful but not indispensable information to all secondary stations. Class BAC,2,8 might be typical of a point-to-point configuration involving two computers where balanced, equal control by both parties is important, where the two-way simultaneous data transfer requirements are such that

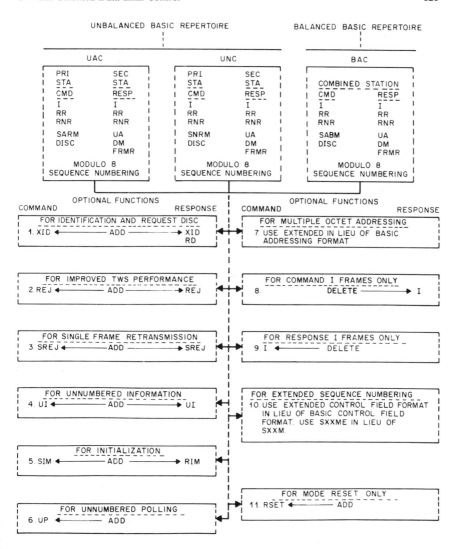

Fig. 2. Basic classes of procedures and their optional functions.

the ability to request retransmission on the fly is desirable, and where the transfer of I frames is used as a continual data link level check that an inadvertent loopback has not occurred somewhere in the system. (BAC,2,8 is the class of procedure with optional functions that is used as the basis for the CCITT Recommendation X.25 Level 2 LAPB procedures described in Chapter 8.)

IV. Examples of Typical Operations

Figures 3–6 illustrate some typical on-line operations, with and without transmission errors, for the two classes of procedure identified above. Figures 3 and 4 depict an unbalanced, two-way alternate, multipoint operation (UNC,3,4) involving three secondary stations (B, C, and D). Figures 5 and 6 illustrate a balanced, two-way simultaneous, point-to-point operation (BAC,2,8) involving two combined stations (L and M). The examples cover normal, error-free operation as well as various error recovery situations. Many of the principles of the bit-oriented data link control concept are illustrated by these examples. The vertical scale has been dramatically reduced so as to allow illustration of various frame exchanges during operations such as link setup, data transfer, recovery from transmission errors, and disconnect. The shorthand used in the figures should be interpreted in the following way:

Consider frame X, Ysr, Z.

X represents the address associated with the frame. Primary station transmissions will use the address of the secondary station for whom the frame is intended. Secondary station transmissions will include the address of the secondary station that is transmitting. Combined station transmissions will use the remote station address when a command frame is sent and will use the local station address when a response frame is sent.

Y represents the abbreviation for the command or response (for example, I, RNR, SNRM, UP, etc.). The "sr" following Y represents the send and receive sequence number values $N(S)$ and $N(R)$, respectively, that are an integral part of I and S format frames. If only a single number is present, it represents the receive sequence number $N(R)$.

Z, when present, indicates that the P or F bit is set to "1" in that frame. When not shown, it means that the value of the P or F bit is set equal to "0."

A. Unbalanced, Two-Way Alternate Multipoint (UNC,3,4)

1. Example of Error-Free Operation

As indicated in Fig. 3, the primary station activates the secondary stations by addressing a SNRM command to each of them individually with the P bit set to "1." Each secondary station acknowledges the set mode command by returning the UA response with the F bit set to "1." (The P bit set to "1" grants the addressed secondary station the right to transmit an appropriate response, and the F bit set to "1" identifies the final frame in the corresponding response transmission.) As each secondary station (B, C, and D) is activated, its data link to the primary station is

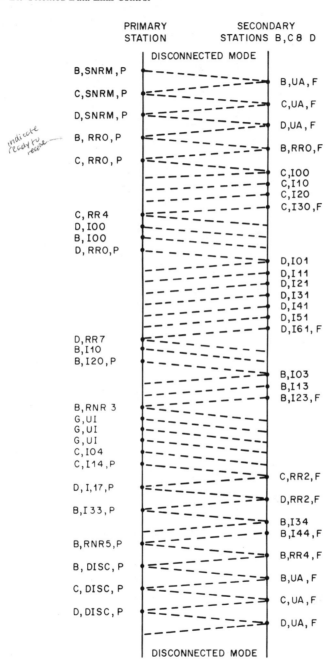

Fig. 3. Class UNC,3,4—examples of error-free operation.

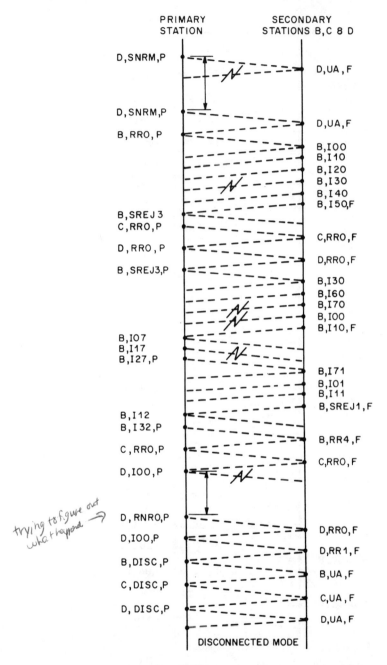

Fig. 4. Class UNC,3,4—examples of recovery procedures.

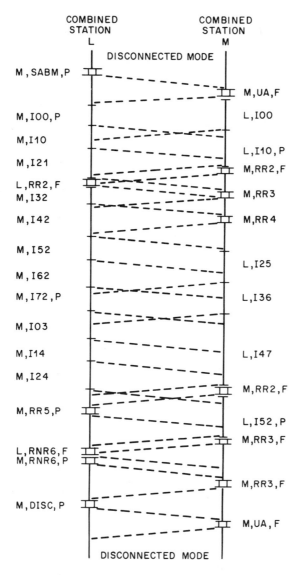

COMBINED
STATION
L

COMBINED
STATION
M

DISCONNECTED MODE

M , SABM , P

M , UA , F

M , I00 , P   L , I00

M , I10

L , I10 , P

M , I21   M , RR2 , F

L , RR2 , F
M , I32   M , RR3

M , I42   M , RR4

M , I52

L , I25

M , I62

M , I72 , P   L , I36

M , I03

M , I14   L , I47

M , I24

M , RR2 , F

M , RR5 , P

L , I52 , P

M , RR3 , F

L , RNR6 , F
M , RNR6 , P

M , RR3 , F

M , DISC , P

M , UA , F

DISCONNECTED MODE

Fig. 5. Class BAC,2,8—examples of error-free operation.

activated and that primary–secondary relationship is in the normal response mode (NRM).

As depicted, after setting up the entire link, the primary station polls B
for information (traffic). B responds that it has no traffic to send by
returning the RR response frame. The primary station then polls C, and
receives four I frames from C. (A basic characteristic of these bit-oriented

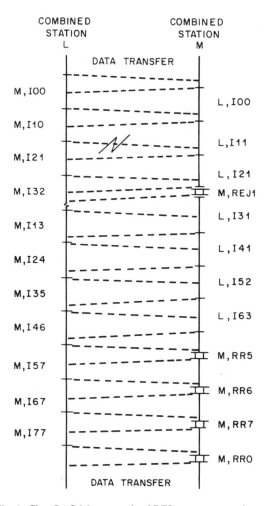

Fig. 6. Class BAC,2,8—example of REJ recovery procedures.

procedures is that a station may send more than one I frame before an acknowledgement is required. In fact, up to the modulus minus one I frames may be sent before an acknowledgment frame is required.) Each I frame has a different in-sequence send sequence number (first digit), plus a receive sequence number (second digit) that identifies the next I frame expected from the other (primary) station. Upon receipt of C,I30,F from C, the primary station acknowledges the receipt of four frames by issuing frame C, RR4. The 4 indicates that I frames numbered 0–3 were received correctly. Because the P bits is *not* set to "1," this frame does not grant C permission to send. It only acknowledges I frames received.

Before continuing the polling cycle with D, the primary station finds that it has received information for D from some other link and so delivers it to D. Prior to completion of the I frame transmission to D, information is received for B. Consequently, the delivery to B is performed before resuming the polling cycle with D (P bit set to "1"). The primary station will check the responses to subsequent polls for acknowledgments from B and D for the I frames delivered to B and D. (If the information for B and D had both been present when the C,RR4 frame was sent, the primary station could have sent B,I00 first and then combined the poll of D with the delivery to D by sending a D,I00,P frame.) D responds to the poll with the maximum number of response I frames (the modulus minus one = seven). When the primary station receives the seventh frame with the F bit set to "1," it acknowledges the transmission by sending D,RR7.

The primary station delivers two I frames to B, setting the P bit to "1" in the second I frame to affect the polling function. B responds with three I frames, identifying the final I frame with an F bit set to "1." The primary station returns B,RNR3, acknowledging the I frames received, but also indicating that the primary station is "not ready" to receive additional I frames from B.

At this point, the typical operation depicted indicates that the primary station has three frames worth of information that should be sent to all of the secondary stations on the link, but that the information is not so important to the operation of each secondary station that it warrants individual delivery, with individual acknowledgments. The nature of the information is such that should it be lost in transit, it will not cause a serious problem at any of the secondary stations. Examples of such information might include (1) some sort of updated hourly production report (missing one out of a series of updated reports may not pose a problem), or (2) periodic time checks or weather reports. The use of the unnumbered information (UI) frame (option 4) plus the global address G (all ones) allows the information to be sent to all three secondary stations at the same time without impacting the send or receive state variables at any of the stations. Following the UI frames, the primary station sends two I frames to C. A series of exchanges with D and B follow. After receipt of B,I44,F from B, the primary station decides that it wishes to go into the disconnected mode. The primary station will check that all secondary station transmissions have been properly acknowledged. The check indicates that the B,I34 and B,I44,F frames have not been acknowledged, so the primary station issues a B,RNR5,P frame. The 5 acknowledges I frames up through 4. The P bit set to "1" will result in a response frame from B acknowledging receipt of the RNR frame. The RNR frame serves to inhibit any additional I frame transmissions from B. Upon receipt of B,RR4,F from B, the primary station initiates a DISC-UA exchange with each of the sec-

ondary stations, resulting in the link being returned to the normal disconnected mode.

2. Examples of Error Recovery Procedures

Figure 4 illustrates some typical error recovery actions that are possible
with the unbalanced normal class of operation. Assume that secondary
stations B and C have already been set up. The primary station sends
D,SNRM,P to activate D. As noted, D responds with a UA response, but
the response is subject to a transmission hit causing it to fail the FCS check
at the primary station and, consequently, to be discarded. To determine
when sufficient time has elapsed while waiting for an expected response, the
primary station would probably activate a time-out function when it sends
the D,SNRM,P frame. When this time-out function runs out, the primary
station may initiate an appropriate recovery action. In the case shown, the
recovery action is to issue the D,SNRM,P frame again and activate the
time-out function again. Although D is already in the data transfer mode as
a result of the first D,SNRM,P frame that it acknowledged, D is reactivated
and returns another D,UA,F frame to the primary station. The arrival of
this second acknowledgment at the primary station without error before the
time-out function runs out, stops the time-out function, and completes the
setup procedure. Normal polling and delivery functions then follow.

As depicted, the primary station polls B, and B responds with six I
frames, the last of which contains an F bit set to "1." To guard against a
"no response," the primary station would activate the response time-out
function when it sent the B,RR0,P frame. The time-out function could then
have been stopped with the reception of each I frame and then restarted in
the case of the first five I frames when it was determined that the F bit was
not set to "1" in that I frame. As indicated, I frame B,I30 set by B was
received in error and was discarded. The two following I frames were
received free of error, and so only B,I30 is needed by the primary station to
complete the data transfer of the six I frames. The first three I frames
received can be passed along to the higher level. The last two I frames
received are held, awaiting correct reception of the fourth I frame, B,I30.
The primary station sends a selective reject frame (B,SREJ3) to acknowledge reception of I frames up through I frame numbered 2. B can then free
up the buffers holding I frames 0–2. The SREJ also tells B that I frame
numbered 3 will have to be retransmitted when permission to transmit is
again granted.

As illustrated, the primary station has decided to complete the polling
cycle before giving B permission to retransmit its I frame numbered 3.
(Alternately, the primary station could have initiated retransmission immediately if it had wished to do so.) When the polling of C and D results in
"no traffic" responses from each station, the primary station requests

retransmission of the missing I frame numbered 3 from B by sending the B, SREJ3,P frame.

B responds by retransmitting B,I30, but does not retransmit I frames numbered 4 and 5 that were transmitted originally. Since its last transmission, B has obtained some additional I frames for transfer to the primary station. The SREJ command provided B with the opportunity to transmit new I frames up to the point where it would have the modulus minus one unacknowledged I frames outstanding. Consequently, B follows the I30 frame with I frames numbered 6, 7, 0, and 1. The I frame numbered 1 has the F bit set to "1" to identify it as the final frame in the transmission. As indicated, frames numbered 7 and 0 are subjected to transmission errors.

When the primary station receives these I frames, it treats them in the following manner. Upon correct receipt of B,I30, the primary station passes it plus I frames B,I40 and B,I50 from the original transmission up to higher level. When B,I60 is received, it is passed to higher level also. Frames B,I70 and B,I00 are identified as missing when B,I10,F is received out of sequence. In this instance, there are two frames requiring retransmission. The primary station will evaluate this situation to determine the type of recovery action to initiate.

If the SREJ function is used, it will have to be used twice. First, it would be used to have B,I70 retransmitted. Then, after B,I70 is received correctly, it would be used to acknowledge B,I70 and to have B,I00 retransmitted. After the successful transmission of B,I00, I frame B,I10 would *not* have to be retransmitted since it was received correctly in the original transmission.

On the other hand, if the SREJ function is not used, the primary station can utilize the checkpointing mechanism that is available with I, RR, and RNR frame transmissions with the P/F bit set to "1" to indicate which I frames have been acknowledged and where in the numbering sequence retransmission of I frames should begin. With this approach, only a single control exchange is required, but any already successfully transmitted I frames (such as B,I10,F) will be retransmitted. When this approach is chosen, the retransmitted I frame (B,I11) should be passed up to higher level, not the originally received I frame (B,I10,F). There is no link level assurance that the contents of an I frame are not altered (updated or made more current) at the time of its retransmission. The link control ensures sequence integrity at the data link level, but depending upon implementation, may not have complete control concerning the contents of I frames ready for transmission or retransmission.

In the case being considered here, the primary station detects that I frames 7 and 0 are missing, chooses not to use the SREJ recovery action, and has I frames of its own to deliver to B. The primary station sends three I frames to B, acknowledging receipt of I frames numbered 3–6 from B. (The middle frame is subject to a transmission error, resulting in an out-of-sequence condition at B as well.) The primary station gives B permission to transmit by setting the P bit equal to "1" in the third I frame sent to B. This causes a checkpoint to take place at B. The P bit is complementary to the earlier F bit sent by B. Since the $N(R)$ associated with this P bit does not acknowledge all of the I frames sent by B since B last sent an F bit set to "1," it is interpreted by B as an indication of where in B's send sequence numbering the retransmission of I frames should begin. The subsequent transmission from B includes all of the I frames from the number identified (7) and acknowledges only the first of the three I frames sent by the primary station. B concludes its transmission with an SREJ frame that requests the retransmission of primary station I frame numbered 1. The primary station retransmits frame B,I12 and adds frame B,I32,P. All of B's frames are acknowledged and a response from B has been requested. B acknowledges the primary station I frames and indicates "no traffic" to send.

The primary station polls C. C responds "no traffic." The primary station sends an I frame to D and requests in the same frame that D respond (P bit set to "1"). The time-out function at the primary station runs out waiting for the response. In a situation like this, the primary station does not know if it was its transmission or D's response that was lost (hit by a transmission error). The primary station could assume that D did not receive the I frame and simply retransmit it, or as is indicated here, the primary station could send a supervisory command with the P bit set to "1" to find out the number of the I frame that D next expects to receive. By sending the RNR command, the primary station restricts D's response to a supervisory frame. Hence, with the exchange of two short frames, the primary station is able to determine if retransmission is required or not. In this case, the primary station has to retransmit the I frame. However, in many instances, the primary station will determine that retransmission is not required (that is, the acknowledging response frame was lost), and occasionally, will discover that D is in a condition in which it would not be able to accept an I frame anyway (for instance, in a disconnected mode because of a power failure or equipment failure at the secondary station). As a general rule, it is considered to be a wise decision to first inquire as to the status of a secondary station before retransmitting unacknowledged I frames. This is even truer if the transmission consisted of multiple I frames.

After receiving the acknowledgment for frame D,I00,P, the primary station in this example initiates the disconnect procedure with each secondary station. Unlike the situation in Fig. 3 where the primary station needed to initiate an exchange of supervisory frames first in order to complete the acknowledgment cycle with a secondary station, in Fig. 4 the primary station knows that all of the secondary stations are in step with the primary station and so proceeds directly with the disconnect procedure.

B. Balanced, Two-Way Simultaneous Point-to-Point (BAC,2,8)

1. Examples of Error-Free Operation

In order to illustrate the interplay of two independent transmission flows in a balanced, two-way simultaneous operation, a slightly different representation of frames is used in Figs. 5 and 6. I frames are shown as having random lengths (solid vertical bars between horizontal boundary lines) and S and U format frames are shown as very short, fat vertical bars. In this way the interactions can be more reasonably noted. The angling lines between the columns serve to indicate where a frame begins or ends at the receiver relative to the receiver's transmissions as well as to its beginning and end at the sender. Many of the operational characteristic cited for the unbalanced case described above also apply here, but are activated in a continuous manner instead of only at the point of P/F exchanges.

As indicated in Fig. 5, combined station L activates the data link by sending an SABM command to combined station M. (In the general case, either station may initiate this action.) M responds with a UA response, acknowledging receipt of the mode setting command. Because of the asynchronous nature of these procedures, a P bit set to "1" is not necessary in order for the receiving station to respond. In this example, however, the F bit is set to "1" in the UA response because the P bit was set to "1" in the SABM command. As soon as M returns the UA response, it is in the information transfer state and may initiate transmission of frames to L. In this instance, M has two I frames ready to send and sets the P bit to "1" in the second frame in order to get an immediate response acknowledgment concerning their arrival at L. On the other hand, L enters the information transfer state when it receives the UA response to its SABM command.

Because of the two-way simultaneous nature of the operation, I frames may be flowing in both directions at the same time as shown. Because option 8 restricts I frames to being command frames only, I frames from L to M will always have the address M and I frames from M to L will always have the address L. Hence, the reception of an I frame with the remote station's address is an indication of a fault in the system, probably an inadvertent loopback somewhere between the two stations.

As each station prepares I frames for transmission, it increases the send sequence number $N(S)$ by one on each successive I frame as long as it does not have more than the modulus minus one unacknowledged I frames outstanding, and it sets the receive sequence number $N(R)$ to the value of the I frame next expected, thereby indicating acknowledgment of I frames number $N(R) - 1$ and below. Hence, a continual I frame flow from each station automatically acknowledges I frames received from the other station (freeing buffers at that station for other use) along with exercising its own information transfer. In the absence of errors and as long as both stations have I frames to send, the process is self-perpetuating.

As long as a station does not have an unanswered P bit outstanding, a station may decide to set the P bit equal to "1" in any one of the I frames transmitted. In the example, M has done so on its last of two I frames and L has done so on its first I frame. (Such action might be taken for a variety of reasons, including, in L's case, to check that an operational data link is present, and in M's case, to get an acknowledgment response that is logically associated with the frame that initiated it, or to provide a P/F exchange protected frame for the transfer of a REJ response in the case of errors.)

After the frame with the P bit set to "1" is sent, the station may continue to send I frames to the other station. Since the receiving station cannot send an I frame with an F bit set to "1" (it would be a response I frame), the receiving station must interrupt its sequence of command I frames to insert a supervisory response frame (for example, L,RR2,F) that will convey both the F bit set to "1" and the $N(R)$ indicating the sequence number of the I frame next expected. I frames $N(R) - 1$ and below are acknowledged. The $N(R)$ should acknowledge at least all I frames transmitted up through the I frame sent with the P bit set to "1." In some instances, the $N(R)$ may acknowledge I frames beyond the $N(S)$ at the time that the P bit was set to "1," but less than or equal to the $N(S)$ at the time that the $N(R)$ is received. (The M,I72,P and M,RR2,F exchange shown later in this example illustrates this.)

When one of the stations runs out of I frames to send, then it will generate appropriate supervisory frames (such as M,RR3 and M,RR4 in the example) in order to acknowledge I frames received. When I frames again become available, I frame transmission will resume with the next highest $N(S)$ and the current value of $N(R)$. The $N(S)$ sequence will be ever increasing (except during recovery—Fig. 6), incremented by one with each new I frame. The $N(R)$ value contained in each frame will be the current value of the receive state variable and may be incremented by more than one from frame to frame as a result of the transmission of different length I frames by the two stations involved (illustrated by the transmission of L,I10,P and L,I47).

In the example, L has concluded its transmission of I frames with M,I24 and is interested in disconnecting the link. After having sent M,I24, L receives M,RR2,F that is the reply to the earlier P bit (M,I72,P). Except for M,I24, all of L's transmitted I frames have been acknowledged. Hence, L transmits M,RR5,P to get an acknowledgment report on its last outstanding I frame. In the process, L acknowledges receipt of the I frame numbered 4 from M by setting $N(R)$ equal to 5 in M,RR5,P. While L is cleaning up its records in this fashion, M transmits an I frame L,I52,P with the P bit set to "1." When M receives the P bit set to "1" from L, it generates the M,RR3,F response as a reply. L receives the P bit set to "1" from M, generates an RNR response (L,RNR6,F) in an effort to preclude further I frame transmission by M. In addition, as soon as L receives the F bit frame M,RR3,F in response to its P bit frame (M,RR5,P), L sends M,RNR6,P to stop any further I frame transmission by M. As soon as the F response frame M,RR3,F is received from M, L indicates the disconnect procedure by sending M,DISC,P. When M transmits the M,UA,F frame and L receives it, the link is in the disconnected mode.

2. Examples of Error Recovery Procedures

Many of the error recovery principles that were explained for the primary station in the unbalanced case are applicable to both combined stations in the balanced type of operation as well. Included are time-out functions associated with P bit transmissions, automatic retransmission of S or U format frames, and inquiry of status with a supervisory command before retransmitting I format frames. Consequently, they are not reiterated here.

Because of the two-way simultaneous nature of the operation and the asynchronous mode of operation, the only way to determine when a response received is a response to a command sent is to set the P bit equal to "1" in the command frame for which a response is required. Otherwise, because of propagation delays and offset of possibly different length I frames, it is possible for frame n to be acknowledged in an I or RR frame after I frame numbered $n + x$ has been transmitted. It is also possible for several I frames in a row to have the same $N(R)$ value because of a long frame having been sent in the other direction. If the last I frame in a series is sent without the P bit set to "1," then the lack of a receipt of a response frame from the other station should not of itself be considered grounds for automatic retransmission. An enquiry of the other station's status first (using P/F bit exchange) is the recommended operation. In general, the use of P bits in both directions helps ensure an orderly operation of the two-way simultaneous exchange of data.

The REJ command/response function provides a mechanism for the receiving station to indicate to the sending station that the transmission in progress should be halted and retransmission should begin from the I frame number indicated by the $N(R)$ value in the REJ frame. It provides a mechanism for reporting an error condition prior to the point where a P/F exchange would provide for reporting the same condition from the receiving station. [As mentioned earlier, the repeat of the same $N(R)$ value in non-P/F frames cannot be construed in any way as a request for retransmission of I frames from that specified point.]

Figure 6 shows how the REJ function can be used to initiate retransmission earlier than waiting for the P/F cycle to be activated or for the number of outstanding unacknowledged I frames to equal the modulus minus one. Having the ability to send multiple frames between acknowledgments is a valuable asset of the bit-oriented procedures. Using the REJ function where two-way simultaneous capability exists helps remove what might otherwise be considered to be a negative aspect of the multiple frame transmission capability.

The examples of bit-oriented operation that are given here provide a brief insight into the capabilities of such procedures. Needless to say, many points have not been covered. The ADCCP standard is offered as a more complete description of the general features and capabilities of this new breed of data link control.

V. Related Standards Activities

The American National Standards Institute approved ADCCP in January, 1979.

There are related efforts in the area of bit-oriented data link control standards development that also will likely influence the destiny of bit-oriented procedures. These are the International Organization for Standardization (ISO) HDLC, the International Telegraph and Telephone Consultative Committee (CCITT) Recommendation X.25 Level 2 (LAPB), Recommendation X.75 Level 2 and Recommendation Q.921 (ISDN Level 2), the Federal Government FIPS 71/Fed, Std. 1003, and the Institute of Electrical and Electronics Engineers (IEEE) Project 802.

The ISO bit-oriented activity is known as HDLC, High-Level Data Link Control. The related ISO documents [3]–[5] cover essentially the same material that is covered by the ANSI ADCCP standard. Not included are (1) switched network conventions, and (2) four reserved U frame format commands and responses that ADCCP sets aside for system designer use. The four reserved U frame format code points are reserved for implementor use to provide data link functions that are not included in the standard but that may be required in certain applications, with assurance that the selected code points will not be assigned to a standard function at some later time. Like ADCCP, HDLC is general in scope and broad in possible applications.

The CCITT bit-oriented activities that are consistent with the standards activities take the form of Recommendation X.25 Level 2 LAPB, Recommendation X.75, Level 2 and Recommendation Q.921 Level 2. (The Recommendation X.25 Level 2 LAP procedures are not cited here because the LAP procedures are predicted on a different (nonstandard) definition and use of the SARM command than that recognized by the ADCCP or ISO standards.)

Recommendation X.25 defines a DTE/DCE interface to public data networks. Recommendation X.75 defines the interface between two network signaling terminals (gateways). Recommendation Q.921 defines the DTE/DCE interface on the D-channel for ISDN. In all cases, the bit-oriented procedures employed were adapted from the ISO balanced class of procedures (BAC) with optional functions 2 (REJ) and 8 (I Frames as Commands Only). Future studies of CCITT needs may result in reconsideration of other classes and/or optional functions.

The potentially large scale of applications of the BAC,2,8 class of procedure makes it a front runner when it comes to those classes of procedure that will likely find themselves "engraved in silicon" and hence will become major factors in the design of data link control procedure IC chips for link level operation between "logical equals."

The Federal Government also plays a significant role in the development of bit-oriented data link control procedures. FIPS 71/Fed Std 1003 is essentially the ADCCP standard, with a few minor exceptions. For example, the four reserved U frame format commands and responses are not supported.

FIPS 71/Fed Std 1003 is intended to serve as the basis for all future bit-oriented data link control equipment procurement throughout the government. For that reason, the large number of optional functions that are defined presents a potential nightmare when it comes to the interconnection of stations, or when it is necessary to move a station from one use and application to another. At present, in order for a piece of equipment to satisfy the condition in FIPS 71/Fed Std 1003 that will ensure a certain level of interoperability, it must, as a primary or combined station, be capable of (1) accepting a FRMR response from the other station that indicates that a command/response received was not implemented, (2) resetting the data link, and (3) following a revised data link procedure that does not involve the use of the nonimplemented command or response. A future alternative is being considered wherein the federal standard would specify only a limited set of classes with specified optional functions. If this latter course of action is followed, there would likely be a significant amount of pressure to rethink the subject of bit-oriented data link control procedure standards from the standpoint of standardizing only a minimal set of fairly specifically spelled out classes of procedures instead of standardizing a general framework and the elements of a much larger group of possible combinations of classes and optional functions. Whether such action would be in the best interest of the data communication community at this time is not clear.

Beginning early in 1980 an intense effort to define bit-oriented data link control procedures for local area network applications was initiated by the Institute of Electrical and Electronics Engineers (IEEE) Project 802. In general, these local area network applications covered communications between multiple stations operating in a "peer" relationship in a noncentralized, multiaccess environment on a common medium, running at speeds in the tens-of-megabits range. The "peer" relationship has identified a number of necessary and desirable extensions and enhancements to the present bit-oriented data link control definition. Examples include (1) provision for both "destination" and "spirce" addressing in each frame, and (2) control of station access to the medium. To maximize the likelihood of satisfying the needs of the local area network applications in an evolving bit-oriented data link control definition, close liaison has, and is, being maintained with ANSI, ISO, CCITT, and others. ISO Standard 8802-2 has been approved. It covers the Logical Link Control (LLC) protocols used in LANs.

VI. The Future

Work is still going on in the area of bit-oriented data link control standardization. Additional commands and responses are under consideration. Some of the existing functions (e.g., XID, Initialization Mode) are being examined in greater detail to identify if there are additional aspects of their utility that warrant definition and standardization.

A multilink operating capability has been approved by CCITT and by [6]. Multilink operation provides a mechanism that interfaces the higher level (Level 3) with a multiplicity of individually operating single link data link control procedure packages in such a manner that the group of single links appears to be the higher level as a single data link of greater bandwidth and increased reliability and integrity. The multilink procedures provide for the automatic distribution of information fields to the single link data link controls for transmission to the remote station where the information fields will be returned to their original sequence before delivery to the higher level. The nature of the multilink procedure is predicated on the high level of data transfer integrity of the single link procedures and relies on the capabilities of the higher levels to recover from any error situations that could result from infrequent equipment malfunctions or system errors that might occur above the single link data link level.

There has also been an increasing interest in possible extension to, or other applications of, these bit-oriented procedures. To date, the procedures are designed for application in synchronous systems operating in either a centralized control environment or an equal-party noncentralized control environment (e.g., point-to-point data links, and multiplex data links in local area networks). There is now a proposal for applying the procedural concepts to asynchronous (start–stop) environments. Use of bit-oriented procedures in areas of application like facsimile, TELETEX, text processing, and electronic mail, may also result in additions, extensions, or modifications to the bit-oriented procedures in the future.

The secure nature of the data link frame—its transparency (bit stuffing), its high level of error protection (16-bit or 32-bit CRC), plus its directability (address field assignment)—make it attractive for other applications besides transferring end user data. From a Level 1 maintenance and operations point of view, the bit-oriented frame appears to be an ideal vehicle for use in activating and deactivating Level 1 (physical line level) loopbacks or initiating other Level 1 related functions. Similarly, data link frames seem to be natural vehicles for Level 3 (and higher) use in exchanging user/network information relative to the establishment of a user-to-user (end-to-end) data transfer exchange. Consideration of uses such as these may be subjects of near-future activities. What impact they may have on the present organization of bit-oriented data link procedures is not clear.

VII. Conclusions

After nearly a decade and a half of inventing, evaluating, deliberating, and comprising, a bit-oriented data link control approach has emerged that satisfies the general requirements cited earlier for interactive operation. Although known by many names at the moment (ADCCP, LAPB, SDLC, etc.), this new breed of data link control contains the necessary features, capabilities, characteristics, and growth potential, and has the level of acceptance nationally and internationally to make it *the* Data Link Control approach for use in providing a high-integrity, transparent data transfer mechanism at the data link level that will satisfy today's as well as tomorrow's synchronous data communication needs.

References

[1] ANSI Standard X3.66-1979, "Advanced Data Communication Control Procedures," (Copies obtainable from American National Standards Institute, 1430 Broadway, New York, NY 10018).
[2] R. J. Cypser, *Communications Architecture for Distributed Systems*. Reading, MA: Addison-Wesley, 1978, p. 385.
[3] ISO-3309:1984, "HDLC, Frame structure," (References [3]–[5] and [6] are available from Computer and Business Equipment Manufacturers Assoc., 311 First Street, N.W., Suite 500, Washington, DC 20001).
[4] ISO-4335:1987, "HDLC, Elements of procedures,"
[5] ISO-7809:1984, "HDLC, Consolidation of classes of procedure."
[6] ISO-7478:1987, "Multilink procedures."

6

Multiaccess Link Control

Fouad A. Tobagi

I. Introduction

The need for multiaccess protocols arises whenever a resource is shared (and thus accessed) by a number of independent users. One main reason contributing to such a situation is the need to *share scarce and expensive resources*. An excellent example is typified by time-sharing systems. Time-sharing was developed in the 1960s to make the powerful processing capability of a large computer system available to a large population of users, each of whom has relatively small or infrequent demands so that a dedicated system cannot be economically justified. Two advantages are gained: the smoothing effect of large populations on the demand, an effect resulting from the law of large numbers, and a lower cost per unit of service resulting from the (almost always existing) economy of scale.

A second major reason contributing to the multiaccess of a common resource by many independent entities is the need for communication among the entities; we refer to this as the *connectivity requirement*. An excellent example today is the telephone system, the main purpose of which is to provide a high degree of connectivity among its subscribers. The multiaccess protocol used in the telephone system is conceptually simple; it merely consists of placing a request for connection to one or several parties, a request which gets honored by the system if all the required resources are available.

FOUAD A. TOBAGI • Computer Systems Laboratory, Stanford University, Stanford, California 94305.

A. Packet Communication

Let us now consider data communication systems. Communications engineers have long recognized the need to multiplex expensive transmission facilities and switching equipment. The earliest techniques for doing this were synchronous time-division multiplexing and frequency-division multiplexing. These methods assign a fixed subset of the time-bandwidth space to each of several subscribers and are very successful for stream-type traffic such as voice. With computer traffic, however, usually characterized as *bursty*, fixed assignment techniques are not nearly so successful, and to solve this problem, *packet communication systems* have been developed over the past decade [1]–[7]. Packet communication is based on the idea that part or all of the available resources are allocated to one user at a time but for just a short period of time. Here each component of the system is itself a resource which is multiaccessed and shared by the many contending users. To achieve sharing at the component level, customers are required to divide their messages into small units called packets which carry information regarding the source and the intended recipient.

One type of packet communication network, known as the *point-to-point store-and-forward* network, is one where packet switches are interconnected by point-to-point data circuits according to some topological structure. Packets are transmitted independently and pass asynchronously from one switch to another until they reach their destination. The multiplexing of packets on a channel is done by queueing them at each switch until the outgoing channel is free. Typical examples are the ARPANET [7], the CIGALE subnetwork [8], TELENET [9], and DATAPAC [10].

Another type of packet transmission network is the (single-hop) *multiaccess/broadcast* network typified by the ALOHA network [11], SATNET [12], and Ethernet [5]. Here a *single* transmission medium is shared by all subscribers; the medium is allocated to each subscriber for the time required to transmit a single packet. The inherent single-hop broadcast nature of these systems achieves full connectivity at small additional cost. Each subscriber is connected to the common channel through a smart interface which listens to all transmissions and absorbs packets addressed to it.

Yet a third type of packet network can be identified. It is the (multihop) *store-and-forward multiaccess/broadcast* type which combines the features exhibited (and problems encountered) in the two types just mentioned. The best and perhaps only example of this type is the packet radio network (PRNET) sponsored by the Defense Advanced Research Projects Agency [13], [14]. The concept of the PRNET is an extension of that of the ALOHA network in that it includes many added features such as direct communication by a ground radio network between mobile users over wide geographical areas, coexistence with possibly different systems in the same frequency band and antijam protection. The key requirement of direct communication over wide geographical areas renders store-and-forward switches, called repeaters, integral components of the system. Furthermore, for easy communication among mobile users and for rapid deployment in military applications, all devices employ omnidirectional antennas and share a high-speed radio channel; hence the multiaccess/broadcast nature of the system.

The main issue of concern in this chapter is how to control access to a common channel to efficiently allocate the available communication bandwidth to the many contending users. The solutions to this problem form the set of protocols known as *multiaccess protocols*. These protocols and their performance differ according to the environment in question and the system requirements to be satisfied. We devote the next few paragraphs to summarizing the basic relevant characteristics underlying these environments.

B. Multiaccess Broadcast Environments

Consider first *satellite channels*. A satellite transponder in a geostationary orbit above the earth provides long-haul communication capabilities. It can receive signals from any earth station in its coverage pattern and can transmit signals to all such earth stations (unless the satellite uses spot beams). Full connectivity and multidestination addressing can both be readily accommodated. The many characteristics regarding data rates, error rates, satellite coverage, channelization, and design of earth stations have been fully discussed in a paper by Jacobs *et al.* [12]. Perhaps the most important characteristic relevant to this discussion is the inherent long propagation delay of approximately 0.25 s for a single hop. This delay, which is usually long compared to the transmission time of a packet, has a major impact on the bandwidth allocation techniques and on the error and flow control protocols.

In *ground radio* environments, the propagation delay is relatively short compared to the transmission time of a packet, and as we shall see in the sequel, this can be of great advantage in controlling access to a common channel. It is important, however, to distinguish single-hop environments where direct full connectivity is assumed to prevail, and more complex user environments where, due to geographical distance and/or obstacles opaque to UHF signals, limited direct connectivity is achieved. Clearly, the latter situation is significantly more complex as it gives rise to a multihop system where global control of system operation and resource allocation (whether centralized or distributed) is much harder to accomplish. Another dimension of complexity results from the fact that, unlike satellite environments where earth stations are stationary, ground radio systems must also support mobile users. With mobile users, not only does demand on the system exhibit relatively fast dynamic changes, but the radio propagation characteristics are subject to important variations in received signal strength so that system connectivity is at all times difficult to predict; with these considerations it is important to devise access schemes and system control mechanisms that allow the system to adapt itself to these changes. Furthermore, multipath effects in urban environments can be so disastrous that special signaling schemes, such as spread spectrum, may be in order. Finally, another point of growing concern today is the efficient utilization of the radio frequency spectrum. This is becoming an increasingly predominant factor in determining the structure of radio systems, both in satellite and ground environments. A packet radio system which allows the dynamic allocation of the spectrum to a large population of bursty mobile users needs flexible high-performance multiaccess schemes which can take advantage of the law of large numbers, and which permit coexistence of the system with other (possibly different) systems in the same frequency band.

Finally, we consider *local area communication* systems. These span short distances (ranging from a few meters up to a few kilometers) and usually involve high data rates. The transmission medium can be privately owned and inexpensive, such as twisted pair, coaxial cable, or optical fiber. Local area environments are characterized by a large and often variable number of devices requiring interconnection, and these are often inexpensive. These situations call for communication networks with simple topologies and simple and inexpensive connection interfaces that can provide great flexibility in accommodating the variability in the environment and that achieve the desired level of reliability. With these constraints, we again face the situation in which a high bandwidth channel is to be shared by independent users. Short propagation delays and special physical aspects of the medium are the main characteristics that are exploited in devising multiaccess schemes appropriate to local area environments.

Multiaccess schemes are evaluated according to various criteria. The performance characteristics that are desirable are, first of all, high bandwidth utilization and low message delays. But a number of other attributes are just as important. The ability for an access protocol to simultaneously support traffic of different types, different priorities, with variable message lengths, and differing delay constraints is essential as higher bandwidth utilization is achieved by the multiplexing of all traffic types. Also, to guarantee proper operation of schemes with distributed control, robustness, defined here as the insensitivity to errors resulting in misinformation regarding the status of the system, is also most desirable.

C. A Classification of Multiaccess Protocols

Having so far discussed briefly the basic characteristics and system requirements underlying the various communication environments, we now proceed with a discussion of the multiaccess protocols appropriate to these environments. Besides their various degrees of appropriateness to these environments, these protocols differ by several aspects, namely, the static or dynamic nature of the bandwidth allocation algorithm, the centralized or distributed nature of the decision-making process, and the degree of adaptivity of the algorithm to changing needs. Accordingly, these protocols can be grouped into five classes. The first class, labeled *fixed assignment techniques*, consists of those techniques which allocate the channel bandwidth to the users in a static fashion, independently of their activity. The second class is that of *random access techniques*. In this class the entire bandwidth is provided to the users as a single channel to be accessed randomly; since collisions may result which degrade the performance of the channel, improved performance can be achieved by either synchronizing users so that their transmissions coincide with the boundaries of time slots, by sensing carrier prior to transmission, or both. The third and fourth classes correspond to *demand assignment* techniques. Demand assignment techniques require that explicit control information regarding the users' need for the communication resource be exchanged. A distinction is made between those techniques in which the decision making is centralized (constituting the third class in question), and those techniques in which all users individually execute a distributed algorithm based on control information exchanged among them. The latter constitute the fourth class. The fifth class, labeled *adaptive strategies and mixed modes*, includes those techniques which consist of a mixture of several distinct modes, and those strategies in which the choice of an access scheme is itself adaptive to the varying need, in the hope that near-optimum performance will be achieved at all times.

We describe the various protocols known today, either implemented or proposed, and discuss their performance and applicability to the different environments introduced in this section. For simplicity and clarity in presentation, we consider the (conceptually simplest) situation consisting of M users sharing a common channel over which they communicate. This situation arises typically in satellite communication, in a single-hop ground radio environment, or in a shared bus local network.

II. Fixed Assignment Techniques

Fixed assignment techniques consist of allocating the channel to the users, independently of their activity, by partitioning the time-bandwidth space into slots which are assigned in a static predetermined fashion. These techniques take two common forms: *orthogonal*, such as frequency division multiple access (FDMA) or synchronous time division multiple access (TDMA), and *"quasiorthogonal"* such as code division multiple access (CDMA).

A. FDMA and TDMA

FDMA consists of assigning to each user a fraction of the bandwidth and confining its access to the allocated subband. Orthogonality is achieved in the frequency domain. FDMA is relatively simple to implement and requires no real time coordination among the users.

TDMA consists of assigning fixed predetermined channel time slots to each user; the user has access to the entire channel bandwidth, but only during its allocated slots. Here, signaling waveforms are orthogonal in time.

A number of disadvantages exist for both FDMA and TDMA. FDMA wastes a fraction of the bandwidth to achieve adequate frequency separation. FDMA is also characterized by a lack of flexibility in performing changes in the allocation of the bandwidth and certainly the lack of broadcast operation. The major disadvantages in TDMA are the need to provide A/D converters for analog traffic such as voice, and rapid burst synchronization and sufficient burst separation to avoid time overlap. However, it has been shown that guard bands of less than 200 ns are achievable (as in INTELSAT'S MAT-1 TDMA system, for example) and many operational systems are moving towards the use of TDMA [16]. Timing at an earth station is provided by a global time reference established either explicitly by a reference station, or implicitly by measurement of the propagation delay from the earth station to the transponder. In order to allow the TDMA modems to acquire frequency, phase, bit timing, and bit framing synchronization for each received burst, a preamble is included in front of each burst requiring typically from 100 to 200 bit times. Thus clearly, TDMA is more complex to implement than FDMA, but an important advantage is the connectivity which results from the fact that all receivers listen to the same channel while senders transmit on the same common channel at different times. Accordingly, many network realizations, both in satellite and ground radio environments, are easier to accomplish.

From the performance standpoint it has also been established that TDMA is superior to FDMA in many cases of practical interest. Rubin has shown that the random variable representing packet delay is always larger in FDMA than in TDMA for comparable systems [17]. Lam derived the average message delay for a TDMA system with multipacket messages and a nonpreemptive priority queue discipline [18]. There, too, it was shown that TDMA is superior to FDMA.

For both FDMA and TDMA, the fixed preallocation of the frequency or time resource does not have to be equal for all users, but can be tailored to fit their needs (assumed constant). Kosovych studied two TDMA implementations [19]. In the first, called *contiguous assignment*, the users are cyclically ordered in the time sequence in which they have access to the channel. Each user is periodically assigned its *own* fixed time duration. In the second implementation, called *distributed allocation*, all access periods are of equal time duration, but the frequency of accesses can be different from one user to the other. It was shown that for situations in which the transmission overhead (defined as guard time and synchronization preamble time) is large, the contiguous fixed assignment implementation is better suited and provides substantially better performance than distributed fixed assignments, while when the transmission overhead is small, distributed fixed assignments provide slightly better performance.

Finally we note that, even though the allocation can be tailored to the relative need of each user, fixed allocation can be wasteful if the users' demand is highly bursty, as we shall explicitly see in the sequel. Given these limitations, one may increase the channel utilization beyond FDMA and TDMA by using asynchronous time division multiple access (ATDMA), also known as statistical multiplexing [70]. Basically the technique consists of switching the allocation of the channel from one user to another only when the former is idle and the latter is ready to transmit data. Thus the channel is *dynamically* allocated to the various users according to their need. The performance of ATDMA in packet communication systems corresponds to that of a work-conserving single server queueing system, and is the best we can achieve under unpredictable demand. Unfortunately, it is not always possible to accomplish the necessary coordination among the users. This mode of multiplexing is possible only when several collocated users (such as the same earth station) are sharing a single point-to-point channel.

B. CDMA

Unlike FDMA and TDMA, code division multiple access allows overlap in transmission both in the frequency and time coordinates. It achieves orthogonality by the use of different signaling codes in conjunction with matched filters (or equivalently, correlation detection) at the intended receivers. Multiple orthogonal codes are obtained at the expense of increased bandwidth requirements (in order to spread the waveforms); this also results in a lack of flexibility in interconnecting all users (unless, of course, matched filters corresponding to all codes are provided at all receivers). However, CDMA has the advantage of allowing the coexistence of several systems. Moreover, it is also possible to separate, by "capture," time overlapping signaling waveforms with the same code, thus achieving connectivity and efficient spectrum utilization. This interesting possibility falls into the class of random access techniques and is addressed in the following section.

III. Random Access Techniques

In computer communication, much data traffic is characterized as bursty (e.g., interactive terminal traffic). Burstiness is a result of the high degree of randomness seen in the generation time and size of messages and of the relatively low-delay constraint required by the user. If one were to observe the user's behavior over a period of time, one would see that the user requires the communications resources rather infrequently; but when he does, he requires a rapid response. That is, there is an inherently large peak-to-average ratio in the required data transmission rate. If fixed sub-channel allocation schemes are used, then one must assign enough capacity to each subscriber to meet his peak transmission rates, with the consequence that the resulting channel utilization is low. A more advantageous approach is to provide a single sharable high-speed channel to the large number of users. The strong law of large numbers then guarantees that, with a very high probability, the demand at any instant will be approximately equal to the sum of the average demands of that population. As stated in the introduction, packet communication is a natural means to achieve sharing of the common channel. When dealing with shared channels in a packet-switched mode, one must be prepared to resolve conflicts which arise when more than one demand is placed upon the channel. For example, in narrowband radio channels, in the absence of any form of capture, whenever a portion of one user's transmission overlaps with another user's transmission, the two collide and "destroy" each other; similarly for baseband signaling over a coaxial cable. The existence of some positive acknowledgment scheme permits the transmitter to determine if his transmission is successful or not. The problem is how to control the access to the common channel in a fashion which produces, under the physical constraints of simplicity and hardware implementation, an acceptable level of performance. The difficulty in controlling a channel which must carry its own control information has given rise to the so-called random-access protocols, among others. We describe these here by considering again single-hop environments.

A. ALOHA

Historically, the *pure* ALOHA protocol was first used in the ALOHA system, a single-hop terminal access network developed in 1970 at the University of Hawaii, employing packet-switching on a radio channel [11], [20]. The simplest of its kind, pure ALOHA permits users to transmit any time they desire. If within some appropriate time-out period following its transmission, a user receives an acknowledgment from the destination (the central computer), then it knows that no conflict occurred. Otherwise it assumes that a collision occurred and it must retransmit. To avoid continuously repeated conflicts, the retransmission delay is randomized across the transmitting devices, thus spreading the retry packets over time. Assume all packets are of the same size, and let T denote the transmission time of a packet. The transmission of a tagged packet will be successful if no other packets begin transmission T seconds before and T seconds after the start of transmission of the tagged packet. Thus we say that in pure ALOHA a packet transmission has a vulnerable period equal to twice its transmission time. A slotted version, referred to as *slotted* ALOHA, is obtained by dividing time into slots of duration equal to the transmission time of a single packet (assuming constant-length packets) [21], [22]. Each user is required to synchronize the start of transmission of its packets to coincide with the slot boundary. When two packets conflict, they will overlap completely rather than partially. This decreases the vulnerability period of a packet to only one transmission time, providing an increase in channel efficiency over pure

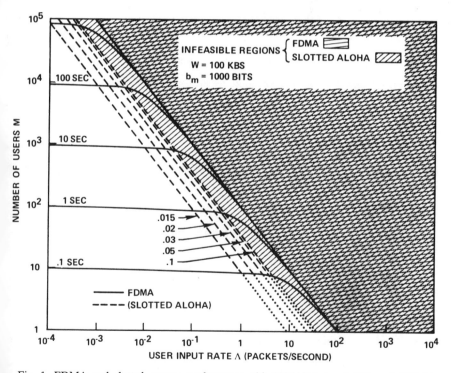

Fig. 1. FDMA and slotted ALOHA: performance with 100-kbit/s bandwidth and 1000-bit packets. Contours are for constant delay [23].

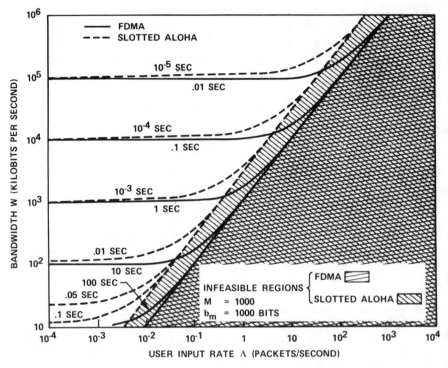

Fig. 2. FDMA and slotted ALOHA bandwidth requirements for 1000 terminals and 1000-bit packets. Contours are for constant delay [23].

ALOHA. Owing to conflicts and idle channel time, the maximum channel efficiency available using ALOHA is less than 100 percent, 18 percent for pure ALOHA and 36 percent for slotted ALOHA. Both schemes are theoretically applicable to satellite, ground radio, and local bus environments. The slotted version has the advantage of efficiency, but it has the disadvantage that synchronization may be hard to achieve, especially in multihop ground radio.

Although the maximum achievable channel utilization is low, the ALOHA schemes are superior to fixed assignment schemes when there is a large population of bursty users. This point is illustrated in comparing for example the performance of FDMA with that of slotted ALOHA when M users, each of which generates packets at a rate of Λ packets per second, share a radio channel of W Hz [23]. Figures 1 and 2 display the constant delay contours in the (M, Λ) and (W, Λ) planes, respectively. These figures clearly show the important improvement gained in terms of bandwidth required, population size supported and delay achieved when the users are bursty.

B. Carrier Sense Multiple Access (CSMA)

In certain environments such as ground radio and local area, the propagation delay between any source–destination pair may be small compared to the packet transmission time. In such an environment one may attempt to avoid collisions by listening to the carrier due to another user's transmission before transmitting, and inhibiting transmission if the channel is sensed busy. This feature gives rise to a random access scheme known as carrier sense multiple access (CSMA) [24], [25]. While in the ALOHA scheme only one action could be taken by the terminals, namely, to transmit, here many strategies are possible so that many CSMA protocols exist differing according to the action that a terminal takes to transmit a packet after sensing the channel. In all cases, however, collisions cannot be totally avoided owing to the nonzero propagation delay among the devices; thus, when a terminal learns that its transmission had incurred a collision, it reschedules the transmission of the packet according to the randomly distributed delay. At this new point in time, the transmitter senses the channel again and repeats the algorithm dictated by the protocol. There are two main CSMA protocols known as *nonpersistent* and *p-persistent* CSMA depending on whether the transmission by a station which finds the channel busy is to occur later or immediately following the current one with probability p. (Many variants and modifications of these two schemes have also been proposed.) Thus, in nonpersistent CSMA, a ready terminal (i.e., a terminal with a packet ready for transmission) senses the channel and operates as follows:

1. If the channel is sensed idle, it transmits the packet.
2. If the channel is sensed busy, then the terminal schedules the retransmission of the packet to some later time according to the retransmission delay contribution. At this new point in time, it senses the channel and repeats the algorithm described.

The 1-persistent CSMA protocol, a special case of p-persistent CSMA, was devised in order to (presumably) achieve acceptable throughput by never letting the channel go idle if some ready terminal is available. More precisely, a ready terminal senses the channel and operates as follows:

1. If the channel is sensed idle, it transmits the packet with probability one.
2. If the channel is sensed busy, it waits until the channel goes idle and then immediately transmits the packet with probability one (i.e., persisting on transmitting with $p = 1$).

In the case of a 1-persistent CSMA, we note that whenever two or more terminals become ready during a packet transmission period, they wait for the channel to become idle (at the end of the transmission) and then they all transmit with probability one. A conflict will also occur with probability one. The idea of randomizing the starting time of transmission of packets accumulating at the end of a transmission period seems reasonable for interference reduction and throughput improvement. Thus we have the p-persistent scheme which involves including an additional parameter p, the probability that a ready packet persists $(1 - p$ being the probability of delaying transmission by τ seconds, the propagation delay). If at this new point in time, the channel is still detected idle, the same process is repeated. Otherwise some packet must have started transmission, and the terminal in question schedules the retransmission of the packet according to the retransmission delay distribution (i.e., acts as if it had conflicted and learned about the conflict); alternatively, the algorithm may be defined such that the terminal waits until the channel becomes idle again (at the end of the current transmission) and then operates as above. The parameter p is chosen so as to reduce the level of interference while keeping the idle periods between any two consecutive nonoverlapped transmissions as small as possible.

To best see the improvement due to carrier sensing, consider, for example, the nonpersistent CSMA protocol. Given that a tagged packet has started transmission following an idle channel period, the transmission will be successful if no other transmission begins during the first τ seconds of the tagged packet, where τ is the maximum propagation delay among all pairs of devices. Accordingly, τ represents the vulnerable period of a packet, a fraction that we denote by $a \triangleq \tau/T$.

A slotted version of these CSMA protocols can also be considered in which the time axis is slotted and the slot size is τ seconds. Note that this definition of a slot is different from that used in the description of slotted ALOHA. Here a packet transmission time is equivalent to several slots. We make this distinction by referring to a slot of size τ seconds as a "minislot." All terminals are synchronized and are forced to start transmission only at the beginning of a minislot.

Packet broadcasting technology has also been shown to be very effective in satisfying many local area in-building communication requirements. A prominent example is Ethernet, a local communication network which uses CSMA on a tapped coaxial cable to which all the communicating devices are connected [5]. The device connection interface is a passive cable tap so that failure of an interface does not prevent communication among the remaining devices. The use of a single coaxial cable achieves broadcast communication. The only difference between this and the single-hop radio is that, in addition to sensing carrier, it is possible for the transceivers to detect interference among several transmissions (including their own), and to abort the transmission of colliding packets. This is achieved by having each transmitting device compare the bit stream it is transmitting to the bit stream it sees on the channel. This variation of CSMA is referred to as carrier sense multiple access with collision detection (CSMA-CD).

While until recently most of the concepts described in this section had been realized in experimental systems (namely, the ALOHA system, the PRNET, and Xerox's experimental Ethernet), it is important to note that today contention systems of the Ethernet type are available on the market, and new ones have been announced to be soon available. Examples are the Hyperchannel and the Hyperbus of Network Systems Corporation [73], and Ethernet itself. In fact, the IEEE committee on standardization of local networks has adopted 1-persistent CSMA-CD as one of its standards (the IEEE 802.3 standard) [75]. One aspect that distinguishes this protocol from other CSMA protocols is its contention resolution algorithm, referred to as the binary exponential backoff. The mean rescheduling delay is doubled with each successive collision of the same packet up to the tenth collision. Following the tenth collision, the mean rescheduling delay remains unchanged until the sixteenth collision. After the sixteenth collision transmission attempt is abandoned.

C. Performance of Random Access

Many theoretical studies have been carried out to determine the performance of these random access schemes [20]–[22], [24]–[29]. We summarize here the most important results. Assume all packets are of the same size, and let again T denote the packet transmission time. Assume also (pessimistically) that all users are separated by the maximum propagation delay over all pairs, denoted again by τ. Consider that there exists an infinite population of users, and let S denote the aggregate rate of packet generation from the entire population of users, G the rate of packet transmissions (new and repeated, hence $G \geq S$), and D the packet delay (defined as the time elapsed between the time that the packet is originated and the time it is successfully received at the destination), all normalized to T. Analytic and simulation models provide us, for each random access scheme, with a relationship between S and G (displayed in Fig. 3), and the throughput delay tradeoff (displayed in Fig. 4) for a normalized propagation delay $a = \tau/T = 0.01$. We note that the behavior of these schemes is typical of contention systems, namely, that the throughput increases as the offered channel traffic increases from zero, but reaches a maximum value for some optimum value of G, and then constantly decreases as G increases beyond that optimal value. Maximizing S with respect to the channel traffic rate G for each of the access modes leads to the channel capacity for that mode. From Fig. 4 we clearly note that D increases as the throughput increases, and reaches infinite values as the throughput approaches the channel capacity. These results show the evident superiority of CSMA over the ALOHA scheme. The CSMA channel capacity in some cases may be as

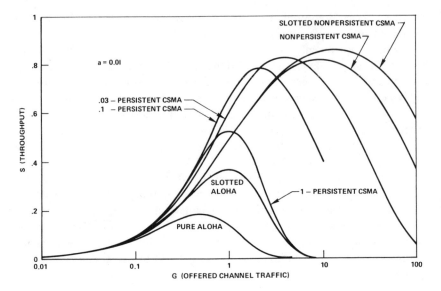

Fig. 3. Throughput for the various random access modes (propagation delay $a = 0.01$) [24].

high as 90 percent of the available bandwidth. It is clear, however, that, as expected, the channel capacity and the throughput-delay tradeoff for the CSMA schemes degrade as the normalized propagation delay ($a = \tau/T$) increases. Figure 5 illustrates the sensitivity of the channel capacity to a.

CSMA-CD offers even more improvement. A system parameter affecting this improvement is the time required to detect collisions and abort ongoing colliding transmissions. The smaller this parameter is, the better the improvement is [26].

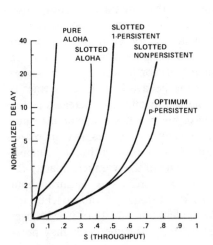

Fig. 4. CSMA and ALOHA: throughput-delay tradeoffs from simulation (propagation delay $a = 0.01$) [24].

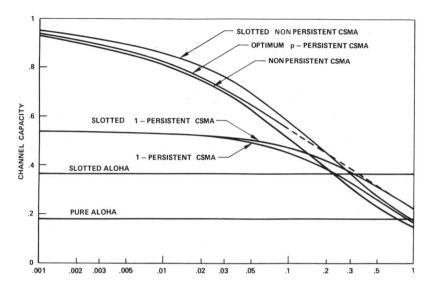

Fig. 5. CSMA and ALOHA: effect of propagation delay a on channel capacity [24].

The results displayed in the above figures have two important assumed conditions, namely (1) acknowledgments are instantaneous, always received correctly and for free (i.e., do not occupy any channel time), and (2) all devices are within range and in line-of-site of each other so that sensing of all transmissions on the channel is perfect. While Condition (1) is relevant to both ALOHA and CSMA, Condition (2) is mostly relevant to CSMA. We discuss these issues in the following.

D. Acknowledgment Procedures and Their Effect

Basically, errors in multiaccess channels are due to two major causes: (1) random noise on the channel and (2) multiuse interference in the form of overlapping packets. A very reliable method ensuring the integrity of the transmitted data is the use of an error-detecting (e.g., cyclic) block code in conjunction with a positive acknowledgment of each correctly received message. Each packet contains a field for the cyclic checksum. Each receiver responds to a complete packet addressed to it with a correct checksum by transmitting an acknowledgment packet back to the originating terminal. This acknowledgment contains (among other things) the unique identification of the originating terminal along with a checksum to ensure the integrity of the acknowledgment packet itself.

It is all too evident that acknowledgments will use part of the total available bandwidth (our limited resource). The amount of overhead introduced, as well as the degradation in delay incurred, varies with the mode of operation. When the available bandwidth is provided as a single channel to be shared by both information and acknowledgment packets, then the channel performance will further suffer from interference between information packets and acknowledgment packets unless some kind of priority scheme is provided. Concerning the degradation in channel capacity due to the overhead created by the error control traffic, it has been shown [30] that, in a common-channel configuration with nonpriority acknowledgment traffic, the channel capacity of slotted ALOHA drops to 14 percent of the channel bandwidth. However, if by some means acknowledgment traffic can be given priority so as to guarantee its transmission free of conflict, then the channel capacity for slotted ALOHA can be maintained at around 26 percent (assuming here that an acknowledgment packet uses an entire slot). The effect of acknowledgment traffic on CSMA channels need not be as dramatic since it is very simple to implement schemes which give priority to acknowledgment packets. One mode of operation is as follows [30, 74].

1. If a terminal with a packet ready for transmission senses the channel idle, then the terminal transmits its packet τ seconds (the propagation delay) later if and only if the channel is still sensed idle.

2. If such a terminal senses the channel busy, then it follows the protocol in question (nonpersistent, 1-persistent, ...) repeating step (1) whenever the channel is sensed idle.

3. All acknowledgment packets are transmitted immediately, without incurring the τ seconds delay.

The capacity of the nonpersistent CSMA protocol with priority acknowledgment and $a = 0.01$ drops gradually from 0.85 to about 0.45 as the acknowledgment packet size increases from 0 to a full packet size.

E. The Hidden Terminal Problem in CSMA and the Busy-Tone Multiple Access (BTMA)

We now relax the assumption that all users are in line of sight and within range of each other. Typically, two terminals can be within range of the intended receiver, but out of range of each other or separated by some physical obstacle opaque to radio signals. The existence of hidden terminals in a radio environment significantly degrades the performance of CSMA. To illustrate this effect, consider a population of users, each of which is communicating with a central station. This station is in line-of-sight communication with the entire population, but this population is divided into two groups (of relative sizes α and $1 - \alpha$) such that the radio connectivity exists only between users in the same group. Figure 6 displays the CSMA channel capacity versus α, showing that the channel capacity drops drastically as α increases from 0 and reaches a minimum at $\alpha = 0.5$ [28].

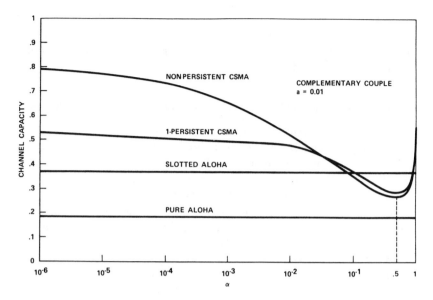

Fig. 6. An example of the hidden terminal situation: complementary couple configuration. Channel capacity versus α, the relative sizes of the two decoupled populations [28].

The hidden-terminal problem can be eliminated by frequency-dividing the available bandwidth into two separate channels, a busy-tone channel and a message channel, and by requiring a device to emit a (sine-wave) busy tone signal on the busy tone channel whenever it is receiving, thus blocking user terminals in its vicinity from using the message channel and interfacing with its reception. Such a scheme is referred to as *busy-tone multiple access* (BTMA). The action that a terminal takes pertaining to the transmission of the packet is again prescribed by the particular protocol being used.

In CSMA, the difficulty of detecting the presence of a signal on the message channel when this message occupies the entire bandwidth is minor and is therefore neglected. This is not realistic when we are concerned with the (statistical) detection of the (sine wave) busy-tone signal on a narrow-band channel. In BTMA, the system's design involves a more complex set of system variables, namely the window detection time, the false alarm probability F, and the fraction of bandwidth devoted to the busy-tone signal. For a detailed analysis of this scheme, the reader is referred to [28]. The throughput-delay tradeoff for BTMA is slightly degraded in comparison to CSMA with no hidden terminals, but still exhibits relatively good performance. The use of BTMA in multihop packet radio networks, where the hidden node problem prevails, brings about important improvements in performance. Analysis and simulation studies have shown that in a packet radio network with a ring topology (an approximation of a tandem network), the network capacity with CSMA lies between those of pure and

slotted ALOHA, while with BTMA, the capacity is about three times higher
[76]–[78]. This result, however, does not take into account the bandwidth
required for the busy tone channel, which is a relatively small portion of the
total available bandwidth.

F. Dynamic Behavior and Dynamic Control of Random Access Schemes

The performance results reported upon above were based on renewal
theory and probabilistic arguments, assuming that steady-state conditions
exist. If one examines in more detail the (S, G) relationships displayed
above for infinite populations, one can see that the steady state may not
exist because of an inherent *instability* of these random access techniques.
This instability is simply explained by the fact that statistical fluctuations in
the offered traffic increase the level of mutual interference among transmissions which in turn increases G, which increases the frequency of collisions,
and so forth. Such positive feedback causes the throughput to decrease to
very low values (zero for the infinite population case). Extensive simulation
runs performed on a slotted ALOHA channel with an *infinite population* of
users have indeed shown that the assumption of channel equilibrium is not
strictly speaking valid; in fact after some finite period of quasistationary
conditions, the channel will drift into saturation with probability one [31].
Thus a more accurate measure of channel performance must reflect the
tradeoffs among stability, throughput, and delay. To that effect, Markov
models have been formulated to analyze slotted ALOHA and CSMA when a
finite number M of interactive users are contending for the channel
[31]–[34]. These models permit one not only to derive valid analytic
expressions for the average throughput-delay performance, since the models
are then ergodic, but also to understand the dynamic behavior of these
systems. In particular, it was observed that in a finite population of users, if
the retransmission delay is sufficiently large, then the stationary performance attained is high, and the system is said to be in good operating
condition. If, however, the retransmission delay is not sufficiently
large, then the stationary performance attained is significantly degraded
(low throughput, very high delay), so that, for all practical purposes, the
channel is said to have failed; it is then called an unstable channel. For a
given access scheme, the appropriate rescheduling delay is a function
primarily of the population size and the offered load, but it also depends on
the physical characteristics of the environment (mostly the parameter a).
With an infinite population, stationary conditions just do not exist; the
channel is always unstable, thus confirming the results obtained from
simulation, as just discussed. For unstable channels, Kleinrock and Lam
[32] defined a stability measure which consists of the average time the
system takes, starting from an empty channel, to reach a particular state
determined to be critical. In fact, this critical state partitions the state space
into two regions: a safe region, and an unsafe region in which the tendency

is towards degraded performance. The stability measure is the average first exit time (FET) into the unsafe region. As long as the system operates in the safe region, the channel performance is acceptable; but then, of course, it is only usable over a finite period of time with an average equal to FET. For more details concerning the determination of FET and the numerical results, the reader is referred to [32] and [33].

In the above discussion, it was furthermore assumed that the system parameters were all fixed, time invariant, and state independent. These systems are referred to as *static*. It is often advantageous to design systems that dynamically adapt to time-varying input and to system state changes, thus providing improved performance. Dynamic adaptability is achieved via dynamic control consisting of time- and state-dependent parameters. The basic problem then is to find the control functions which provide the best system performance. Markov decision theory has successfully been applied by Lam and Kleinrock to the design and analysis of control procedures suitable to slotted ALOHA in particular and random access techniques in general [35]. Two main types of control are proposed: an *input* control procedure (ICP) consisting of choosing between accepting or rejecting all new packets generated in the current slot, and a *retransmission* control procedure (RCP) consisting of selecting an appropriate retransmission delay; in both cases the action taken is a function of the current system state, defined as the number of active users with outstanding packets. In order to implement such control schemes, each channel user must individually estimate the channel state by observing the channel outcome over some period of time. The control is of a distributed nature, as there is no central station monitoring and broadcasting state information or control actions. In the context of slotted ALOHA, Lam and Kleinrock give some heuristic control-estimation algorithms which prove to be very satisfactory [35]. With appropriate modification and extensions, these algorithms can be applied to CSMA channels as well. These algorithms are best suited to fully connected single-hop environments. The dynamic control problem in multihop environments is more complex and little progress has yet been made in this area.

G. Distributed Tree Retransmission Algorithms in Packet Broadcast Channels [71]

In the protocols examined above, conflict resolution is achieved by retransmitting randomly in the future. Such a rescheduling discipline in slotted ALOHA achieves a 36 percent bandwidth utilization, but exhibits some sort of instability unless the rescheduling is controlled, as discussed above. Tree algorithms are based on the observation that a contention among several active sources is completely resolved if and only if all the sources are somehow subdivided into groups such that each group contains at most one active source. (Such observation is similar to that made in the probing technique discussed later in Section IV.) In its simplest form, the tree algorithm consists of the following. Each source corresponds to a leaf on a *binary* tree. The channel time axis is slotted and the slots are grouped into pairs. Each slot in a pair corresponds to one of the two subtrees of the node being visited. Starting with the root node of the tree, we let all terminals in each of the two subtrees of the root transmit in their corresponding slot. If any of the two slots contains a collision, then the algorithm proceeds to the root of the subtree corresponding to the collision and repeats itself. This continues until all the leaves are separated into sets such that each of them contains at most one packet. This is known to all users, as the outcome of the channel is either a successful transmission or an idle slot. Collisions caused by the left subtree (first slot of a pair) are resolved prior to resolving collisions in the right subtree. This scheme provides a maximum throughput of 0.347 packets/slot, and all moments of the delay are finite if the aggregate packet arrival rate is less than $1/3$ packets/slot [71].

Clearly, a binary tree is not always optimum. If each time we return to the root node we allow the tree to be reconfigured according to the current traffic conditions, it can be shown that the optimum tree is binary everywhere except for the root node whose optimum degree depends on traffic conditions [71]. The dynamic scheme achieves a throughput of 0.430 packets/slot, and all the moments of the delay are finite for $\lambda < 0.430$ packets/slot.

A variant of the tree resolution algorithm divides the users into groups based on the time of arrival of their packets. All stations are supposed to know of a common time window in the past. All stations with a packet generated during that window are allowed to transmit. One of three outcomes may occur: If no transmission takes place, the window is advanced according to some formula in order to allow later packets to be transmitted. If a successful transmission takes place, then it is known that no other packet arrived in the window, and there too the window is advanced. If a collision is detected, then the original window is divided into smaller windows, and the earliest of these would be the one determining packets to be transmitted. The interesting feature in this scheme is the order of service, which is first-come-first-served (FCFS); thus the delay variance is smaller than in networks with random order. Tree algorithms are implementable in both ground radio and satellite channels as long as the broadcast capability is available.

H. Power Capture

In the preceding discussions it was assumed that whenever two packet transmissions overlap in time, these packets destroy each other. This assumption is pessimistic as it neglects *capture* effects in radio channels. Capture can be defined as the ability for a receiver to successfully receive a packet (with nonzero probability) although it is partially or totally over-lapped by another packet transmission. Power capture is mainly due to a discrepancy in receive power between two signals allowing the receiver to correctly receive the stronger; both distance and transmit power contribute to this discrepancy. Clearly power capture improves the overall network performance, and, by the means of adaptive transmit power control, it allows one to achieve either fairness to all users, or intentional discrimination. Some of these effects have been addressed in [27] and [36].

I. Spread Spectrum Multiple Access (SSMA)

Spread spectrum multiple access (SSMA) is the most common form of CDMA whereby each user is assigned a particular code sequence which is modulated on the carrier with the digital data modulated on top of that. Two common forms exist: the frequency-hopped SSMA and the phase-coded SSMA. In the former, as its name indicates, the frequency is periodically changing according to some known pattern; in the latter the carrier is phase modulated by the digital data sequence and the code sequence. SSMA has many applications: it is useful in satellite communications, mobile ground-radio, and computer communication networks [37]. In a recent article Kahn *et al.* addressed many of the issues concerning the use of SSMA in packet radio systems. Security, coexistence with other systems, and ability to counteract the effects of multipath are key factors contributing to the choice of SSMA in the PRNET; however, one main point of interest in this presentation is the benefit of capture in asynchronous SSMA. Even when several users employ the same code, the effect of interference is minimized by the "capture effect," defined here as the ability of the receiver to "lock on" one packet while all other overlapping packets appear as noise. The receiver locks on a packet by correctly receiving the preamble appended in the front of the transmitted packet. As long as the preamble of different packets do not overlap in time, and the signal strength of the late packets is not too high, capture of the earliest packet can be guaranteed with a high probability. In essence SSMA allows a packet to be captured at the receiver, while CSMA allows a user to capture the channel. CSMA can still be used in conjunction with SSMA. This mode will have the benefit of keeping away all users within hearing distance of the transmitter and thus help keep the capture effect and antijamming capability of the system at the desired level. For a complete discussion of all these issues, the reader is referred to [14] and [78].

IV. Centrally Controlled Demand Assignment

We have so far discussed the two extremes in the bandwidth allocation spectrum as far as control over the user's access right is concerned: the tight fixed assignment, which has the most rigid control, is nonadaptive to dynamically varying demand, and can be wasteful of capacity if small-delay constraints are to be met; and random access, which involves no control, is simple to implement, is adaptive to varying demand, but which, in some situations, can be wasteful of capacity due to collisions. In this and the following subsections, we examine demand assignment techniques which require that explicit information regarding the need for the communication resource be exchanged. We distinguish those demand assignments which are controlled by a central scheduler from those which employ a distributed algorithm executed by all users. We address centrally controlled assignments in the present section.

A. Circuit-Oriented Systems

In these systems, the bandwidth is divided into FDMA or TDMA subchannels which are assigned on demand. The satellite SPADE system, for example, has a pool of FDMA subchannels which get allocated on request [38]. It uses one subchannel operated in a TDMA fashion with one slot per frame permanently assigned to each user to handle the requests and releases of FDMA circuits. Intelsat's MAT-1 system uses the TDMA approach [39]. TDMA subchannels are periodically reallocated to meet the varying needs of earth stations.

The Advanced Mobile Phone Service (AMPS), also presently known as the cellular mobile phone system, is yet another example of a centrally controlled FDMA system [40]. The uniqueness of this system, however, lies in an efficient management of the spectrum based on space division multiple access (SDMA). That is, each subchannel in the pool of FDMA channels is allocated to different users in separate geographical areas, thus considerably increasing the spectrum utilization. To accomplish space division, the AMPS system has a cellular structure and uses a centralized handoff procedure (executed by a central office) which reroutes the telephone connections to other available subchannels as the mobile users move from one cell to another.

Given the significant setup times required in allocating subchannels, the above systems are attractive only when applications have stream-type traffic. When traffic is bursty, we again turn to packet-oriented systems, such as in the following.

B. Polling

In packet-oriented systems, polling is one of two modes used to centrally control access to the communication bandwidth, again provided as a single high-speed channel. A central controller sends polling messages to the terminals, one by one, asking the polled terminal to transmit. For this, the station may have a polling list giving the order in which the terminals are polled. If the polled terminal has something to transmit, it goes ahead; if not, a negative reply (or absence of reply) is received by the controller, which then polls the next terminal in sequence. Polling requires this constant exchange of control messages between the controller and the terminals, and is efficient only if (1) the round-trip propagation delay is small, (2) the overhead due to polling messages is low, and (3) the user population is not a large bursty one. Polling has been analyzed by Konheim and Meister [41], and their analysis has been applied to the environment of M users sharing a radio channel in [23]. Denoting by L the ratio of the data message length to the polling message length, and by a the ratio of propagation delay to message transmission time, Fig. 7 displays numerical results corresponding to some typical values of L and a. These curves show that indeed as the population size increases, thus containing more and more bursty users, the performance of polling degrades significantly. Channel utilization can reach 100 percent of the channel if the terminals are allowed to empty their buffers when they are polled. But as a result, the variance of packet delay can become intolerably large.

The primary limitation of polling in lightly loaded systems is the high overhead incurred in determining which of the terminals have messages. In order to decrease this overhead, a modified polling technique, based on a tree searching algorithm, and referred to as *probing*, has been proposed [42]. This technique assumes that the central controller can *broadcast* signals to all terminals. First the controller interrogates all terminals simultaneously, asking if any of them has a message to transmit, and repeats this question until some terminals respond by putting a signal on the line. When a response is received, the central station breaks down the population into subsets (according to some tree structure) and repeats the question to each of the subsets. This can be performed simply, for example by using binary addresses for the terminals and by transmitting as probing signal the common prefix of the addresses of a group of terminals. The process is continued until the terminals having messages are identified. When a single terminal is interrogated, it transmits its message.

Assume that the number of terminals is a power of 2, say $M = 2^n$. Let a cycle be recursively defined as the time required for the polling and transmission of all messages that were generated in the preceding cycle. If a single terminal has a message to transmit, probing requires $2n + 1$ inquiries per cycle as opposed to 2^n for conventional polling; but if all terminals have messages, probing requires $2^{n+1} - 1$ inquiries as opposed to 2^n for conventional polling. To avoid incurring such a penalty when the system is heavily loaded, the probing technique can be made adaptive whereby the controller starts a cycle by probing smaller groups as the probability of terminals having messages increases. In particular, the group size may be considered a function of the duration of the immediately preceding polling cycle. Simula-

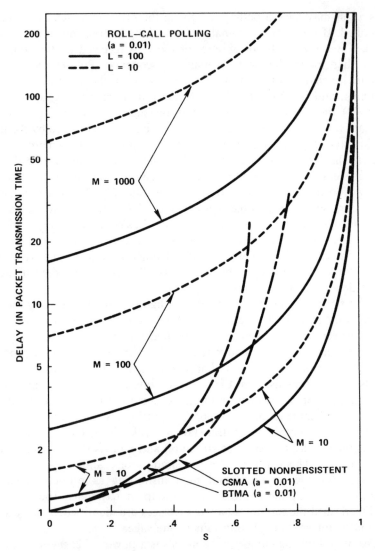

Fig. 7. Packet delay in roll-call polling. L = ratio of data message length to polling message length, a = normalized propagation delay, M = number of stations [23].

tion of the adaptive probing technique has shown that this scheme is always superior to polling. Figure 8 displays the mean cycle time (obtained from simulation) as a function of the message arrival rate for both polling and probing [42]. Reference [42] did not provide any results concerning the message delay, but it is intuitively clear that the smaller the mean cycle time is, the lower is the average delay.

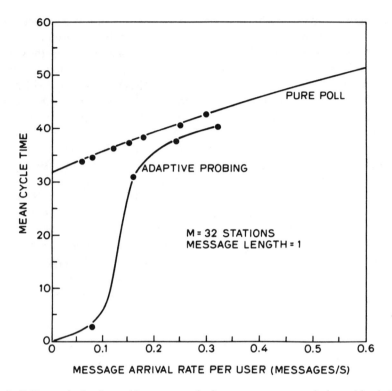

Fig. 8. Polling and adaptive probing: mean cycle time versus message arrival rate (simulation results—32 stations) [42].

C. Reservation-Based Schemes

An attractive alternative to polling is the use of explicit reservation techniques. In dynamic reservation systems, it is the terminal which makes a request for service on some channel whenever it has a message to transmit. The central scheduler manages a queue of requests and informs the terminal of its allocated time.

Since the channel is the only means of communication among terminals, the main problem here is, once again, how to communicate the request to the central scheduler. The contention on the channel of these request packets is of exactly the same nature as the contention of the data packets themselves. Fixed assignment and random access techniques suggest themselves, but it is clear from previous results that random access modes for multiplexing the requests on the channel would be more efficient. Furthermore, in order to prevent collisions between the requests and the actual message packets, the available bandwidth is either time divided or frequency divided between the two types of data. In the split-channel reservation multiple access (SRMA) scheme, frequency division is considered [23]. The available bandwidth is divided into two channels: one used to transmit control information, the second used for the data messages themselves. With this configuration, there are many operational modes. In the request/answer-to-request/message scheme (RAM), the bandwidth allocated for control is further divided into two channels: the request channel and the answer-to-request channel. The request channel is operated in a random access mode (ALOHA or CSMA). Upon correct reception of the request packet, the scheduling station computes the time at which the backlog on the message channel will empty and transmits an answer packet back to the terminal, on the answer-to-request channel, containing the address of the terminal and the time at which it can start transmission. Another version of SRMA, called the RM scheme, consists of having only two channels: the request channel and the message channel. When correctly received by the scheduling station, the request packet joins the request queue. Requests may be serviced on a "first-come first-served" basis (or any other scheduling algorithm). When the message channel is available, an answer packet (containing the ID of a queued terminal scheduled for transmission) is transmitted by the station on the message channel. After hearing its own ID repeated by the station, the terminal starts transmitting its message on the message channel. If a terminal does not hear its own ID repeated by the scheduling station within a certain appropriate time after

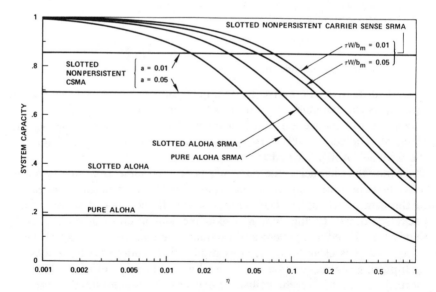

Fig. 9. SRMA: channel capacity versus η, ratio of request packet length to data packet length (normalized propagation delay of 0.01 and 0.05) [23].

the request is sent, the original transmission of the request packet is assumed to be unsuccessful. The request packet is then retransmitted.

We now examine the performance of SRMA. Let η denote the ratio of request packet length to data packet length, this representing a measure of the overhead due to control information. In Fig. 9 we plot the (RAM) SRMA system capacity versus η for the following access modes: pure ALOHA SRMA, slotted ALOHA SRMA, and slotted nonpersistent carrier sense SRMA. In addition, we show the system capacity for both ALOHA and CSMA. We note that the system capacity in SRMA reaches 1 for very small η. Typical values for η fall in the range $(0.01, 0.1)$. Figure 9 shows that a high improvement is gained when the request channel is operated in CSMA

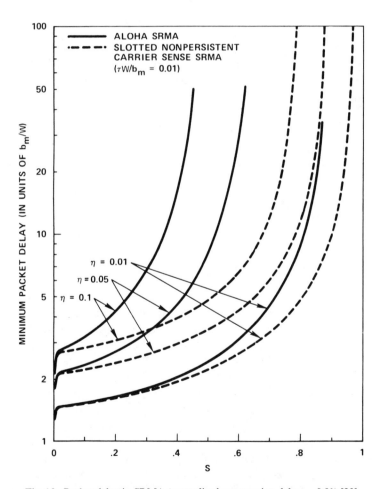

Fig. 10. Packet delay in SRMA (normalized propagation delay = 0.01) [23].

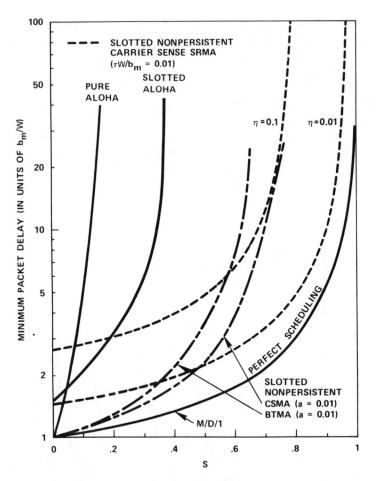

Fig. 11. Comparison of various schemes (parameters defined as in Figs. 9 and 10) [23].

as compared to ALOHA. The delay for ALOHA SRMA and carrier sense SRMA (normalized to b_m/W, where W denotes again the total channel bandwidth, and b_m is the number of bits per packet) is shown in Fig. 10 as a function of S for various values of η. We again note an important improvement in using CSMA for the request channel. Finally, in Fig. 11 we compare carrier sense SRMA with the random access modes ALOHA, CSMA, and BTMA, and with $M/D/1$, representing perfect scheduling with fixed size packets and Poisson sources. We note that unless η is large (0.1 and above), there is a value of S below which CSMA or BTMA performs better than SRMA and above which the opposite is true.

For a high-speed data bus, a conflict-free reservation multiaccess scheme which is based on the time-division concept for reservation is

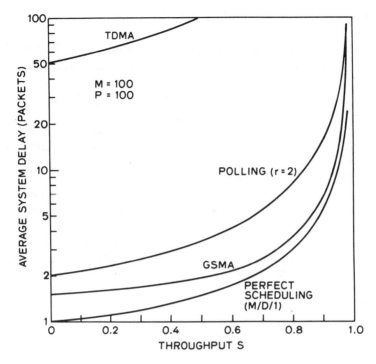

Fig. 12. GSMA: throughput delay performance (M = number of stations, P = number of bits per data packet) [43].

Global Scheduling Multiple Access (GSMA) [43]. Here too a scheduler oversees all scheduling tasks. The users, all connected to the same line, listen for scheduling assignments and transmit in accordance with the slot allocation initiated by the scheduler. The channel time is divided into frames (of variable lengths). A frame is partitioned into two subframes: a subframe of status slots statically assigned to the users (in a fixed TDMA mode) to request data slot allocation, and a subframe of data slots, each sufficient to transmit a data packet of P bits. The fixed assignment of the status slots removes the need to transmit users' IDs and thus reduces the size of these slots. In each frame, a user can be allocated a number of data slots which does not exceed the number of packets generated at the user during the preceding frame or a maximum number specified, whichever is smaller. As a consequence each active user is guaranteed at least one slot per frame. Figure 12 displays the performance of GSMA (with $P = 100$ and number of stations $M = 100$) in comparison to polling (for some typical parameter values regarding the polling overhead r) and $M/D/1$ (the perfect scheduling). This illustrates some improvement gained in GSMA over polling [43].

V. Demand Assignment with Distributed Control

There are two reasons why distributed control is desirable. The first is *reliability*: with distributed control the system is not dependent on the proper operation of a central scheduler. The second is improved *performance*, especially when dealing with systems with long propagation delays, such as those using satellite channels. Indeed, if an earth station were to play the role of a scheduler, the minimum packet delay in a packet reservation scheme would be three times the round-trip propagation delay. (Of course, this can be decreased if on-board processing is available.) With distributed control, this minimum delay can be brought down to twice the round-trip delay or less without affecting the bandwidth utilization. Clearly, in slotted ALOHA, the best random access scheme available for satellite channels, the minimum packet delay is exactly one round-trip delay; but this is guaranteed only for a channel utilization approaching zero! In fact, the inherent long propagation delay in satellite channels is really the nasty characteristic that makes this environment "more distributed" than the single-hop ground radio or local area environments. In the latter, we have seen that efficient random access schemes, such as CSMA, are available; and the shorter the propagation delay, the better the performance. With zero propagation delay, collisions in CSMA can be completely avoided and CSMA's performance then corresponds to that of an $M/D/1$ queue,* the best we can achieve under random demand. In fact, when the propagation delay is zero we no longer have a distributed environment, and the cost of creating a common queue disappears.

The basic element underlying all distributed algorithms is the need to exchange control information among the users, either explicitly or implicitly. Using this information, all users then execute independently the same algorithm resulting in some coordination in their actions. Clearly, it is essential that all users receive the same information regarding the demand placed on the channel and its usage in order to achieve a global optimum, and thus distributed algorithms are most attractive in fully connected systems. This attribute is not always present in ground radio environments, but certainly exists in satellite environments due to their inherent broadcast nature.† The long-delay/broadcast combination of attributes is one of the reasons why many distributed control algorithms have been proposed in the context of satellite environments. We examine in this section distributed control algorithms suitable for each of our three environments (satellite, ground radio, and local area), starting with satellite channels.

---

*This correspondence applies to CSMA and fixed size packets and Poisson sources.
†This is valid unless the satellite uses spot beams, in which case we may lose on the connectivity requirement but gain the benefits of space division multiple access (SDMA).

A. Distributed Demand Assignment Schemes for Satellite Environments

*Reservation-*ALOHA *[45].* Reservation-ALOHA for a satellite channel is based on a slotted time axis, where the slots are organized into frames of equal size. The duration of a frame must be greater than the satellite propagation delay. A user who has successfully accessed a slot in a frame is guaranteed access to the same slot in the succeeding frame and this continues until the user stops using it. "Unused" slots, however, are free to be accessed by all users in a slotted ALOHA contention mode. An unused slot in the *current* frame is a slot which, in the *preceding* frame, either was idle or contained a collision. (Note again the effect of long delays on the control procedure.) Users need to simply maintain a history of the usage of each slot for just one frame duration. Since no request is explicitly issued by the user, this scheme has been referred to as an *implicit reservation* scheme. Clearly Reservation-ALOHA is effective only if the users generate stream-type traffic or long multipacket messages. Its performance will degrade significantly with single-packet messages, as every time a packet is successful the corresponding slot in the following frame is likely to remain empty.

A First-In First-Out (FIFO) Reservation Scheme [46]. In this scheme, reservations are made explicitly. Time division is used to provide a reservation subchannel. The channel time is slotted as before, but every so often a slot is divided into V small slots which are used for the transmission of reservation packets (as well as possibly acknowledgments and small data packets); these packets contend on the V small slots in a slotted ALOHA mode. All other slots are data slots and are used on a reservation basis, free of conflict. The frequency of occurrence of reservation slots can be made adaptive to the load on the channel and the need to make new reservations. This adaptivity can be achieved as a result of the time division of bandwidth between reservations and data packets.

To execute the reservation mechanism properly, each station must maintain information on the number of outstanding reservations (the "queue in the sky") and the slots at which its own reservations begin. These are determined by the FIFO discipline based on the successful reservations received. Each successful reservation can accommodate up to a design maximum of, say, eight packets, thus preventing stations from acquiring exclusivity of the channel for long periods of time. To maintain synchronization of control information at the proper time, and to acquire the correct count of packets in the queue if out-of-sync conditions do occur, each station sends, in its data packet, information regarding the status of its queue. This information is also used by new stations which need to join the queue. The robustness of this system is achieved by a proper encoding of the reservation packets to increase the probability of their correct reception at *all* stations. Furthermore, to limit the effect of errors, a station reacquires synchronization if it detects a collision in one of its reserved slots or an error in a reservation packet.

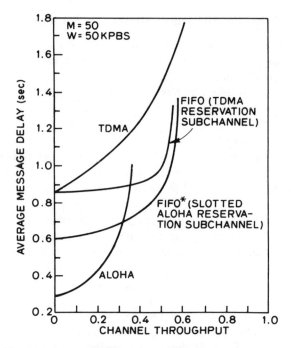

Fig. 13. Slotted ALOHA, TDMA, and FIFO reservation: delay throughput tradeoff for 50 users and single-packet messages in a satellite environment [66].

Figure 13 compares the throughput-delay tradeoff of the FIFO reservation scheme (operated with either a TDMA or a slotted ALOHA reservation subchannel) to that of TDMA and slotted ALOHA [66]. FIFO-Reservation offers delay improvements over TDMA. When compared to ALOHA, we note that higher system capacity is achieved but at the expense of a higher delay at low channel throughputs (due to a higher overhead).

A Round-Robin (RR) Reservation Scheme [47]. The basis of this scheme is fixed TDMA assignment, but with the major difference that "unused" slots are assigned to the active stations on a round-robin basis. This is accomplished by organizing packet slots into equal size frames of duration greater than the propagation delay and such that the number of slots in a frame is larger than the number of stations. One slot in each frame is permanently assigned to each station. To allow other stations to know the current state (used or unused) of its own slot, each station is required to transmit information regarding its own queue of packets piggybacked in the data packet header (transmitted in the previous frame). A zero count indicates that the slot in question is free. All stations maintain a table of all stations' queue lengths, allowing them to allocate among themselves free unassigned slots in the current frame. Round-robin is the discipline proposed by Binder [47], but other scheduling disciplines can be used as well. A station recovers its slot by deliberately causing a conflict in that slot which other users detect. For a station which was previously idle, initial acquisition of queue information is required and is achieved by having one of the stations transmit its table at various times. However, it is interesting to note that in this scheme, while acquiring synchronization, a station can always reclaim and use its own assigned slot.

The above three schemes have been proposed for satellite channels. All of them assumed fixed size slots and thus can be implemented in systems which have been built for synchronous TDMA. The effect of large propagation delay is important. Framing is used in two of the schemes to deal with it, with the frame duration being equal to or longer than the propagation delay. Due to their dynamic nature, these protocols perform better than synchronous TDMA. However, when compared to random access (namely, ALOHA here), they offer higher capacity, but also higher delay at low throughput. If used in systems with small propagation delay, such as ground radio, then they will perform significantly better, and are expected to have a performance comparable to SRMA. In fact, due to the inherent small propagation delay in ground radio environments, other access modes with distributed control are also possible if all devices are in line-of-sight and within range of each other. We describe these in the following.

B. Demand Assignment Schemes Suitable for Ground Radio Environments

Minislotted Alternating Priorities (MSAP) [48]. MSAP is a conflict-free multiple access scheme suitable for a *small* number of data users. In essence, MSAP is a "carrier-sense" version of polling with distributed control. The time axis is slotted with the minislot size again equal to the maximum propagation delay. All users are synchronized and may start transmission only at the beginning of a minislot. Users are considered to be ordered from 1 to M. When a packet transmission ends, the alternating priorities (AP) rule assigns the channel to the same user who transmitted the last packet (say user i) if he is still busy; otherwise the channel is assigned to the next user in sequence [i.e., user $(i, \mod M + 1)$]. The latter (and all other users) detects the end of transmission of user i by sensing the absence of carrier over one minislot. At this new point in time, either user $(i \mod M + 1)$ starts transmission of a packet (which will be detected by all other users) or he is idle, in which case a minislot is lost and control of the channel is handed to the next user in sequence. The overhead at each poll in this scheme is simply one minislot.

Scheduling rules other than AP are also possible, namely, round-robin (RR) or random order (RO). MSAP, however, exhibits the least overhead incurred in switching control between users. On the other hand, MSRR may be more suitable to environments with unbalanced traffic since then smaller users will be guaranteed more frequent access than with MSAP. These scheduling rules have also appeared in the literature as BRAM, the broadcast recognizing access method. For details, see [72] and Section V D below.

The Assigned-Slot Listen-before-Transmission Protocol [49]. MSAP, being a "carrier sense" version of polling, behaves like polling. In particular, as the system load decreases, the overhead incurred in locating a nonidle user increases, and so does the delay. The assigned-slot listen-before-transmission protocol has been proposed to improve on MSAP by allowing several users to share common minislots. In such a case, there exists a tradeoff between the time wasted in collisions and the time wasted in control overhead. Time is divided into frames, each containing an equal number of minislots (say, L). To each minislot of a frame is assigned a given subset of M/L users. A user with a packet ready for transmission in a frame can sense the channel only in his assigned minislot. If the channel

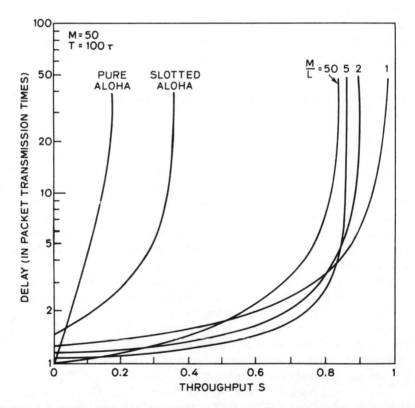

Fig. 14. Assigned-slot listen-before-transmission protocol: throughput delay tradeoff for 50 users and $T = 100$ (propagation time $a = 0.01$) [49].

is sensed idle, transmission takes place; if not, the packet is rescheduled for transmission in a future frame. A packet transmission spans T slots. The parameter M/L is adjusted according to the load placed on the channel. For high throughput, $M/L = 1$ is found to be optimum. In fact, with $M/L = 1$, the scheme becomes a conflict-free one which approaches MSAP and gives nearly identical results [49]. For very low throughput, $M/L = M$ (i.e., $L = 1$) is found to be optimum; this corresponds to pure CSMA. In between the two extreme cases intermediate values of M/L are optimum. Figure 14 displays the throughput-delay performance of this scheme for various values of M/L when $M = 50$ and $T = 100$. It also shows how this scheme (and thus, MSAP) compare to CSMA.

C. Distributed Control Algorithms in Ring Local Area Networks

In addition to the random access schemes described previously in Section III, the above two algorithms are also applicable to local area (broadcast) *bus* networks as these exhibit the required characteristics of small propagation delay and full connectivity. But in local area communication, a slightly different topology has also been widely considered, namely, the *ring* (or loop). In the ring topology, messages are not broadcast but rather passed from node to node along unidirectional links until they reach their destination or, if required by the protocol, until they return to the originating node. Each subscriber is attached to the cable by means of an active tap which allows the information to be examined before it proceeds on the cable. To avoid excessive transit delays, messages are not stored in their entirety, but rather forwarded onto the cable as soon as possible. The delay incurred at each intermediate node can thus be limited to a small number of bit times. Messages are removed from the cable by the receiver (or the originator if the receiver is inactive).

A simple access scheme suitable for a ring consists of passing the access right sequentially from node to node around the ring. (Note that in a ring, the physical location of the nodes defines a natural ordering among them.) One implementation of this scheme is exemplified by the Distributed Computing System's network where an eight-bit control *token* is passed sequentially around the ring. Any node with a ready message may, upon receiving the control token, remove the token from the ring, send the message and then pass on the control token [50]. Note that this token ring scheme is an IEEE 802 Standard, namely 802.5 [79]. Another implementation consists of providing a fixed number of message slots which are continuously transmitted around the ring. A message slot may be empty or full; a node with a ready message waits to see an empty slot pass by, marks it as full, and uses it to send its message [51]–[53]. A still different strategy is known as the *register insertion* technique [3], [54], [55]. Here a message to be transmitted is first loaded into a shift register. If the ring is idle, the

shift register is just transmitted. If not, the register is inserted into
the network loop at the next point separating two adjacent messages: the
message to be sent is shifted out onto the ring while the incoming message
is shifted into the register. The shift register can be removed from the
network loop when the transmitted message has returned to it. The insertion of a register has the effect of increasing the transport delay of messages
on the ring.

D. Demand Assignment Multiple Access in Broadcast Bus
Local Area Networks

Recently, a number of new demand assignment multiple access
(DAMA) schemes have been proposed for broadcast bus networks. These
schemes provide conflict-free transmission using distributed access protocols with *round-robin scheduling functions*. Contrary to random access
schemes, these lead to bounded delay. The stations that are "alive" are
ordered so as to form what is called a *logical ring*, according to which they
are given their chance to transmit. In some of these schemes, such as the
Token-Passing Bus Access Method, an *explicit message* gets sent around the
logical ring to provide the required scheduling; the station holding the
token at any instant is the one that has access to the channel at that instant.
It relinquishes its right to access the channel by transmitting the token to
the next one in turn. The token-bus access method constitutes yet another
standard of IEEE 802, namely, 802.4 [80]. As in rings, the robustness of the
token-bus depends on the integrity of the token and on the proper operation of the involved stations. As in random access networks, the performance degrades significantly with larger values of a [81].

In contrast to those schemes where a station transmits an explicit
token to the next in turn, in others the stations rely on various events due to
activity on the channel to determine when to transmit. Since the token
passing operation is *implicit*, the overall robustness of the network is
improved over token bus networks. Here too, packet delay is bounded; but
in addition both throughput and delay can be made much less sensitive to
a, thus rendering these schemes particularly suitable to networks with high
bandwidth, small-size packets (such as those arising from real-time applications), and long distances.

It is possible to identify three basic access mechanisms according to
which these schemes can be classified. These are the *scheduling-delay access*
mechanism, the *reservation access* mechanism, and the *attempt-and-defer
access* mechanism. We give here a brief description for each of these
mechanisms and list a number of schemes that belong to the classes thus
defined. For more detail, the reader should consult [82].

In presenting these basic mechanisms, three distinct broadcast bus
network configurations can be identified. The first is the *bidirectional bus*

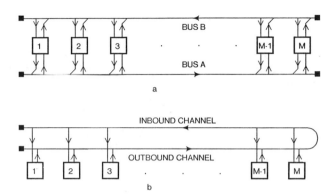

Fig. 15. The unidirectional bus system.

system (BBS), in which, as with Ethernet, the signal transmitted by a station propagates in both directions to reach all other stations on the bus. The second is the *unidirectional bus system* (UBS), in which the transmitted signal propagates in only one direction. In this case, broadcast communications can be achieved in various ways. One way is by means of two unidirectional busses with signals propagating in opposite directions as shown in Fig. 15a, so as to provide each station with a direct path to every other station. Another way is to fold the cable onto itself (or to use a separate frequency channel in the case of broadband signaling) so as to create two channels, an outbound channel onto which the users transmit packets and an inbound channel from which users receive packets, and such that all signals transmitted on the outbound channel are repeated on the inbound channel (see Fig. 15b). The third configuration is the *bidirectional bus with control* (BBC), which consists of a bidirectional bus along with an auxiliary control wire used to control the allocation of the bus.

Consider that there are M stations in the network. Assume the stations to be numbered $1-M$. For some schemes this numbering is a requirement and is explicitly made use of in the algorithm, while for other schemes it merely serves the purpose of clarity in presentation. Denote by S_i the station with index i. Furthermore, a station that has a message to transmit is said to be backlogged. Otherwise, it is said to be *idle*.

The scheduling delay access mechanism is suitable for the BBS configuration where the only means for coordinating the access of the various users following the end of a transmission is by staggering the potential starting time of these users. More specifically, each station is assigned a unique index number. These indexes form a logical ring which determines the order in which stations are allowed to transmit. Included with each transmission is a field for the index number of the sending station. Let S_i be the station currently transmitting. Let $EOC(i)$ denote its *end-of-carrier*. Following the detection of EOC(i), station S_j assigns itself a *scheduling delay* $H_j(i)$, function of both i and j, according to which it schedules its potential transmission. $H_j(i)$ is sufficiently long such that, if at least one of the stations with indexes between S_i and S_j is backlogged, then that backlogged station that is the next in sequence following S_i would have begun to transmit its packet and would have been detected by S_j before the scheduled transmission time of S_j, thus resulting in a round-robin scheduling. Network-schemes that use this access mechanism are BRAM [72] and MSAP [48], [83] presented above, sosam [84], [85], BID [86], Silentnet [87], and L-Expressnet [88].

The reservation access mechanism is mainly suitable for the BBC configuration in which the stations use the control wire to place reservations and to reach a consensus on the next station to transmit, prior to transmission on the bus. The next station to transmit is determined according to some measure, such as the relative positions of the stations on the network, or their addresses. Examples of such schemes are DSMA [89], [90], and the control wire system [91], [92]. The reservation access mechanism can also be implemented on a UBS configuration. For robustness purposes, reservations consist of unmodulated bursts of carrier. These are transmitted on the same bus interleaved with packet transmissions. Consensus here can be reached due to the ordering of the stations, implied by the unidirectionality in transmission and the stations' positions on the bus. An example of this is UBS-RR [93], [94].

The attempt-and-defer access mechanism can only be implemented on UBS configurations where there is an implicit ordering of the stations. Using this access mechanism, a station wishing to transmit waits until the channel is idle. It then begins to transmit, thus establishing its desire to acquire the channel. However, if another transmission from upstream is detected, then this station aborts its transmission and defers to the one from upstream. The upstream transmission is therefore allowed to continue conflict free. In slotted systems, a station wishing to transmit waits for the next slot to arrive, and subsequently asserts its desire to transmit in that slot by marking the slot full. However, if the station finds that the slot has already been marked full, it defers and waits for the next slot. Examples of network schemes that use the attempt-and-defer access mechanism are Expressnet [95], [96], D-Net [97], Fasnet [98], U-Net [100], token-less protocols [101], MAP [102], CSMA-DCR [103], and Buzznet [104].

VI. Adaptive Strategies and Mixed Modes

We have so far examined quite a large number of multiaccess schemes and compared their performance. One thing is clear: each of these schemes has its advantages and limitations. No one scheme performs better than all others over the entire range of system throughput (except, of course, the hypothetical perfect scheduling, which is clearly unachievable in a distributed environment). If a scheme performs nearly as well as perfect scheduling at low input rates, then it is plagued by a limited achievable channel capacity. Conversely, if a scheme is efficient when the system utilization is high, the overhead accompanying the access control mechanism becomes relatively large at low utilization. Although some characteristics of a system (propagation delay, channel speed, etc.) are unlikely to vary during operation, it is certain that the load placed upon the system will be time varying. In the case of a single subscriber type (say with periodic traffic, stream-type traffic, or bursty traffic) the volume of the traffic may be varying; if several subscriber types are simultaneously present, the volume of traffic introduced by each, and therefore the proportional mix of traffic types, may also be time varying.

We have discussed at several points in this chapter the dynamic control of a specific access scheme which improved its performance to a certain extent; but such an adaptive control did not change the nature of the access scheme nor the nature of its limitation. Dynamically controlled random access schemes provide improved packet delay over uncontrolled versions, but still exhibit channel capacity less than 1. The adaptive polling technique decreased the overhead at low throughput but only to a certain extent Actually, what one really needs is a strategy for choosing an access mode which is itself adaptive to the varying need so that optimality is maintained at all times. Clearly, in order to accomplish adaptivity, a certain amount of information is needed by the distributed decision makers. The type and amount of information required by an adaptive strategy, as well as the implementation of the information acquisition mechanism are among the most crucial factors in determining the performance and robustness of the strategy. A great deal of effort has been spent in recent years on such adaptive strategies. We devote this section to schemes which fall into this category.

A. The Urn Scheme [56]

We start with this more recent scheme because of its simplicity, elegance, and the smoothness by which it adapts to varying loads. It has been proposed for fully connected ground radio environments. The time axis is divided into packet slots, and all users are synchronized. Assuming that all users know the exact number n of busy users, the scheme consists of giving full access right (i.e., the right to transmit with probability 1) to some subset of k users. A successful transmission will result if there is exactly one busy user among these k. The probability of such an event is maximized when $k = \lfloor M/n \rfloor$, where $\lfloor M/n \rfloor$ denotes the integer part of M/n.

This is in contrast to the controlled slotted ALOHA scheme, where all users are given the same partial access right: the right to transmit with probability $p = 1/n$. If the system is lightly loaded then a large subset of users is given access right, but only a few, and hopefully only one user, will make use of it. As the load increases, k decreases and the access right is gradually restricted. If $n = 1$, for example, then $k = M$ and a successful transmission takes place. For the extreme case of $n = M$, $k = 1$ and the scheme converges to TDMA. If the sampling of the k users is random, the Urn scheme converges to random TDMA; if the sampling is without repetitions from slot to slot until all users have been sampled once, the Urn scheme converges to round-robin TDMA.

Two important questions remain: how to estimate n, and how to reach a consensus on who the k users are. One possible means for estimating n with good accuracy is to include a single reservation minislot at the beginning of each data slot. An idle user who turns busy sends a standard reservation message of few bits. All users are able to detect the following three events: no new busy users, one new busy user, and more than one new

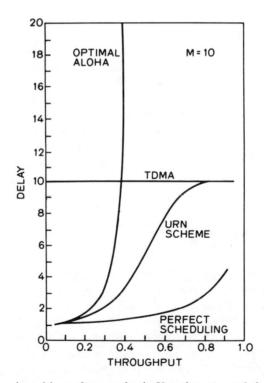

Fig. 16. Throughput-delay performance for the Urn scheme (example for 10 users) [56].

busy user (termed an erasure). As it is impossible with this minimal overhead to estimate the exact number of new busy users when the latter is greater than one, errors in estimation result; however, analysis and simulation have shown that this error is negligible, and that the scheme is insensitive to small perturbations in n. This last statement is even more important with respect to the robustness of the scheme since it means that all users need not have exactly the same estimate for n. As for coordinating the selection of the k users, an effective mechanism is the use of synchronized pseudorandom generators at all users which allow them to draw the same k pseudorandom numbers. Another mechanism, referred to as a round-robin slot-sharing window mechanism, consists of having a window of size k move over the population space. When a collision occurs, the window stops and decreases in size. When there is no collision, the tail of the window is advanced to the head of the previous window, and the size is again set to k as determined by n.

The improvement obtained by this scheme over slotted ALOHA and TDMA can be seen in Fig. 16, where the throughput-delay performance of all these schemes is displayed for a population size $M = 10$ [56].

B. Another Adaptive Strategy for the Dynamic Management of Packet Radio Slots [57]

Another way to achieve adaptivity is as follows. The time axis is again slotted with the slot size equal to a packet transmission time. Slots are divided into k classes or subchannels. Slots are furthermore grouped into frames of m slots, $m \geq k$, each containing at least one slot for every class. Let M be again the number of users. Each user is at any one time assigned to one of the k classes. All stations in a given class use a random access mode to access slots assigned to their class. If CSMA is used as the contention scheme, then time slots are minislots of size τ, assigned to the k classes just as before. By dynamically varying the size of the frame and the assignment of slots within the frame to classes of users, one can vary the access mode to best fit the situation. At low load, for example, choosing $k = m = 1$ with all users in the same class leads to a pure random access mode of low delay. Choosing $k = m = M$ with each user constituting a separate class leads to TDMA. Increasing the parameter k has the effect of decreasing the rate of collisions among users of the same class. The frame size m can be used to allow a smooth changeover between the schemes. By partitioning the frame into two subframes, both contention and pure TDMA can coexist simultaneously. The information used in adapting to the situation is the collision rate and the rate of empty slots (or minislots) for the randomly accessed slots, and the rate of empty slots for the TDMA assigned slots. For example, when one minislot of a TDMA slot goes

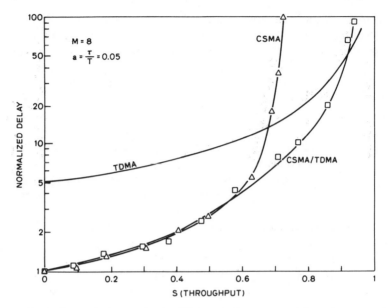

Fig. 17. Simulation results for an adaptive CSMA/TDMA strategy (eight stations, normalized propagation delay of $a = 0.05$) [57].

empty, the remainder of the TDMA slot may be canceled and reassigned to some other groups (then to be used via CSMA).

Schemes other than CSMA and TDMA can be combined by this adaptive strategy. One may, for example, mix CSMA with MSRR. In [57], Ricart and Agrawala studied, via simulation, some typical adaptation algorithms of this type. Some of their simulation results for a CSMA/TDMA combination are shown in Fig. 17. These results exhibit clearly the improvement gained over the entire throughput range by using the adaptive strategy.

C. The Reservation upon Collision Schemes (RUC) [58]

The basic concept in these schemes is to switch back and forth between contention mode and reservation mode. The channel time is divided into slots of fixed length, which in turn are divided into two parts: a data subslot SS0 for transmission of information packets and a subslot SS1 for the transmission of (signaling) information regarding the transmitting user(s). The data subchannel can be in one of two states: the contention state or the reserved state. It is normally in the contention state and users can access the slots in a slotted ALOHA mode as long as no collisions occur. When a collision is detected, then the data subchannel switches to the reserved state and remains in that state until the queue of reservations is cleared, at which time it switches back to the contention state. That is, if a collision is detected, reservations are automatically implied for the colliding users. To accomplish this, the signaling information identifying the users must be received by all users free of interference, and thus a conflict-free use of the SS1 subslots must be devised. CDMA and TDMA have been proposed in [58]. When the number of users is large, a particularly suitable approach is to consider grouping the slots into a frame of, say, L slots. Each of the L SS1 subslots is assigned to a group of size M/L users instead of M users, thus decreasing the degree of multiplexing signaling information over the SS1 subslots. TDMA or CDMA still needs to be used. In this approach, users need not transmit their identification as this is implied from the position of the SS1 subslot. However, each user has to send the number of packets transmitted in the frame, and this information requires at most $\log_2(L + 1)$ bits. This scheme is referred to as the split reservation upon collision (SRUC).

Figure 18 shows the performance of SRUC in a satellite environment as compared to slotted ALOHA and pure reservation for two values of the overhead Ψ required per frame for the signaling information. Clearly, this performance degrades as Ψ increases. More detailed results can be found in [58].

Since slotted ALOHA and reservations are both suitable for satellite channels, RUC schemes are also particularly suitable for these as well as ground radio channels.

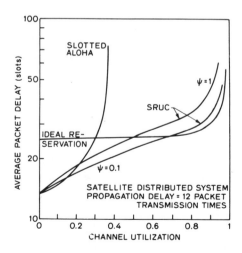

Fig. 18. Split reservation upon collision: throughput-delay performance for various values of the overhead Ψ [58].

D. Priority-Oriented Demand Assignment (PODA) [12]

In the context of a satellite channel, PODA has been proposed as the ultimate scheme which attempts to incorporate all the properties and advantages seen in many of the previous schemes. It has provision for both implicit and explicit reservations, thus accommodating both stream and packet-type traffic. It may also integrate the use of both centralized and distributed control techniques thus achieving a high level of robustness.

To accomplish this flexibility, channel time is divided into two basic subframes, an information subframe and a control subframe. The information subframe contains scheduled packets and packet streams, with the packets also containing, piggybacked, control information such as reservations and acknowledgments. The control subframe is used exclusively to send reservations that cannot be sent in the information subframe in a timely manner. In order to achieve integration of centralized and distributed assignments, the information subframe is further divided into two sections, one for each type.

Access to the control subframe (which is divided into slots accommodating fixed size control packets) can take any form that is suitable to the environment. It can be by *fixed assignment* (TDMA) if the number of stations is small (giving rise to the so-called FPODA), or by *contention* as in ALOHA if the stations have a low-duty cycle (giving rise to CPODA), or a combination of both. The boundary between the control subframe and the information subframe is not fixed, but varies with the demand placed on the channel. As in the FIFO and RR reservation schemes, distributed control is achieved by having all stations involved in this type of control keep track of their queue length information. Priority scheduling can thus be achieved. For stream traffic, a reservation is made only once, and is retained by each station in a stream queue. Centralized assignment may be used when delay is not the crucial element. This scheme has been proposed in the context of a satellite channel but may be applied to other environments as well.

E. More on Mixed Modes

Other studies have appeared in the literature that also deal with integrating several different access modes into the same system.

The *Mixed ALOHA Carrier Sense* (MACS) scheme consists of allowing a large user to steal, by carrier sensing, slots which are unused by a large population of small users accessing the channel in a slotted ALOHA mode [59]. Analysis has shown that the total channel utilization is significantly increased with MACS, and that the throughput-delay performance of both the large user and the background ALOHA users is better with MACS than with a split-channel configuration in which the larger user and the ALOHA users are each permanently assigned a portion of the channel [59].

Group Random Access (GRA) procedures consist of using only certain channel time periods to allow some network terminals to transmit their information-bearing packets on a random access basis. The channel can then be utilized at other times to grant access to other terminals or other message types, by applying, as appropriate, group random access, reservation procedure, or fixed assignment. The idea is simply a fixed time-division assignment among groups utilizing different access schemes. For more details on and analysis of GRA, the reader is referred to [60] and [61].

Finally, we consider satellite systems with on-board processing capability. These have recently received increased attention and are being considered as a means to increase the capacity of packet satellite channels [62–65]. One example is typified by the integration of slotted ALOHA on several uplink channels, with TDMA on one or several downlink channels. The on-board processing capability is used to filter out all collisions and thus improve the utilization of the downlink channels. The overall spectrum efficiency is also improved especially if the ratio of uplink channels to downlink channels is properly chosen. Analysis of these disciplines is given in [62] and [63]. Additional improvement over these disciplines is possible by providing buffering capability on board the satellite to smooth the input and more completely fill the downlink channels.

VII. Conclusion

Tremendous advances have been made in the past two decades in devising multiaccess schemes suitable to a variety of data communication environments. In this chapter, we have briefly reviewed a large number of these protocols which we have grouped into five categories according to (1) the degree of control exercised over the users' access; (2) the (centralized or distributed) nature of the decision-making process; and (3) the degree of adaptivity of the algorithm to the changing need. We have seen that these link level protocols have a great impact on the utilization of the communication resource in particular and the overall system performance in general. We have also briefly discussed their suitability to various traffic characteristics.

Although an attempt has been made to render the presentation complete, it is by no means exhaustive of all existing schemes, and the field is still so wide open that new schemes are constantly being introduced. Throughout the chapter, an emphasis was placed on that class of packet communications that service very many bursty users, since this has been a major concern for many years. It is important, however, to note that there is a growing interest in the support of applications which lend themselves to stream-type traffic (such as packetized voice, facsimile, and video data for remote conferencing) and which may also require real-time communications service on the part of the network. Moreover, with an even greater interest in integrating the many different applications onto the same network structure, it is becoming important to devise multiaccess protocols which can provide all the capabilities and features required for this integration. The adaptive strategies and the demand assignment schemes in local area networks discussed in this chapter provide an attempt at solving this problem.

Another point of great importance is the impact that these link level protocols have on the design of higher-level protocols. Indeed, owing to the basically different nature and behavior of some of these multiaccess schemes, one is faced with the necessity to find new ways to deal with many of the higher-level functions. To briefly illustrate this point, we consider for example store-and-forward multiaccess/broadcast systems. The routing problem in these systems is significantly different from the well-known routing algorithms devised for point-to-point store-and-forward networks; here the transmitted packet should carry, at each transmission, the next node's address, and each *receiving* node has to decide as to whether to relay or ignore the packet. A discussion of routing schemes appropriate to these systems can be found in [14]. Clearly, in single-hop broadcast systems, and in local area ring architectures, the routing problem is absent.

Acknowledgment procedures may also have to be handled differently in broadcast networks. In the PRNET, for example, hop-by-hop acknowledgments can be passive, in the sense that, due to the broadcast nature of transmission, the relaying of a packet over a hop constitutes the acknowledgments for the transmission over the previous hop. Acknowledgments may also be active in the sense that an acknowledgment packet is actually created and transmitted. If acknowledgment packets are given priority, the active acknowledgment procedure has the benefit of minimizing buffering requirements at the repeaters since the acknowledgments are sent at the earliest opportunity, and possibly minimizing channel overhead since the additional transmissions beyond success resulting from delayed acknowledgments can then be kept to a minimum [67]. (In fact, it was found that if acknowledgments were instantaneous, then a few buffers in each packet radio unit appear to be sufficient to handle the storage requirements, indicating that the system becomes more channel bound than storage bound [68, 69].) In satellite environments, PODA achieves the same objective by piggybacking acknowledgments, whenever possible, on pending reservation requests which are heard by all users including the sender.

To conclude, we can say that despite the many advances already accomplished, this area still presents many challenging open problems, and that to best make use of the progress already achieved in link level protocols, one also needs to turn one's attention to the many unresolved issues concerning higher-level protocols.

References

[1] D. W. Davies, K. A. Bartlett, R. A. Scantlebury, and P. T. Wilkinson, "A digital communication network for computers giving rapid response at remote terminals," preented at ACM Symp. Operating System Principles, Gatlinburg, TN, Oct. 1–4, 1967.

[2] W. D. Farmer and E. E. Newhall, "An experimental distributed switching system to handle bursty computer traffic," in *Proc. ACM Conf.*, Pine Mountain, GA, Oct. 1969.

[3] M. T. Liu and C. C. Reames, "Communication protocol and network operating system design for the distributed loop computer network (DLCN), in *Proc. 4th Annu. Symp. Computer Architecture*, Mar. 1977, pp. 193–200.

[4] M. T. Liu, "Distributed loop computer networks," in *Advances in Computer Networks*, M. Rubinoff and M. C. Yovitts, Eds. New York: Academic, 1978.

[5] R. M. Metcalfe and D. R. Boggs, "ETHERNET: Distributed packet switching for local computer networks," *Commun. Ass. Comput. Mach.*, vol. 19, pp. 395–403, 1976.

[6] L. Pouzin, "Presentation and major design aspects of the CYCLADES computer network," presented at Datacom 73, ACM/IEEE, 3rd Data Commun. Symp., St. Petersburg, FL, Nov. 1973, pp. 80–87.

[7] L. G. Roberts ad B. D. Wessler, "Computer network developments to achieve resource sharing," in *1970 Spring Joint Comput. Conf.*, *Proc. AFIPS Conf.*, vol. 36, 1970, pp. 543–549.

[8] L. Pouzin, "CIGALE, The packet switching machine of the CYCLADES computer network," presented at IFIP Congress, Stockholm, Sweden, Aug. 1974, pp. 155–159.

[9] H. Opderbeck and R. B. Hovey, "Telenet—Network features and interface protocols," in *Proc. NTG-Conf. Data Networks*, Baden-Baden, West Germany, Feb. 1976.

[10] W. W. Clipshaw and F. Glave, "Datapac network review," in *Int. Comput. Commun. Conf. Proc.*, Aug. 1976, pp. 131–136.

[11] N. Abramson, "The Aloha system," in *Computer Communication Networks*, N. Abramson and F. Kuo, Eds. Englewood Cliffs, NJ: Prentice-Hall, 1973.

[12] I. M. Jacobs, R. Binder, and E. V. Hoversten, "General purpose packet satellite networks," *Proc. IEEE*, vol. 66, Nov. 1978.

[13] R. E. Kahn, "The organization of computer resources into a packet radio network," in *Nat. Comput. Conf., AFIPS Conf. Proc.*, vol. 44, Montvale, NJ: AFIPS Press, 1975, pp. 177–186; also in *IEEE Trans. Commun.*, vol. COM-25, Jan. 1977.

[14] R. E. Kahn, S. A. Gronemeyer, J. Burchfiel, and R. C. Kunzelman, "Advances in packet radio technology," *Proc. IEEE*, vol. 66, Nov. 1978.

[15] D. Clark *et al.*, "An introduction to local area networks," *Proc. IEEE*, vol. 66, Nov. 1978.

[16] W. G. Schmidt, "Satellite time-division multiple access systems: Past, present and future," *Telecommun.*, vol. 7, pp. 21–24, Aug. 1974.

[17] I. Rubin, "Message delays in FDMA and TDMA communication channels," *IEEE Trans. Commun.*, vol. COM-27, May 1979.

[18] S. Lam, "Delay analysis of time-division multiple access (TDMA) channel," *IEEE Trans. Commun.*, vol. COM-25, Dec. 1977.

[19] O. Kosovych, "Fixed assignment access technique," *IEEE Trans. Commun.*, vol. COM-26, Sept. 1978.

[20] N. Abramson, "The ALOHA system—Another alternative for computer communications," in *1970 Fall Joint Comput. Conf. AFIPS Conf. Proc.*, vol. 37. Montvale, NJ: AFIPS Press, 1970, pp. 281–285.

[21] L. G. Roberts, "ALOHA packet system with and without slots and capture," *Comput. Commun. Rev.*, vol. 5, pp. 28–42, Apr. 1975.

[22] L. Kleinrock and S. Lam, "Packet-switching in a slotted satellite channel," *Nat. Computer Conf.*, *AFIPS Conf. Proc.*, vol. 42. Montvale, NJ: AFIPS Press, 1973, pp. 703–710.

[23] F. A. Tobagi and L. Kleinrock, "Packet switching in radio channels: Part III—Polling and (dynamic) split channel reservation multiple access," *IEEE Trans. Commun.*, vol. COM-24, pp. 832–845, Aug. 1976.

[24] L. Kleinrock and F. A. Tobagi, "Packet switching in radio channels: Part I—Carrier sense multiple access modes and their throughput-delay characteristics," *IEEE Trans. Commun.*, vol. COM-23, pp. 1400–1416, Dec. 1975.

[25] F. Tobagi, "Random access techniques for data transmission over packet switched radio networks," Ph.D. dissertation, Comput. Sci. Dep., School of Eng. and Appl. Sci., Univ. California, Los Angeles, Rep. UCLA-ENG 7499, Dec. 1974.

[26] F. Tobagi and V. B. Hunt, "Performance analysis of carrier sense multiple access with collision detection," in *Proc. Local Area Commun. Network Symp.*, Boston, MA, May 1979; also *Computer Networks*, vol. 4, No. 5, Oct./Nov. 1980.

[27] N. Abramson, "The throughput of packet broadcasting channels," *IEEE Trans. Commun.*, vol. COM-25, pp. 117–128, Jan. 1977.

[28] F. Tobagi and L. Kleinrock, "Packet switching in radio channels: Part II—The hidden terminal problem in carrier sense multiple access and the busy tone solution," *IEEE Trans. Commun.*, vol. COM-23, pp. 1417–1433, Dec. 1975.

[29] F. A. Tobagi, M. Gerla, R. W. Peebles, and E. G. Manning, "Modeling and measurement techniques in packet communication networks," *Proc. IEEE*, vol. 66, pp. 1423–1447, Nov. 1978.

[30] F. Tobagi and L. Kleinrock, "The effect of acknowledgment traffic on the capacity of packet-switched radio channels," *IEEE Trans. Commun.*, vol. COM-26, pp. 815–826, June 1978.

[31] S. S. Lam, "Packet switching in a multiaccess broadcast channel with application to satellite communication in a computer network," Ph.D. dissertation, Dep. Comput. Sci., Univ. California, Los Angeles, Mar. 1974; also in Univ. California, Los Angeles, Tech. Rep. UCLA-ENG-7429, Apr. 1974.

[32] L. Kleinrock and S. S. Lam, "Packet switching in a multiaccess broadcast channel; Performance evaluation," *IEEE Trans. Commun.*, vol. COM-23, pp. 410–423, Apr. 1975.

[33] F. Tobagi and L. Kleinrock, "Packet switching in radio channels: Part IV—Stability considerations and dynamic control in carrier sense multiple access," *IEEE Trans. Commun.*, vol. COM-25, pp. 1103–1120, Oct. 1977.

[34] G. Fayolle, E. Gelembe, and J. Labetoule, "Stability and optimal control of the packet-switching broadcast channels," *J. Ass. Comput. Mach.*, vol. 24, pp. 375–386, July 1977.

[35] S. S. Lam and L. Kleinrock, "Packet switching in a multiaccess broadcast channel: Dynamic control procedures," *IEEE Trans. Commun.*, vol. COM-23, pp. 891–904, Sept. 1975.

[36] J. Metzner, "On improving utilization in ALOHA networks," *IEEE Trans. Commun.*, vol. COM-24, Apr. 1976.

[37] Special Issue on Spread Spectrum Communications, *IEEE Trans. Commun.*, vol. COM-25, Aug. 1977.

[38] B. Edelson and A. Werth, "SPADE system progress and appolication," *COMSAT Tech. Rev.*, vol. 2, pp. 221–242, Spring 1972.

[39] W. Schmidt *et al.*, "Mat-1: INTELSAT's Experimental 700-channel TDMA/DA system," in *Proc. INTELSAT/IEEE Int. Conf. Digital Satellite Commun.*, Nov. 1969.

[40] N. Erlich, "The advanced mobile phone service," *IEEE Commun. Mag.*, vol. 17, Mar. 1979.

[41] A. G. Konheim and B. Meister, "Service in a loop system," *J. Ass. Comput. Mach.*, vol. 19, pp. 92–108, Jan. 1972.

[42] J. F. Hayes, "An adaptive technique for local distribution," *IEEE Trans. Commun.*, vol. COM-26, Aug. 1978.

[43] J. W. Mark, "Global scheduling approach to conflict-free multiaccess via a data bus," *IEEE Trans. Commun.*, vol. COM-26, Sept. 1978.

[44] L. Kleinrock, "Performance of distributed multiaccess computer communication systems," in *Proc. IFIP Congress*, 1977.

[45] W. R. Crowther, R. Rettberg, D. Walden, S. Ornstein, and F. Heart, "A system for broadcast communication: Reservation-ALOHA," in *Proc. 6th Hawaii Int. Syst. Sci. Conf.*, Jan. 1973.

[46] L. Roberts, "Dynamic allocation of satellite capacity through packet reservation," in *Proc. AFIPS Conf.*, vol. 42, June, 1973.

[47] R. Binder, "A dynamic packet switching system for satellite broadcast channels," in *Proc. ICC'75*, San Francisco, CA, June 1975.

[48] L. Kleinrock and M. Scholl, "Packet switching in radio channels: New conflict-free multiple access schemes for a small number of data users," in *ICC Conf. Proc.*, Chicago, IL, June 1977, pp. 22.1-105–22.1-111.

[49] L. W. Hansen and M. Schwartz, "An assigned-slot listen-before-transmission protocol for a multiaccess data channel," *IEEE Trans. Commun.*, vol. COM-27, pp. 846–857, June 1979.

[50] D. C. Loomis, "Ring communication protocols," Univ. California, Dep. Inform. and Comput. Sci., Irvine, CA Tech. Rep. 26, Jan. 1973.

[51] J. R. Pierce, "Network for block switching of data," *Bell Syst. Tech. J.*, vol. 51, pp. 1133–1143, July/Aug. 1972.

[52] A. Hopper, "Data ring at computer laboratory, University of Cambridge," *Computer Science and Technology: Local Area Networking*. Washington DC: Nat. Bur. Stand., NBS Special Publ. 500-31, Aug. 22–23, 1977, pp. 11–16.

[53] P. Zafiropoulo and E. H. Rothauser, "Signalling and frame structures in highly decentralized loop systems," *Proc. Int. Conf. on Comput. Commun.* (Washington, DC), IMB Res. Lab., Zurich, Switzerland, pp. 309–315.

[54] E. R. Hafner *et al.*, "A digital loop communication system," *IEEE Trans. Commun.*, p. 877, June 1974.

[55] M. V. Wilkes, "Communication using a digital ring, "in *Proc. PACNET Conf.*, Sendai, Japan, Aug. 1975, pp. 217–255.

[56] L. Kleinrock and Y. Yemini, "An optimal adaptive scheme for multiple access broadcast communication," *ICC Conf. Proc.*, Chicago, IL, June 1977.

[57] G. Ricart and A. Agrawala, "Dynamic management of packet radio slots," presented at *Third Berkeley Workshop on Distributed Data Management and Comput. Networks*, Aug. 1978.

[58] F. Borgonovo and L. Fratta, "SRUC: A technique for packet transmission on multiple access channels," in *Proc. Int. Conf. Comput. Commun.*, Kyoto, Japan, 1978.

[59] M. Scholl and L. Kleinrock, "On a mixed mode multiple access scheme for packet-switched radio channels," *IEEE Trans. Commun.*, vol. COM-27, pp. 906–911, June 1979.

[60] I. Rubin, "A group random-access procedure for multi-access communication channels," in *NTC'77 Conf. Rec. Nat. Telecommun. Conf.*, Los Angeles, CA, Dec. 1977, pp. 12 : 5-1–12 : 5-7.

[61] I. Rubin, "Integrated random-access reservation schemes for multi-access communication channels," School Eng. Appl. Sci., Univ. California, Los Angeles, Tech. Rep. UCLA-ENG-7752, July 1977.

[62] J. K. DeRosa and L. H. Ozarow, "Packet switching in a processing satellite," *Proc. IEEE*, vol. 66, pp. 100–102, Jan. 1978.

[63] R. E. Eaves, "ALOHA/TDM systems with multiple downlink capacities," *IEEE Trans. Commun.*, vol. COM-27, pp. 537–541, Mar. 1979.

[64] S. F. W. Ng and J. W. Mark, "A multiaccess model for packet switching with a satellite having some processing capability," *IEEE Trans. Commun.*, vol. COM-25, pp. 128–135, Jan. 1977.

[65] S. F. W. Ng and J. W. Mark, "Multiaccess model for packet switching with a satellite having processing capability: Delay analysis," *IEEE Trans. Commun.*, vol. COM-26, pp. 283–290, Feb. 1978.

[66] S. S. Lam, "Satellite multi-access schemes for data traffic," in *Proc. Int. Conf. Commun.*, Chicago, IL, 1977, pp. 37.1-19–37.1-24.

[67] F. Tobagi, S. Lieberson, and L. Kleinrock, "On measurement facilities in packet radio systems," in *Nat. Comput. Conf. Proc.*, New York, NY, June 1976.

[68] F. Tobagi, "Analysis of a two-hop centralized packet radio network: Part I—Slotted ALOHA," *IEEE Trans. Commun.*, vol. COM-28, pp. 196–207, Feb. 1980.

[69] F. Tobagi, "Analysis of a two-hop centralized packet radio network: Part II—Carrier sense multiple access," *IEEE Trans. Commun.*, vol. COM-28, pp. 208–216, Feb. 1980.

[70] W. W. Chu, "A study of asynchronous time division multiplexing for time-sharing computer systems," in *1969 Spring Joint Comput. Conf. AFIPS Conf. Proc.*, vol. 35, 1969, pp. 669–678.

[71] J. I. Capetanakis, "Tree algorithms for packet broadcasting channels," *IEEE Trans. Inform. Theory*, vol. IT-25, pp. 505–515, Sept. 1979.

[72] I. Chlamtac, W. Franta, and K. D. Levin, "BRAM: The broadcast recognizing access method," *IEEE Trans. Commun.*, vol. COM-27, pp. 1183–1190, Aug. 1979.

[73] J. E. Thornton, "Overview of Hyperchannel," *18th IEEE Comp. Soc. Int. Conf. (CompCon 79 Spring)*, San Francisco, February 1979, pp. 262–265.

[74] M. Tokoro and K. Tamaru, "Acknowledging Ethernet," *Proc. CompCon 77*, pp. 320–325, Sept. 1977.

[75] ANSI/IEEE std 802.3-1985, *Carrier-Sense Multiple Access with Collision Detection*, New York: The Institute of Electrical and Electronics Engineers, 1985.

[76] J. M. Brázio, "Capacity analysis of multihop packet radio networks under a general class of channel access protocols and capture modes," Ph.D. dissertation, Department of Electrical Engineering, Stanford University, Stanford, June 1986.

[77] F. A. Tobagi and D. H. Shur, "Performance evaluation of channel access schemes in multihop packet radio networks with regular structure by simulation," Stanford Electronics Laboratories, Technical Report No. 85-278, June 1985; and, SURAN Temporary Note No. 33, June 1985.

[78] F. A. Tobagi, "Modeling and performance analysis of multihop packet radio networks," *Proc. IEEE*, Special Issue on Packet Radio Networks, January 1987.

[79] ANSI/IEEE std 802.5-1985, *Token Ring Access Method*, New York: The Institute of Electrical and Electronics Engineers, 1985.

[80] ANSI/IEEE std 802.4-1985, *Token-Passing Bus Access Method and Physical Layer Specifications*, New York: The Institute of Electrical and Electronics Engineers, 1985.

[81] W. Stallings, "Local network performance," *IEEE Commun. Mag.*, vol. 22, No. 2, Feb. 1984.

[82] M. Fine and F. A. Tobagi, "Demand assignment multiple access schemes in broadcast bus local area networks," *IEEE Trans. Computers*, vol. C-33, No. 12, 1984.

[83] L. Kleinrock and M. Scholl, "Packet switching in radio channels: New conflict-free multiple access schemes," *IEEE Trans. Commun.*, vol. COM-28, July 1980.

[84] Y. I. Gold and W. R. Franta, "An efficient collision-free protocol for prioritized access-control of cable or radio channels," *Comput. Networks*, 1983.

[85] Y. I. Gold and W. R. Franta, "An efficient scheduling function for distributed multiplexing of a communication bus shared by a large number of users," in *Proc. Int. Conf. Commun.*, Philadelphia, June 13–17, 1982.

[86] M. E. Ulug, G. M. White, and W. J. Adams, "Bidirectional token flow system," in *Proc. 7th Data Commun. Symp.*, Mexico City, Oct. 1981.

[87] E. D. Jensen, M. Tokoro, and L. Sha, "Bus allocation scheme for distributed real time systems," Carnegie-Mellon University, Pittsburgh, Rep. Dec. 1980.

[88] F. Borgonovo, L. Fratta, F. Tarini, and P. Zini, "L-Express-net: A communication protocol for local area networks," in *Proc. INFOCOM '83*, San Diego, April 1983.

[89] J. W. Mark, "Distributed scheduling conflict-free multiple access for local area communications networks," *IEEE Trans. Commun.*, vol. COM-28, pp. 1968–1976, 1980.

[90] T. D. Todd and J. W. Mark, "Waterloo experimental local network (Welnet) physical level design," in *Proc. NTC'80*, Houston, November/December 1981, pp. 41.4.1–41.4.5.

[91] K. P. Eswaran, V. C. Hamacher, and G. S. Shedler, "Collision-free access control for computer communication bus networks," *IEEE Trans. Software Eng.*, vol. SE-7, No. 6, 1981.

[92] K. P. Eswaran, V. C. Hamacher, and G. S. Shedler, "Asynchronous collision-free distributed control for local bus networks," IBM, San Jose, CA, Research Report RJ2482, 1979.

[93] F. A. Tobagi and R. Rom, "Efficient round-robin and priority schemes for unidirectional broadcast systems," in *Proc. IFIP 6.4 Zurich Workshop Local Area Networks*, Zurich, Aug. 27–29, 1980.

[94] R. Rom and F. A. Tobagi, "Message-based priority functions in local multiaccess communications systems," *Comput. Networks*, vol. 5, No. 4, pp. 273–286, July 1981.

[95] L. Fratta, F. Borgonovo, and F. A. Tobagi, "The Express-net: A local area communication network integrating voice and data," in *Proc. Int. Conf. Perform. Data Commun. Syst. Applications*, Paris, Sept. 14–16, 1981.

[96] F. Tobagi, F. Borgonovo, and L. Fratta, "Express-net: A high-performance integrated-services local area network," *IEEE J. Select. Areas Commun.*, vol. SAC-1, No. 5, 1983.

[97] C. Tseng and B. Chen, "D-Net, A new scheme for high data rate optical local area networks," *IEEE J. Select. Area Commun.*, vol. SAC-1, No. 3, 1983.

[98] J. O. Limb and C. Flores, "Description of Fasnet, A unidirectional local area communications network," *Bell Syst. Tech. J.*, Sept. 1982.

[99] J. O. Limb, "Fasnet: A proposal for a high speed local network," in *Proc. Office Inform. Syst. Workshop*, St. Maximin, France, Oct. 1981.

[100] M. Gerla, C. Yeh, and P. Rodrigues, "A token protocol for high speed fiber optics local networks," in *Proc. Opt. Fiber Commun. Conf.*, New Orleans, Feb. 1983.

[101] P. Rodrigues, L. Fratta, and M. Gerla, "Token-less protocols for fiber optics local area networks," in *Proc. ICC'84*.

[102] M. A. Marsan and G. Albertengo, "Integrated voice and data network," *Comput. Commun.*, vol. 5, No. 3, June 1982.

[103] A. Takagi, S. Yamada, and S. Sugawara, "CSMA/CD with deterministic contention resolution," *IEEE J. Select. Areas Commun.*, vol. SAC-1, p. 5, 1983.

[104] M. Gerla, P. Rodrigues, and C. Yeh, "BUZZ-NET: A hybrid random access/virtual token local network," in *Proc. GLOBECOM '83*, San Diego, Dec. 1983.

IV

Network Layer

Circuit-Switched Network Layer

Harold C. Folts

I. Introduction

CCITT Recommendation X.21 has been developed as "The General Purpose Interface between Data Terminal Equipment (DTE) and Data Circuit-terminating Equipment (DCE) for Synchronous Operation on Public Data Networks" [1]. The only "general purpose" part, however, is the designation of the physical elements which include the electrical (X.26/X.27), functional (X.24), and mechanical (ISO 4903) characteristics described in Chapter 3 (Bertine). Additionally, the basic family of quiescent signals and states for the interface is specified. These provide the fundamental components of X.21 which will apply to all modes of operation in new data communications applications for circuit-switched, packet-switched, and general purpose integrated services [2].

The remainder of X.21 includes procedures for leased circuit service (both point-to-point and multipoint) and for circuit-switched services. In relation to the OSI Reference Model of the ISO architecture, discussed in Chapter 2, the leased circuit procedures are a specific application at the Physical Layer, while the circuit-switched procedures involve the Data Link Layer and the Network Layer. The focus of this paper will be on the Network Layer call establishment procedures of Recommendation X.21.

Harold C. Folts • Omnicom, Inc., Vienna, Virginia 22180.

II. Background

Extensive activity by telecommunications administrations around the world is taking place implementing public data networks which will provide tailored data communication services to the user community. In recognition of this new evolution, the International Telegraph and Telephone Consultative Committee (CCITT) established a study program in 1968 by forming a Joint Working Party for New Data Networks (JWP/NRD) to set the basis for international standardization. In 1972, the resulting first X-series of Recommendations, including the original version of X.21, was approved by the Fifth CCITT Plenary Assembly. These Recommendations dealt primarily with circuit-switching technology.

To further refine and expand this work, CCITT then established Study Group VII, Public Data Networks. The main thrust of the work continued toward circuit switching with only a minor question directed toward the emerging packet-switching technology. In 1976, however, a major diversion in direction started to take place with the sudden appearance of the X.25 Virtual Call packet-switched service, described in Chapter 8.

As a result, the emphasis on the circuit switching in public data networks has been subsequently overshadowed by the fascination for packet-switching technology. This does not mean, however, that circuit switching has passed into oblivion, but circuit switching is, in fact, being actively pursued by the Nordic countries (Sweden, Norway, Denmark, and Finland), Japan, the Federal Republic of Germany, Italy, Hungary, and Canada (Infoswitch). Experience and proven technology may very likely lead in the future to an expansion of circuit-switched services in public data networks.

III. Architecture

The architecture of Recommendation X.21 has been a subject of considerable misunderstanding and controversy in the ISO and CCITT work developing the OSI Reference Model, which was introduced in Chapter 2. While X.21 provides the essential physical elements of an interface, it also provides the circuit-switched network control procedures. Some have argued that these procedures are also within the Physical Layer because they result in the establishment of a physical circuit which is the used for data transfer.

By analyzing the X.21 call establishment procedures in comparison with the call establishment procedures of X.25, it will become clear that the basic functionality of each is identical. It is technically possible to use the

X.21 call establishment for a packet-switched virtual circuit as well as to use the X.25 call establishment for a circuit-switched physical circuit. Both are Network Layer control procedures.

The necessity for consistent distribution of functionality among layers for all switched network services was set forth by the United States to ISO. This consistency is essential if the goal of a universal interface for integrated services is to be realized [2]. ISO has now endorsed this determination and includes the provisions at the Network Layer in the OSI Reference Model for establishment of connections through a switched network regardless of the implemented technology: circuit-switched for physical circuits or packet-switched for virtual circuits.

IV. CCITT Recommendation X.21

When work on Recommendation X.21 began in 1969, during the early days of JWP/NRD, it was recognized that use of any CCITT V-series interfaces (equivalent of RS-232-C and RS-366) would not be satisfactory for the new generation of digital public data networks. Therefore, an initial objective was established to develop a new interface that is compatible with advancing technology and tailored for circuit-switched networks providing full transparency (bit sequence and protocol independence) for the transfer of user data.

For call control purposes, use of International Alphabet Number 5 (IA5) was adopted to maintain consistency with the character-oriented data link layer basic mode control procedures of ISO 1745 and ANSI X3.28, discussed in Chapter 4.

The first version of X.21 approved by the CCITT Fifth Plenary Assembly in 1972, was little more than an outline of procedures. It was not complete enough at that time for practical implementation. During 1973–1976, however, substantive work was completed to produce a usable Recommendation [3]. This version was approved by the CCITT Sixth Plenary Assembly in 1976 and appears in the Orange Book [4].

Subsequently, work continued to further refine and expand the Recommendation, as well as to include adjustments resulting from implementation experience. The new version of X.21 was completed at the CCITT Study Group VII meeting in February 1980. In addition to significant technical and editorial enhancements, the new revision of X.21 has been completely reorganized to track with the work in developing the standard architecture for Open System Interconnection (OSI). The presentation in this chapter will relate to the latest revision which was approved by the CCITT VIIth Plenary Assembly, November 1980 [1].

A. General Purpose Physical Layer

1. Basic Elements

The physical elements for X.21 as discussed in Chapter 3 include application of the X.26 and X.27 electrical characteristics, together with functional circuits defined by X.24. The mechanical element of the interface is the 15-pin connector specified by ISO 4903 which is from the same family of connectors as the commonly known 25-pin connector used for RS-232-C and the CCITT V-series interfaces. The physical configuration of the DTE/DCE interface for X.21 consists of six circuits as shown in Fig. 1.

Circuits T and R convey data and control information, while circuits C and I provide control functions similar to "OFF/ON hook" indications. This simple out-of-band control provides an effective mechanism for maintaining full transparency during data transfer. Circuit S provides signal element (bit) timing from the network, and optionally in some networks, circuit B provides an octet byte alignment with the network.

2. Quiescent Phase Signals

The signals during the quiescent phase indicate the ability of the DTE and the DCE to enter the operational phases such as the call control phase. The two basic signals used indicate READY and NOT READY.

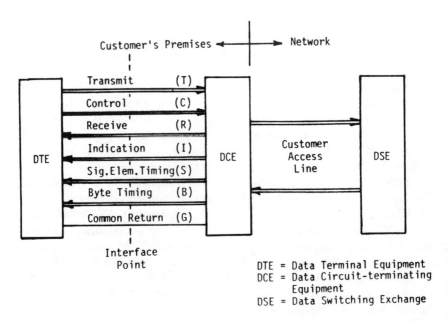

Fig. 1. X.21 interface.

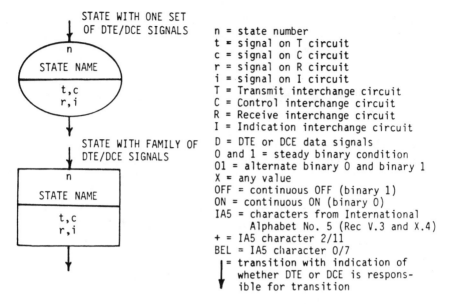

STATE WITH ONE SET
OF DTE/DCE SIGNALS

STATE WITH FAMILY OF
DTE/DCE SIGNALS

n = state number
t = signal on T circuit
c = signal on C circuit
r = signal on R circuit
i = signal on I circuit
T = Transmit interchange circuit
C = Control interchange circuit
R = Receive interchange circuit
I = Indication interchange circuit
D = DTE or DCE data signals
0 and 1 = steady binary condition
01 = alternate binary 0 and binary 1
X = any value
OFF = continuous OFF (binary 1)
ON = continuous ON (binary 0)
IA5 = characters from International
         Alphabet No. 5 (Rec V.3 and X.4)
+ = IA5 character 2/11
BEL = IA5 character 0/7
= transition with indication of
   whether DTE or DCE is respons-
   ible for transition

Fig. 2. Conventions for X.21 state diagrams.

B. Circuit-Switching Procedures

For circuit-switched operation, X.21 defines four phases—quiescent, call control, data transfer, and clearing. Within the phases there are a number of states which are defined by the signals appearing on circuits T, C, R, and I. Each state is essentially a "snapshot" in time of the interface signals presented by the DTE and the DCE as described in Fig. 2. The interface procedures are then illustrated by state diagrams to present a coherent picture of the operations.

1. Data Link Layer Elements

X.21 does not support the full richness of the Data Link Layer functions of the OSI Reference Model but provides only the minimum necessary elements for basic operation. These include character synchronization and error detection.

As X.21 is intended for synchronous operation, the first Data Link Layer function provides for correct alignment of the IA5 character sequences used during the call control phase. The actual method of achieving character alignment was an issue of intense debate for several years [3], but was finally resolved in 1976. One proposal was to provide for character alignment as typically used for synchronous character-oriented operation. This provided for use of two or more contiguous SYN characters preceding each sequence of call control characters. The alignment for each direction of transmission would be independent. The other proposal was to use a separate byte alignment interchange circuit (circuit *B*, Fig. 1) from the DCE to the DTE. Circuit *B* provides the indication of the last bit of an 8-bit byte which represents an IA5 character with parity. The byte alignment information is used both to align characters received on circuit *R* and to align characters transmitted on circuit *T*. Each direction of transmission is then dependent on the byte alignment information provided by the network (DCE).

The compromise which resulted in agreement essentially recognized that either method of operation could be provided, but it requires two or more contiguous SYN characters to be present before each call control sequence in all cases, even when byte timing is provided by circuit *B*. Where byte alignment with the network is required, the DTE must still align transmitted call control characters to the synchronization of either circuit *B*, when used, or received SYN characters from circuit *R*.

This compromise now makes it practical to design a new DTE which can work with all X.21 network implementations where the provision is included in the DTE for alignment of transmitted characters to the synchronization of the received characters. The use of the byte timing circuit *B*, when offered by a network, then becomes a purely optional matter, and operation with a nonbyte aligned network is therefore possible.

Another provision of the compromise agreement allows ready adaptation of existing designs of synchronous character-oriented DTEs to X.21. This requires, for an intermediate period, that all networks accommodate conventional SYN character alignment independent of direction of transmission. The intermediate period is to be determined by customer demand and other relevant factors as interpreted by the network provider.

The other Data Link Layer element of X.21 provides an elementary means of error checking using odd parity according to CCITT Recommendation X.4 [1]. Before the decision was made to employ parity, a thorough study was made as to how powerful an error control was needed. The conclusion showed that with the low error rates expected in public data networks, the use of parity is quite adequate and cost effective.

2. Network Layer Procedures

The character-oriented procedures used during the call control phase establish a connection to one or more distant subscribers through a circuit-switched public data network. To clearly define the procedures, a state diagram, Fig. 3, is used to show the relationship among the various call control phase states which are defined by the text. Only the recognized transitions among the states under normal operating conditions are shown by Fig. 3. As further clarification of the procedures, illustrative time

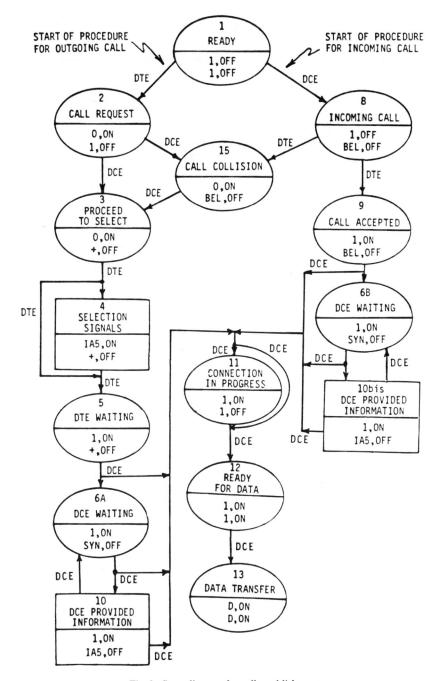

Fig. 3. State diagram for call establishment.

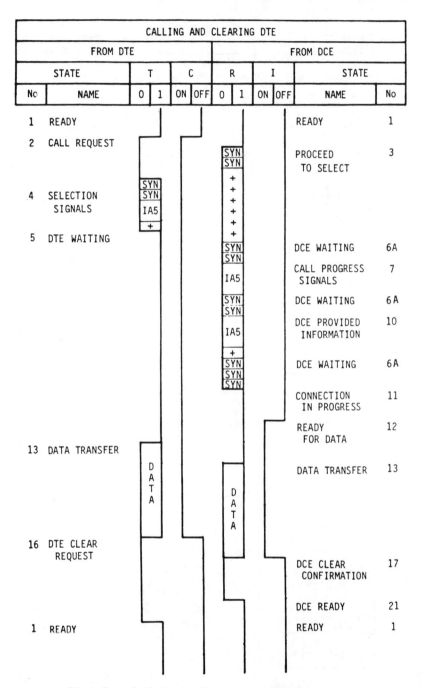

Fig. 4. Example of sequence of events: successful call and clear.

sequence diagrams are also provided in the X.21 documentation. One of these examples is shown in Fig. 4.

Call establishment can begin only from the READY state (state 1). Both the DTE and DCE must be READY before either INCOMING CALL (state 8) or CALL REQUEST (state 2) can be signaled across the interface. It was proposed that the DTE and DCE be allowed to enter the call establishment phase directly either from or toward a NOT READY state to allow more flexibility in operation. Some proposed network designs, however, precluded these additional state transitions.

The process for the calling DTE starts with the signaling of CALL REQUEST with $t = 0$ and $c = $ ON. The simple steady-state signal rather than a character sequence was used to alert the DCE of the request. As a result, only a minimum of intelligence for detection of the signal is needed. Next, in response to CALL REQUEST, the DCE signals PROCEED TO SELECT (state 3), $r = +$, $i = $ OFF.

It is possible for an INCOMING CALL and CALL REQUEST to be inadvertently signaled at the same time. Therefore the CALL COLLISION (state 15) has been included. There was considerable debate as to how a CALL COLLISION should be resolved. In following the principal of simplicity, only one means of resolution was desired. It was finally agreed that CALL REQUEST should always win because a DTE preparing for an outgoing call may not be able to readily reallocate its internal resources to handle an incoming call. Accordingly, the only exit transition from state 15 is toward PROCEED TO SELECT where the network continues to process the outgoing call and abandons the incoming call.

The DTE can then proceed with the SELECTION SIGNAL SEQUENCE for the specific call. During the SELECTION SIGNAL SEQUENCE, the DCE continues to signal $r = +$, $i = $ OFF, while the DTE sends a family of signals containing facility and address information. The formats for these signals are specified in detail in X.21 using the Backus Normal Form. For a simplistic description, Table I presents an example illustration of the format.

The FACILITY REQUEST enables selection of special service features for each call. It consists of a FACILITY REQUEST CODE followed by one or more FACILITY REQUEST PARAMETERS all separated by the "/" character. Multiple FACILITY REQUESTS are separated by "," characters. The last FACILITY REQUEST is ended with a "−" character. The list of recognized international facilities is given in CCITT Recommendation X.2, while the actual coding is specified in Annex 7 of X.21. A further discussion on optional user facilities is presented in a later section of this chapter.

The full address signals are in accordance with the format of the International Numbering Plan for public data networks of CCITT Recommendation X.121 [1]. ABBREVIATED ADDRESS signals can be used to represent, with a reduced number of characters, designated FULL ADDRESS signals

Table I. Simplified Example of Selection Signals

```
− = End of facility request
+ = End of selection signal
. = Beginning of abbreviated address
/ = Parameter separator
, = Facility request and address separator
Selection signal sequence = facility requests − addresses +
Facility requests = facility request code/facility parameter
   / ... / ... , ... −
Addresses = full address signal, ... , ... or
   .abbreviated address signal, ... ,
      ... +
Facility registration or cancellation = facility request code/
   parameter/parameter/
   parameter − +
```

as established by agreement with the specific network. A single ABBREVI-
ATED ADDRESS code may represent either a single address or a group of
multiple addresses. Each ABBREVIATED ADDRESS signal is preceded by the
"." character. Multiple FULL or ABBREVIATED ADDRESSS, which can be
intermixed, are separated by "," characters. The last ADDRESS signal is
followed by the " + " character as the "end of selection."

If there is no FACILITY REQUEST in the SELECTION SIGNAL SEQUENCE,
the sequence will start immediately with the ADDRESS signals without any
" − " character. If there is no ADDRESS signal, but there is a FACILITY
REQUEST, the sequence is ended by the " − " followed by the " + " character.

As shown in Fig. 3, the SELECTION SIGNAL SEQUENCE may be bypassed.
This provides for a direct calling feature similar to an "OFF-HOOK" or "hot
line" service which may be used as either a fixed mode of operation or on a
dynamic per-call basis. After receiving the PROCEED TO SELECT signal, the
DTE signals DTE WAITING (state 5); then the DCE proceeds to establish a
connection to a predesignated address or group of addresses. If the choice
of direct call or addressed call is allowed dynamically on a per-call basis,
the DTE can enter either state 5 or state 4 depending on the service desired.

Once the DCE has the request and necessary information to establish a
connection through the network, the DCE signals r = SYN, i = OFF (state
6a) as it processes the call. If establishment of the call is successful, there
will normally not be any CALL PROGRESS SIGNALS (state 7), and in the
absence of any special facilities, there will not normally be any DCE
PROVIDED INFORMATION (state 10).

Depending on how fast the connection is made, the DCE may bypass
state 6a and proceed directly to CONNECTION IN PROGRESS (state 11) or
READY FOR DATA (state 12). The difference will be whether the connection is
made to a subscriber within the same switching center, the same network,

or through an international connection to another network where the processing time would be greater. The procedure allows a great deal of flexibility in this respect.

The term CALL PROGRESS SIGNALS in state 7 is perhaps a misnomer because they primarily indicate the reasons for "nonprogress" or unsuccessful completion of the call. The CALL PROGRESS SIGNALS are defined by CCITT Recommendation X.96. The 1980 revision of X.96 has now established a great deal of commonality with the CALL PROGRESS SIGNAL definitions used for packet-switching operation. The codings for the two applications, however, are quite different. Figure 5 gives a list of the CALL PROGRESS SIGNALS applicable to X.21, together with the respective coding.

CODE GROUP	CODE	SIGNIFICANCE	CATEGORY
0	00	RESERVED	WITHOUT CLEARING
	01	TERMINAL CALLED	
	02	REDIRECTED CALL	
	03	CONNECT WHEN FREE	
2	20	NO CONNECTION	WITH CLEARING DUE TO SHORT TERM CONDITIONS
	21	NUMBER BUSY	
	22	SELECTION SIGNALS PROCEDURE ERROR	
	23	SELECTION SIGNAL TRANSMISSION ERROR	
4&5	41	ACCESS BARRED	WITH CLEARING DUE TO LONG TERM CONDITIONS
	42	CHANGED NUMBER	
	43	NOT OBTAINABLE	
	44	OUT OF ORDER	
	45	CONTROLLED NOT READY	
	46	UNCONTROLLED NOT READY	
	47	DCE POWER OFF	
	48	INVALID FACILITY REQUEST	
	49	NETWORK FAULT IN LOCAL LOOP	
	51	CALL INFORMATION SERVICE	
	52	INCOMPATIBLE USER CLASS OF SERVICE	
6	61	NETWORK CONGESTION	WITH CLEARING DUE TO NETWORK SHORT TERM CONDITIONS
7	71	LONG TERM NETWORK CONGESTION	WITH CLEARING DUE TO NETWORK LONG TERM CONDITIONS
	72	RPOA OUT OF ORDER	
8	81	REGISTRATION/CANCELLATION CONFIRMED	WITH CLEARING DUE TO DTE-NETWORK PROCEDURE
	82	REDIRECTION ACTIVATED	
	83	REDIRECTION DEACTIVATED	

Fig. 5. Coding of call progress signals.

Initially, a two-digit code is applied where the first digit indicates a general category of signal. This enables a relatively simple terminal to translate only the basic category of the CALL PROGRESS SIGNAL. The second digit indicates the more specific reason which can be translated by more intelligent terminals. In the future, it will be possible to expand the number of digits if further enrichment is needed.

The CALL PROGRESS SIGNAL SEQUENCE must be preceded by at least two "SYN" characters as described earlier for the character synchronization. These "SYN" characters will be sent during state 6a. If there is more than one block of signals, the period between them will be filled by additional "SYN" characters during state 6a.

In the 1976 issue [4] of X.21, state 10 was named CALLED LINE IDENTIFICATION. Further study, however, showed that more flexibility will be needed for future enhancements providing a family of signals that may be provided to the DTE from the network. Therefore, the name was changed to DCE PROVIDED INFORMATION. The only signal presently designated for state 10 is the original CALLED LINE IDENTIFICATION. In effect, CALL PROGRESS SIGNALS are really a subset of the more general DCE PROVIDED INFORMATION. This logically suggests a possible merger of states 7 and 10 for a simplification of the state diagram, which was done at the very last minute during the meeting of Study Group VII in February 1980 and therefore is part of the 1980 revised Recommendation.

While the above actions have been occurring at the calling DTE/DCE interface, the state diagram also shows the procedures unique at the called DTE/DCE interface in states 8, 9, and 10bis. The INCOMING CALL signal (state 8) with r = BEL, i = OFF is presented to a READY DTE where $t = 1$, c = OFF. The DTE answers the call by signaling the steady-state conditions of $t = 1$, c = ON for CALL ACCEPTED (state 9).

At this point, the network may wish to provide the called DTE additional information relating to the call (state 10bis). Similar to state 10 as earlier described, state 10bis was originally named CALLING LINE IDENTIFICATION in the previous version of X.21 [4], but further work has now changed the name to DCE PROVIDED INFORMATION. The new state 10bis includes the original CALLING LINE IDENTIFICATION and the new addition of CHARGING INFORMATION which will be described later. Consideration in the future will be given to further enhancing state 10bis to include additional capabilities such as subaddressing and a means for acceptance of reverse charging calls.

Another feature under future consideration will be a means for positive and negative acknowledgment of the CDE PROVIDED INFORMATION. As presently defined, negative acknowledgment due to error or rejection is only possible with a complete clearing of the call. It is felt that this may be too drastic where a simple retransmission could solve the problem.

Upon acceptance of the call, DCE WAITING (state 6b) is signaled. Then after at least two SYN characters being signaled in state 6b, DCE PROVIDED

INFORMATION (state 10bis) may be signaled. As with state 6a, state 6b may be bypassed when connection time is very fast and no DCE PROVIDED INFORMATION will be sent.

The transition to states 11 and 12 is the process known as "connect-through" in the original version of X.21 in 1972. This was a issue of great confusion and debate which resulted in a carefully constructed agreement. The concerns were related to the danger of losing bits of user data and the possible presence of spurious bits during the "connect-through" process.

As the "connect-through" procedure is very complex to describe, the following extracted text from X.21 is presented to assist understanding:

> All bits sent by a DTE after receiving READY FOR DATA and before sending DTE CLEAR REQUEST will be delivered to the corresponding DTE after that corresponding DTE has received READY FOR DATA and before it has received DCE CLEAR INDICATION (provided that the corresponding DTE does not take the initiative of CLEARING).

> All bits received by a DTE after receiving READY FOR DATA and before receiving DCE CLEAR INDICATION or receiving DCE CLEAR CONFIRMATION were sent by the corresponding DTE. Some of those may have originated as DTE WAITING before that corresponding DTE has received READY FOR DATA; those bits are binary 1.

In effect, the result of the process of the transition on circuit R from "SYN" in state 6, or from " + " of state 5, to "1" of states 11 and 12 is the completion of the end-to-end connection. The "SYN" (or " + ") is generated internally within the network, while the "1" originates from the distant DTE on circuit T and is carried through the network and presented to the local DTE on circuit R. Because the transition on circuit I may not be concurrent with the transition on circuit R, owing to network signaling differences, state 11 has been included but, as shown, may be bypassed. The significant state is READY FOR DATA (state 12) where a guaranteed transparent end-to-end path is established and ready for transfer of user data in state 13.

3. Clearing

In the proposed OSI Reference Model, a disconnection function is defined for each layer to terminate operational phases. In the case of X.25, there is a disconnection function at each of the first three layers, each of which serves a specific purpose. X.21, being a greatly simplified procedure for circuit-switched applications, does not provide for any disconnection function at either the Data Link or the Network Layer. Instead, the basic Physical Layer NOT READY functions of $t = 0$, $c = $ OFF, and $r = 0$, $i = $ OFF serve to terminate the operational phases of a call.

Figure 6 shows the state diagram for clearing a circuit-switched connection and return to the READY state. Clearing can be initiated at any time by either the DTE or the DCE from any state in Fig. 3 except READY. A

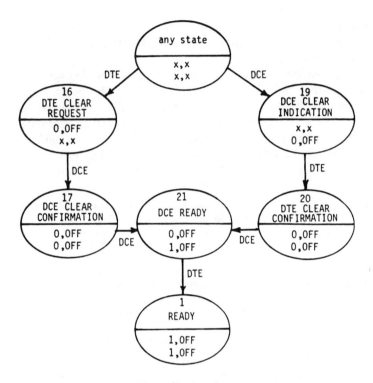

Fig. 6. Clearing phase.

DTE initiates clearing by sending DTE CLEAR REQUEST (state 16) and the DCE responds with DCE CLEAR CONFIRMATION (state 17) followed at least 24 bit times later by DCE READY (state 21). The DCE initiates clearing by sending DCE CLEAR INDICATION (state 19), and the DTE responds with DTE CLEAR CONFIRMATION (state 20). The DCE then responds with DCE READY (state 21). In a normal clearing sequence, regardless of whether the DTE or the DCE initiates clearing, the DCE must first indicate DCE READY (state 21) $r = 1$, i = OFF before the DTE can signal READY $t = 1$, c = OFF to enter state 1. This was necessary due to the operation of the network signaling system in CCITT Recommendation X.60. Once READY (state 1) is reached, a new call can then be processed.

4. DTE Time-Limits and DCE Time-Outs

In order to detect error or fault situations and provide a recovery mechanism, a family of DTE time-limits and DCE time-outs has been specified. Each timer is started by a transition into a particular state. For normal operation within the specified time, the timer stops when the designated next state is entered. If the timer expires before the recognized normal transition, then a recovery action can be initiated. As a result, lock-up or endless loop operations are avoided so when the problem clears, normal operation can resume.

It should be noted that these time-outs and time-limits are not an indication of typical response times, but are used to determine when most probably a failure in operation has occurred in either the DTE or DCE. Much faster response times under normal operation are expected for efficient network operation.

5. Optional User Facilities

There are a number of optional user facilities (special service features) defined for circuit-switched service by CCITT Recommendation X.2. These may be selected on a per-call basis by a facility request in the SELECTION SIGNAL SEQUENCE. A list of these facilities is given in Table II.

The Closed User Group provides for communication only among a designated group of subscribers. A subscriber may belong to more than one such group and therefore, a calling DTE must then designate to which group the subscriber being called belongs. A particular closed user group can be designated by the DTE as preferential to enable the network to process a call to the requested called subscriber in the preferential group without having to receive a facility request. Calls within nonpreferential groups would then need a facility request to specify the applicable closed user group desired.

Multiple Address Calling is allowed for circuit-switched service. This enables establishment of conference or broadcast types of communication. Additionally a centralized multipoint connection can be arranged on a addressed call basis.

Charge Advice is a new facility established to provide a calling DTE with the charging information related to an immediately preceding call.

Table II. Optional User Facilities

Closed user group
Multiple address calling
Charge advice
Calling line identification
Called line identification
DTE inactive
Redirection of call
Abbreviated addressing
Direct call
Facility registration/cancellation

Upon clearing of a call for which the charge advice has been requested by a facility request, the network will, within 200 ms, return the charging information to the DTE by means of an INCOMING CALL (state 8). When the call is accepted by the DTE (state 9), the DCE will provide the charging information in state 10bis. At the present time, there is no generalized error recovery defined if the DTE fails to receive the information correctly. One means under consideration is to repeat the information two or more times or until the DTE clears.

Called Line Identification can be requested by a DTE on a per-call basis so the network will verify the called number during state 10. Additionally, Calling Line Identification can be provided on a continuing basis to called DTEs as part of state 10 bis. This feature is intended to facilitate screening of incoming calls by a DTE to avoid unauthorized access.

A newer facility to appear in X.21 is the DTE Inactive facility. It is invoked when a subscriber is to be out of operation for a period of time. The DTE notifies the network of certain information indicating the reason and when normal operation will resume. The network forwards this information to a calling DTE through the DCE PROVIDED INFORMATION state 10.

Another facility that has been defined is Redirection of Call. This enables a subscriber to have incoming calls rerouted to an alternate number when desired, such as during nonbusiness hours. Other facilities that are also included in X.2 for circuit-switching service are Abbreviated Addressing and Direct Call. These have been discussed in detail earlier.

Facilities can be initiated on a per-call basis using a facility request or continuously on a subscription basis. Additionally, there is a procedure defined where a DTE can dynamically change or modify a particular facility. This is the Facility Registration/Cancellation procedure. It can be applied to reallocate or change full X.121 addresses for assigned abbreviated address codes. It can also be used to add and delete subscribers from closed user groups.

6. Test Loops

The new CCITT Recommendation X.150 has been developed to define a family of test loops to assist in the location of faults in an interconnection. These are shown in Fig. 7.

The DTE test loop 1 is implemented in the DTE and is under the full control of the DTE.

The local test loop 3 types are located in the DCE and provide a loop toward the DTE. This enables the DTE to verify the operation of the DTE/DCE interface. Loop 3 can only be activated by a switch on the

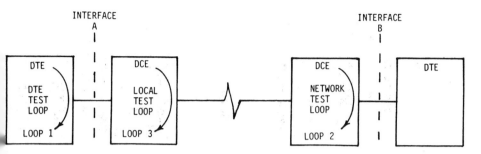

Fig. 7. Test loops.

DCE, although a means for automatic activation across the interface is being studied.

Finally, network test loop 2 types are implemented in the DCE and provide a loop toward the network. This can also be activated manually by a switch on the DCE. A provisional procedure was proposed for use in some networks but was rejected before the new version of X.21 got approved. Automatic activation of loops through the network is very controversial with many of the nations and will probably not be universally agreed on.

V. Future Evolution

As circuit-switched networks commence operation, practical experience will be gained as to the efficacy of this technology for data communications applications. As a result, the question can then be answered within the next few years as to whether an efficient, fast circuit-switched operation will prove to be more effective than the popularized packet-switched service of X.25 for a number of applications.

One significant issue that must be dealt with in future work is the convergence toward common protocols to satisfy all modes of operation. It is not a practical matter on a continuing basis to have two very different protocols satisfying identical functions, viz. X.21 call establishment and X.25 call establishment. As the general purpose physical elements of X.21 become the established universal Physical Layer interface for all data communications applications in the future, universal Data Link and Network Layer protocols should also be established accordingly. HDLC appears appropriate for the Data Link Layer, but considerable study remains to be done for establishment of a universal Network Layer standard [2].

References

[1] CCITT Blue Books, CCITT NINTH Plenary Assembly, Vols. VIII.2 and VIII.3, November 1988, Geneva, Switzerland.
[2] H. C. Folts, "Evolution toward a universal interface for data communications," in Proceedings on International Conference of Computer Communications, Kyoto, Japan, Sep. 1978, pp. 675–680.
[3] H. C. Folts, "X.21—The international interface for new synchronous data networks," in Conference Record of the International Conference on Communications, IEEE, Vol. 1, San Francisco, June 1975, pp. 15–19.
[4] H. C. Folts, ed., *McGraw-Hill's Compilation of Data Communications Standards*, New York: McGraw-Hill, Third Edition, 1986.

X.25 Packet-Switched Network Layer

Antony Rybczynski

I. Introduction and General Description

A. Introduction

The 1970s heralded the beginning of the development of public data networks (PDNs), offering packet switching services, e.g., Canada: Datapac, France: Transpac, USA: Telenet. The commercial viability of these networks hinged largely on the development and adoption of standard access protocols. These standards would facilitate the connection of varying types of data terminal equipments (DTEs) to the various public networks being developed as well as facilitate international internetworking.

The International Telegraph and Telephone Consultative Committee (CCITT), a permanent organ of the International Telecommunications Union, is responsible for establishing recommendations applicable to various aspects of international communications, including public data networks. A number of recommendations related to PDN services have been approved, most notably X.25 for packet-mode DTEs accessing packet-switching PDNs. The equivalent ISO international standard is 8208. Without these standards, users would almost definitely not be benefiting from the expansion of X.25 public data networks on a worldwide basis. Furthermore, compatible worldwide availability of X.25 networks combined with high value to end users has resulted in a broad spectrum of subscriber X.25 products. These include computer and front-end processor software and

ANTONY RYBCZYNSKI • Data Networks Division, Northern Telecom, Ottawa, Ontario K2C 3T1, Canada.

hardware, multiterminal controllers, intelligent terminals, LAN gateways, and X.25 intelligent modem plug-ins for personal computers.

PDNs offering X.25 packet-mode services are available in over 40 countries, typically growing at 30% per year. X.25 is also embedded in ISDN standards and network products and evolving services. Furthermore, international services have been established among these countries. X.25 is also widely used in private packet networks. Within LANs, connectionless network layer protocols are commonly used, though X.25, which is connection oriented, can be used on a LAN following ISO standard 8881. Furthermore, through gateways, X.25 is used for remote PC access to LANs as well as for inter-LAN applications. Finally, X.25 is also finding application in Very Small Aperture Terminal (VSAT) and mobile satellite-based systems.

CCITT Recommendation X.25 was first approved in March 1976. The next formal revision took place in 1977 with the addition of data link control procedures that are compatible with the High-Level Data Link Control (HDLC) procedures standardized by ISO. The last formal revision of X.25 took place in 1988, maintaining full alignment with the OSI network service definitions. With the previous revision in 1984, connectionless operations (i.e., datagrams) were deleted from X.25, owing to absence of any PDN implementation. However, connectionless services can be carried over X.25 networks with interprocess routing provided outside of the network. Concurrently, X.32 was defined for X.25 access via circuit-switching networks. This chapter describes the important characteristics of X.25 that are currently contained in the text of the recommendation.

B. General Description of the X.25 Interface

CCITT Recommendation X.25 is titled: "Interface between Data Terminal Equipment (DTE) and Data Circuit-terminating Equipment (DCE) for Terminals Operating in the Packet Mode on Public Data Networks." However, applying the concepts of the standard layer model introduced in Chapter 2, X.25 is not strictly speaking an interface. In fact, X.25 is a set of three peer protocols, as follows (see Fig. 1):

1. A peer protocol between Physical Level entities in the DTE and the DCE.
2. A peer protocol between Link Control Level entities in the DTE and the network node.
3. A peer protocol between Packet-Switched Network Level entities in the DTE and the network node.

Each of these levels functions independently of the other levels, with the exception that failures at a level may affect the operation of higher

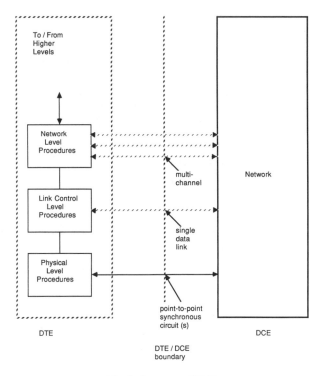

Fig. 1. Structure of X.25.

levels. It is this property that allows the X.25 Network Level Procedures to operate over various Link Control Level Procedures, whether as specified in X.25 or elsewhere; this will be elaborated later.

The Physical Level originally specified the use of a duplex, point-to-point synchronous circuit, thus providing a physical transmission path between the DTE and the network. It also specified the use of Recommendation V.24 (i.e., the EIA RS-232-C standard) between the DTE and a data set or modem. Therefore, no changes to the communications hardware of the DTE are required from those commonly used on a simple direct link. For more robust higher throughput applications, multiple physical connections (of various types and speeds) can be concurrently used. In this case, an additional Link Control Level Procedure is required; this multilink procedure is briefly described below. The Physical Level also specifies the use of Recommendation X.21. Physical level protocols have been discussed in Chapter 3. X.25 is standardized to operate at speeds up to 64 kbps, though megabit rates are foreseen in the early 1990s.

The Link Control Level specifies the use of data link control procedures which are compatible with HDLC and with the Advanced Data

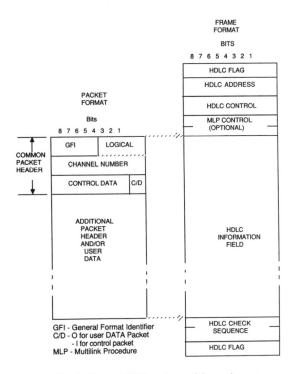

Fig. 2. General X.25 packet and frame formats.

Communications Control Procedure (ADCCP) standardized by the U.S. American National Standards Institute (ANSI) (see Chapter 5). The Link Control uses the principles of an ISO Class of Procedures for a point-to-point balanced system; in X.25, these procedures are referred to as the Link Access Procedures Balanced (LAPB). The use of this data link control procedure ensures that packets provided by the Packet-Switched Network Level and contained in HDLC information frames (see Fig. 2) are accurately exchanged between the DTE and the network. The functions performed by the Link Control Level include the following:

1. The transfer of data in an efficient and timely fashion.
2. The synchronization of the link to ensure that the receiver is in step with the transmitter.
3. The detection of transmission errors and recovery from such errors.
4. The identification and reporting of procedural errors to higher layers for recovery.

The multilink procedure performs the function of distributing frames across a number of physical links, each operating under LAPB, and of resequencing packets received over these links. This is accomplished by inserting an additional two-octet header (immediately after the LAPB header) containing a 12-bit sequence number. This procedure is further described in Chapter 5.

The major significance of the Link Control Level is that it provides the Packet-Switched Network Level with an error-free, variable delay link between the DTE and the network. The Packet-Switched Network Level is the highest level in X.25 and specifies the manner in which control information and user data are structured into Network Protocol Data Units called packets. The control information, including addressing information, is contained in the packet header field and allows the network to identify the DTE for which the packet is destined. It also allows a single physical circuit to support communications to numerous other DTEs concurrently.

The characteristics of the Packet-Switched Network Level Peer Protocol are further described in Section III.

C. Packet-Switched Network Level Services Available to X.25 DTEs

A distinction must be made between the X.25 access protocol and the Network Level services provided on a network operating in the packet mode and accessed by the DTE via X.25. Recommendation X.25 defines a set of three peer protocols to be used between the packet-mode DTE and the common-carrier equipment, generally referred to as the DCE. The X.25 Recommendation provides access to the following network services that may be provided on public or private data networks:

1. Switched virtual circuits (SVCs), also called virtual calls.
2. Permanent virtual circuits (PVCs).

A virtual circuit (VC) is a bidirectional transparent, flow-controlled path between a pair of logical or physical ports. A switched virtual circuit is a temporary association between two DTEs and is initiated by a DTE signaling a call request to the network. A permanent virtual circuit is a permanent association existing between two DTEs which does not require call setup or call clearing action by the DTE.

The characteristics of virtual circuits are now presented.

II. X.25 End-to-End Virtual Circuit Service Characteristics

A. Introduction

The Network Level virtual circuit service characteristics are specified in a nonsystematic way in various sections of Recommendation X.25. This section consolidates the specification of VC characteristics. The perspective for this discussion is a view of the DTE-to-DTE services provided by VCs rather than a view of the signaling performed between the DTE and the network.

B. Establishment and Clearing of a Virtual Circuit

A switched virtual circuit is established when the call request issued by the calling DTE is accepted by the called DTE (see Fig. 3). A permanent virtual circuit is always established and therefore no establishment procedures are required. The call request identifies the called and calling addresses (including address information relevant to private networks, if any) and facilities requested for the call, and may include user data. The user data sent during the call establishment phase is available for use by the higher layers (e.g., system passwords). Up to 16 octets may be sent in a normal call, and up to 128 octets can be sent via the fast select facility.

During the call establishment phase, the calling DTE may request certain optional user facilities (e.g., reverse charging) to be associated with the VC. In some cases (e.g., quality of service options such as throughput class and transit delay), the called DTE can alter the facility values requested by the caller. Thus, the VC service provides mechanisms for facility negotiations during call setup. Optional user facilities are discussed in Section IV.

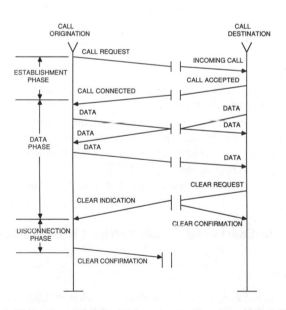

Fig. 3. Illustration of Call Establishment, DATA Transfer, and Call Clearing.

Table I. Clearing Call Progress Signals

Call progress signal	Explanation
DTE originated	Called DTE has refused the call or remote DTE has cleared it; a 7-bit user-to-user signal can be included.
Number busy	The called DTE is engaged in other calls and cannot accept the incoming call.
Out of order	The remote number is out of order. (X.25 Physical and/or Link Control Level not in operation).
Remote procedure error	An X.25 procedure error has occurred at the remote DTE/network boundary.
Reverse charging acceptance not subscribed	Network has blocked the call because the called DTE does not accept reverse charged calls.
Incompatible destination	The remote DTE does not support a user or facility requested.
Fast select acceptance not subscribed	The network has blocked the call because the called DTE does not support fast select calls.
Ship absent	The destination ship is not accessible; used in conjunction with the mobile maritime service.
Invalid facility request	Facility request invalid (e.g., a request for a facility which has not been subscribed to or is not available in the local network).
Access barred	The calling DTE is not permitted the connection to the called DTE (e.g., incompatible closed user group).
Local procedure error	A procedure error is detected at the local DTE/network boundary (e.g., incorrect format).
Network congestion	Temporary network congestion or a temporary fault condition has occurred within the network.
Not obtainable	Called number not assigned.
RPOA out of order	The RPOA nominated by the calling DTE is unable to forward the call.

If the call is refused by the called DTE, the DTE can signal the reason for call clearing to the calling DTE. If the call attempt fails for some other reason, a call progress signal is transmitted across the network indicating one of the causes specified in X.25 and given in Table I. As will be seen in the next section, the diagnostic code is also used by the network to provide extra information to the DTE when it has made a local procedure error.

This latter use is a characteristic of the X.25 Network Level protocol rather than of the virtual circuit itself.

Once the call has entered the data transfer phase, either DTE can clear the call using the diagnostic code to signal to the remote DTE the reason for the clearing. If the call is cleared by the network, it will signal this fact and indicate a call progress signal (Table I). When a call is cleared, data may be discarded by the network since the clear is not sequenced in respect to user data. All data generated by the DTE before initiation of a clear procedure will either be delivered to the remote DTE before completion of the clearing procedure at the remote DTE, or be discarded by the network. When a DTE initiates a clear, all data that were generated by the remote DTE before it has received the corresponding indication will be either delivered to the initiating DTE before the clear procedure is completed locally, or discarded by the network.

C. Facility Registration

Certain types of facilities (e.g., reverse charge calls not accepted) are provided to DTEs on a subscription basis. These may also be provided via an on-line registration procedure using REGISTRATION packets.

D. Data Transfer

In the data transfer phase, user data, which are conveyed in DATA and INTERRUPT packets, are passed transparently through the network. DTEs wishing universal operation on all networks should transmit all packets with data fields containing only an integral number of octets.

Virtual circuit flow control is a mechanism provided to ensure that the transmitting DTE does not generate data at a rate that is faster (on average) than that which the receiving DTE can accept. This is achieved by the receiving DTE controlling the rate at which it accepts DATA packets, noting that there is an upper limit on the number of DATA packets that may be in the network on a virtual circuit. Thus, flow control is performed on an end-to-end basis in that back-pressure exerted by a receiving DTE is reflected back to the sending DTE.

The network assigns resources to VCs based on information available to it (e.g., call routing) and on performance (specifically throughput) characteristics associated with the service. The network determines the maximum number of DATA packets that can be on a VC and the DTE need not be concerned with this aspect.

In some cases, there is a need to allow DTEs to select the maximum number of DATA packets that can be on a VC and thus be able to ascertain whether certain DATA packets have been delivered to the remote DTE. This

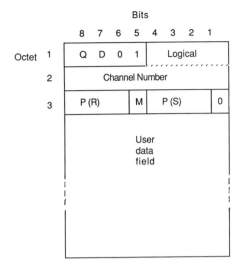

Fig. 4. DATA packet format: Q, data qualifier bit; D, delivery confirmation bit; M, more data bit; $P(S)$, packet send sequence number; and $P(R)$, packet receive sequence number.

information can be used in conjunction with a higher-level DTE-to-DTE error control protocol.

DTE-to-DTE acknowledgment of delivery via the Delivery Confirmation procedure in X.25 is available as a standard characteristic of X.25 virtual circuits. Specifically, if a DTE wishes to receive end-to-end acknowledgment for data it is transmitting, it uses an indicator called the Delivery Confirmation or D bit contained in the header of DATA packets (see Fig. 4). The D bit is always associated with the last octet in the DATA packet in which it is set by the DTE; this relationship is preserved by the network on an end-to-end basis even when the maximum packet lengths at each end of a VC are not the same. The acknowledgment is signaled via the packet receive sequence number $P(R)$, discussed in Section III.

If a DTE does not wish to receive end-to-end acknowledgment for data it is transmitting, it sets the D bit to zero. In this case, the network determines the maximum number of DATA packets that it is willing to accept, bearing in mind the throughput requirements of the communicating DTEs.

The enhancement of X.25 by the addition of the Delivery Confirmation procedure increases the robustness of virtual circuits by providing a DTE-to-DTE acknowledgment scheme. The communicating DTEs can maintain strict control on the amount of unconfirmed data, thus facilitating error recovery in the event of failure.

Since networks may perform packet-length conversion (though few do), X.25 defines a "complete packet sequence," which is a sequence of DATA packets that may be combined by the network. The only DATA packets that can be combined with subsequent DATA packets are those that are full, have the D bit set to zero, and have an indication set by the sending DTE that More Data is to follow (see Fig. 4).

For example, if a long record needs to be transmitted, it could be sent as a sequence of full packets each with the M bit set to 1. The last packet of the record might be full with $M = 0$ or more likely is partially full. Even if packet length conversion is provided, the record would be received as a sequence of full packets ($M = 1$) and possibly a partially full packet. The start and end of record information will be preserved on an end-to-end basis.

The D bit has priority over the More Data indication in packet combination so that a DATA packet with the D bit set to one is never combined with a subsequent packet. This is required to avoid possible deadlocks in the event that a network, which is providing packet length conversion, is awaiting data to assemble a full packet while the sending DTE is withholding these same data pending delivery confirmation. The More Data Indication may only be set by the DTE in full DATA packets or in partially full DATA packets that also have the D bit set to one. A sequence of DATA packets each carrying a More Data indication except for the last one will be delivered as an equivalent sequence of DATA packets.

Two independent mechanisms are provided to transfer user control information between a pair of DTEs outside the normal flow of data on a VC. The first mechanism transfers user control data within the normal flow control and sequencing procedures on a virtual circuit. This is called the Data Qualifier procedure, uses the Q bit, and applies to "complete packet sequences." This mechanism is used extensively to convey control information in the support of non-X.25 terminals on the network.

The second mechanism bypasses the normal DATA packet transmission sequence and provides an out-of-band (nonsequenced) signaling channel on VCs. The INTERRUPT packet, which is used in this case, may contain 32 octets of user data and is always delivered at or before the point in the stream of DATA packets at which it was generated, even when DATA packets are being flow controlled.

The maximum attainable throughput of a virtual circuit may vary owing to the statistical sharing of transmission and switch resources, and is constrained by the following:

1. The access line speed, local flow control parameters, and traffic on other calls at the local DTE/network boundary.
2. The access line speed, local flow control parameters, and traffic on other calls at the remote DTE/network boundary.
3. The maximum throughput achievable through the network independent of access line characteristics; the limit may differ for national and varying types of international calls.

The above throughput will generally be reached if the following conditions hold:

1. The DTE access data links at both ends of the VC are traffic engineered properly.
2. The receiving DTE is not flow controlling the DCE.
3. The transmitting DTE is sending DATA packets that have the maximum data field length; in addition, excessive use of the D bit will constrain VC throughput since, in this case, the rate of packet transfer will be determined by the rate of packet delivery confirmation by the receiving DTE. Table IV and associated text in Section III D will expand on the above points.

E. Error Recovery

The reset procedure is used to reinitialize the virtual circuit and in so doing removes in each direction all user data that may be in the network. When the reset is initiated by the DTE, it may convey to the remote DTE the reason for the resetting via a diagnostic code. If it is a network-generated reset, the reason is conveyed to both DTEs. Table II lists call progress signals associated with resetting in X.25.

All data generated by a DTE before initiation of a reset will either be delivered to the remote DTE before the corresponding indicator or discarded by the network; all data generated after local completion of a reset procedure will be delivered after completion of the corresponding reset procedure at the remote end. When a DTE initiates a reset procedure, all data that were generated by the remote DTE before its receipt of the corresponding indication are either delivered to the initiating DTE before the procedure is completed locally, or discarded by the network. Multiple and simultaneous resets are handled at the local interface as defined by the procedures for single resets.

The maximum number of packets that may be discarded when the clearing or resetting procedure has been invoked is a function of network end-to-end delay and network resources assigned in conjunction with the provided throughput. The maximum number of packets with the D bit set is a parameter of an X.25 interface (i.e., the local DTE transmit window size discussed in Section III D).

F. Alignment with OSI Network Service

The X.25 virtual circuit service described above is identical to the service specified in Recommendation X.213 for the OSI Network Service of the Model for Open System Interconnection (X.200 and IS 7498). A number of enhancements were approved in X.25 (1984) to achieve this

Table II. Resetting Call Progress Signals

Call progress signal	Explanation
DTE originated	Remote DTE reset the VC; a 7-bit user-to-user signal can be conveyed.
Out of order (PVC only)	The remote DTE is out of order (e.g., X.25 Physical and/or Link Control Level not in operation).
Remote procedure error	The call is cleared because of a procedure error at the remote DTE/network boundary.
Local procedure error	A procedure error is detected at the local DTE/network boundary (e.g., incorrect format, expiration of time-out).
Network congestion	Temporary network congestion has occurred within the network.
Remote DTE operational (PVC only)	Remote DTE is ready to resume normal operation after a temporary failure or out of order condition.
Network operational (PVC only)	Network is ready to resume normal operation after a temporary failure or congestion.
Incompatible destination	The remote DTE does not support a function used.
Network out of order (PVC only)	Temporary network fault condition has occurred.

alignment, most notably the following:

1. The fast select option was extended to allow 128 octets of user data during both call setup and clearing.
2. An address extension facility, providing up to 32 digit calling and called extended addresses, was introduced to allow private network addressing.
3. The INTERRUPT data field was extended to a maximum of 32 octets to match the requirement of the expedited data of the OSI network service.
4. The transit delay facility was introduced to enhance the quality of service negotiation capabilities inherent in the service.

III. X.25 Packet-Switched Network Peer Protocol Characteristics

A. Introduction

Recommendation X.25 specifies the peer protocol to be used by DTEs in establishing, maintaining, and clearing virtual circuits. This section now

Table III. Usage of X.25 Packet Fields[a]

Packet type	Common[b]	Cause	Diagnostic	Addresses and address length	Facilities and facility length	Data
CALL REQUEST	REQ	—	—	REQ	REQ	OPT (16) FS (128)
CALL CONNECTED	REQ	—	—	OPT	OPT	FS (128)
DATA	REQ	—	—	—	—	OPT
INTERRUPT	REQ	—	—	—	—	OPT (32)
RR/RNR	REQ	—	—	—	—	—
CLEAR REQUEST	REQ	REQ	OPT	FS	FS	FS (128)
CLEAR CONFIRMATION	REQ	—	—	FS	FS	—
RESET/RESTART REQUEST	REQ	REQ	OPT	—	—	—
INTERRUPT/RESET/RESTART CONFIRMATION	REQ	—	—	—	—	—
DIAGNOSTIC	REQ	REQ	OPT	—	—	—
REGISTRATION	REQ	REQ	—	REQ	REQ	—
REGISTRATION CONF'N	REQ	REQ	REQ	REQ	REQ	—

[a] REQ, required; REQ (X), required, of maximum length X octets; OPT, optional (only if all subsequent optional fields are not present); OPT (X), optional, of maximum length X octets; —, not applicable; FS, Fast Select extended format.
[b] Three-Octet Common Packet Header Field.

discusses this packet-switched network level protocol used at the user–network interface.

Packet formats for the various types of packets are introduced during the discussion, while Table III summarizes the usage of various packet fields.

B. Multiplexing at the X.25 Interface

In order to allow a DTE to establish concurrent virtual circuits with a number of DTEs over a single physical access circuit, the X.25 Packet-Switched Network Level employs packet-interleaved statistical multiplexing. This multiplexing technique is used to exploit the fact that a typical virtual circuit to a remote DTE may actually be carrying data for only a small percentage of the time. Each packet contains a logical channel number, which identifies the packet with a switched or permanent virtual circuit, for both directions of transmission.

A logical channel is a conceptual access path between a DTE and network which can be dynamically assigned for a new call either originated by the local DTE or by the local switch (to indicate an incoming call). A logical channel, assigned to a call, is busy until the call is cleared.

The range of logical channel numbers that can be used for virtual circuits is established at subscription time by agreement between the DTE and the network. If the DTE can only support a single VC, then logical channel number 1 will be used. If both PVCs and SVCs are used, then individual PVCs are statically assigned logical channel numbers in a range starting from number 1, while logical channels for calls are assigned a range above this. Logical channel numbers for SVCs are dynamically assigned during call establishment and identify all packets (i.e., control and data) associated with the VC. The logical channel numbers are only significant for a particular DTE.

Logical channel number 0 is reserved for non-call-related signaling between the user terminal and the switch. For example, the total interface can be restarted via the restart procedure on logical channel 0. Restarts result in clearing of all active calls and resetting of all PVCs. A second use of logical channel 0 is for on-line registration of facilities.

Every packet consists of a three-octet common packet header field as shown in Fig. 2, identifying the logical channel number and the packet type.

C. Establishing and Clearing a Virtual Circuit

A signaling method is provided to allow a DTE to establish switched virtual circuits to other DTEs, using logical channel numbers at each end to locally identify these switched virtual circuits, once they are established.

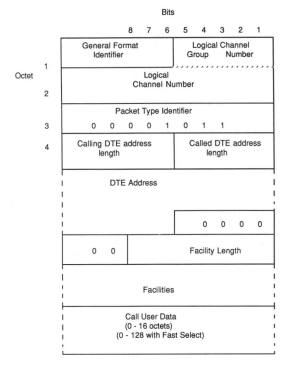

Fig. 5. CALL REQUEST and INCOMING CALL packet format.

A DTE initiates a call by sending a CALL REQUEST packet (Fig. 5) to the network. The CALL REQUEST packet includes the logical channel number chosen by the DTE to be used to identify all packets associated with the call. It also includes the network address of the called DTE and optionally (in the facility field) the calling and called terminal private network addresses. A facility field (to be discussed in Section IV) is present only when the DTE wishes to request an optional user facility requiring some indication at call setup. Reverse Charging is an example of such a facility. User data may follow the facility field and may contain up to a maximum of 16 octets. This field can be extended to 128 octets with invocation of the Fast Select facility.

The calling DTE will receive a CALL CONNECTED packet as a response indicating that the called DTE has accepted the call (Fig. 6). With the Fast Select option, up to 128 octets of user data can be added.

If the call is refused by the called DTE or if the attempt fails, the calling DTE will receive a CLEAR INDICATION (Fig. 7) indicating the appropriate call progress signal, and a one-octet diagnostic field, generated by the DTE and by the network in the former and latter cases, respectively. Up to 128 octets of user data can be transmitted in the fast select mode.

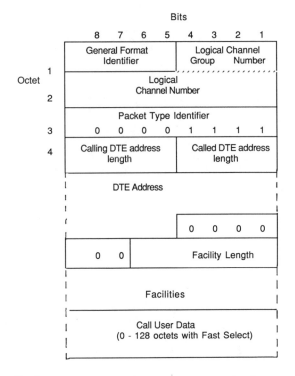

Fig. 6. CALL ACCEPTED and CALL CONNECTED packet format.

Call clearing, once the call enters the data phase, may be initiated by either DTE (or by the network in case of failure).

In any event, the logical channel number can be used again for another call when the clearing procedure is completed, normally by the transfer of a CLEAR CONFIRMATION packet. The CLEAR CONFIRMATON packet is normally three octets long and identifies the logical channel for which the clear procedures is completed.

Figure 8 illustrates the signaling states and transitions associated with call setup and clearing on a particular logical channel.

D. Data Transfer

DATA packets, illustrated in Fig. 4, can only be transferred across a logical channel after the virtual circuit has been established and if flow control constraints are not violated. $P(R)$ is the packet send sequence number of the packet. Only DATA packets are numbered, the numbering normally being performed modulo 8. The maximum number of sequentially numbered DATA packets that the DTE (or DCE) may be authorized to

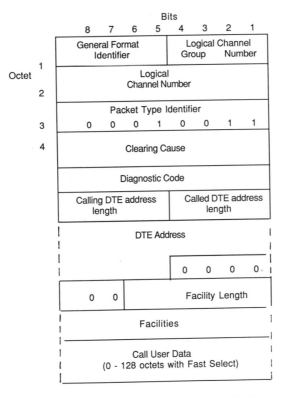

Fig. 7. CLEAR REQUEST and CLEAR INDICATION packet format.

transmit, without further authorization from the network (or DTE), may never exceed the modulus minus one, to ensure unique identification of packets (e.g., for D bit usage). The modulus can be optionally set to 128 (but is normally 8). The actual maximum value, called the window size W, is set for the logical channel either at subscription time or at call setup time (using the facility described in Section IV B 2). The default value for W is 2.

Each DATA packet also carries a packet receive sequence number, $P(R)$, which authorizes the transmission of W DATA packets on this logical channel starting with a send sequence number equal to the value of $P(R)$. If the DTE or the network wishes to authorize the transmission of one or more DATA packets, but there is no data flow on a given logical channel in the reverse direction on which to piggyback this information, it can transmit a RECEIVE READY (RR) Packet. If, on the other hand, the DTE or the network wishes to confirm the acceptance of a DATA packet with the D bit set to 1, but does not wish to authorize the transmission of any more data, it can transmit a RECEIVE NOT READY (RNR) packet. RR and RNR packets are three octets long and identify the logical channel number and a $P(R)$

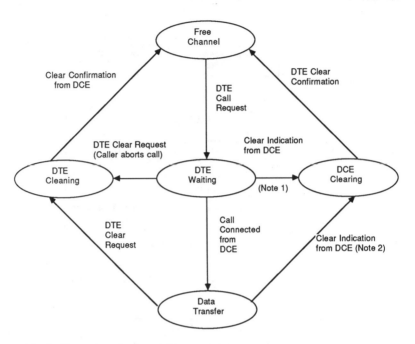

Fig. 8. Illustration of call establishment and clearing over a logical channel.

value. Flow control based on the conveyance of $P(R)$ numbers across a logical channel ensures that a sending DTE does not transmit data at an average rate that is greater than that at which the receiving DTE can accept those data.

The data field of a DATA packet may be any length up to some maximum value. The latter may be established independently at each end to a virtual circuit. Every network will support a maximum value of 128 octets and optionally maxima from 16 to 4096 octets in powers of 2.

When the Delivery Confirmation or D bit is set to 0, the $P(R)$ number is used to locally convey flow control information. When the D bit in Fig. 4 is set to 1, the corresponding $P(R)$ is used to convey delivery confirmation information and therefore has DTE-to-DTE significance.

For example, if the D bit is set to 1 in a DATA packet numbered p [i.e., $P(s) = p$] that it is transmitting, then a $P(R)$, which is received in a DATA, RR, or RNR packet and which is greater than or equal to $p + 1$, confirms acceptance by the remote DTE of the DATA packet. The receiving DTE indicates acceptance of a DATA packet with the D bit set to 1 by transmitting the corresponding $P(R)$ value to the network.

In order to allow two communicating DTEs to each operate at their locally selected packet sizes, the user may indicate, in a full DATA packet or

Table IV. Treatment of Data Packets with *M* and *D* Bits

DATA packet sent by source DTE			Combining with subsequent packet(s) is performed by the network when possible	DATA packet received by destination DTE[a]	
M	D	Full		M	D
0 or 1	0	No	No	0	0
0	1	No	No	0	1
1	1	No	No	1	1
0	0	Yes	No	0	0
0	1	Yes	No[b]	0	1
1	0	Yes	Yes	1	0
1	1	Yes	No	1	1

[a] Refers to the delivered DATA packet whose last bit of user data corresponds to the last bit of user data, if any, that was present in the DATA packet sent by the source DTE.
[b] If the DATA packet sent by the source DTE is combined with other packets, the *M* and *D* bit settings in the DATA packet received by the destination DTE will be according to that given in the two right-hand columns for the last DATA packet sent by the source DTE that was part of the combination.

any DATA packet with the *D* bit set to 1, that there is a logical continuation of his data in the next DATA packet on a particular logical channel. This is done with the More Data *M* bit contained in the DATA packet header as indicated in Fig. 4. Only a full packet is treated as if it had the *M* bit off.

Table IV defines the network treatment of DATA packets with various settings of the *M* and *D* bits.

The procedures used in conjunction with the DATA Qualifier procedure are identical to those that apply to DATA packets. The format used in this procedure is identical to that of the DATA packet except that the *Q* bit is set in the DATA packet header (see Fig. 4).

INTERRRUPT packets (Fig. 9), on the other hand, may be transmitted by the DTE even when DATA packets are being flow controlled. They contain neither send nor receive sequence numbers. Only one unconfirmed INTER-RUPT may be outstanding at a given time.

E. Error Recovery

1. Reset Procedure

The reset procedure is used to reinitialize the flow control procedure on a given logical channel to the state it was in when the virtual circuit was

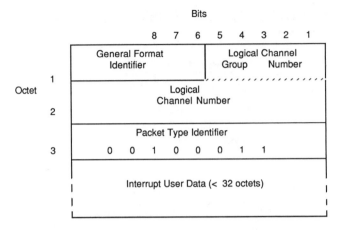

Fig. 9. INTERRUPT packet format.

established (i.e., all sequence numbers equal to zero and no data in transit). To reach this state, all DATA and INTERRUPT packets that may be in transit at the time of resetting are discarded. RESET REQUEST and RESET CONFIRMATION packets are used in the reset procedure.

2. Restart Procedure

The restart procedure provides a mechanism to recover from major failures. The issuance of a RESTART REQUEST packet is equivalent to sending a CLEAR REQUEST on all logical channels for switched virtual circuits and a RESET REQUEST on all logical channels for permanent virtual circuits. Thus, the restarting procedure will bring the DTE and the network to the state they were in when service was initiated.

3. Error Handling

X.25 treats packet level errors by applying the following principles:

1. Procedural errors during call establishment and clearing are reported to the DTE by clearing the call.
2. Procedural errors during the data transfer phase are reported to the DTE by resetting the VC.
3. A diagnostic field provides additional information to the DTE.
4. Timers are specified for resolving some deadlock conditions.

5. DIAGNOSTIC packets are used to indicate to the DTE conditions that may arise due to misalignment of DTE and network subscription options—for example, unauthorized use of logical channels.
6. Error tables define the action of the DCE on receiving various packet types in various states of the interface.

A number of special error cases (e.g., packet received on unassigned logical channel) have been identified in X.25 for which it is inappropriate to inform a DTE of a procedural error by resetting or clearing the logical channel. In this case, a DIAGNOSTIC packet may be used. The DIAGNOSTIC packet is nonprocedural in nature and solely for DTE logging. The DIAGNOSTIC packet identifies the logical channel number on which an error condition has been detected, and includes a diagnostic code and the first three octets of any erroneous packet received by the network.

Two areas associated with timeouts have been addressed. The first area relates to the length of time the DTE has to respond to an incoming call. On one hand, the network wishes to control its resources. On the other hand, short timeout values are not reasonable owing to the interaction between calls on a single interface, and between the Network and Link Control Levels, and owing to user processing within higher layers. A minimum value of three minutes has been agreed.

The second area relates to the action of the DCE when no confirmation has been received to an indication packet (i.e., during resetting, clearing, and restarting). In order to avoid long looping conditions, the DCE will take the following actions:

1. On expiry of a 60-s timer after issuing a RESET INDICATION, the DCE will retransmit the RESET, and on timing out a second time will clear the call, indicating the reason for clearing via the diagnostic code and call progress signal. On a PVC, a DIAGNOSTIC packet may be sent.
2. On expiry of a 60-s timer after issuing a CLEAR INDICATION, the DCE will retransmit the CLEAR, and on timing out again may issue a DIAGNOSTIC packet and will enter the ready state. In this state, the DCE does not ignore any packets sent by the DTE.
3. On expiry of a 60-s timer after a RESTART INDICATION has been issued, the DCE may issue a DIAGNOSTIC packet after one retry.

Diagnostic codes have been defined for reset, clear, and restart packets. The contents of the diagnostic code field provide nonprocedural information that does not alter the meaning of the call progress signal also provided. A DTE is not required to undertake any action on the content of the diagnostic to facilitate the correction of the problem.

Network-generated diagnostic codes are hierarchical. That is, for any specific diagnostic code, there is always a code that is of a more general nature. The specific codes provide information allowing the DTE implementor to quickly diagnose problems. The more general codes are used when relatively uncommon or unanticipated problems occur. To accelerate trouble resolution, the X.25 error tables indicate the diagnostic code generated under each error condition.

F. Interrelationship between Levels

Changes of operational states of the Physical and Link Control Levels do not implicitly change the state of each logical channel at the Network Level; such changes when they occur are explicitly indicated by the use of Network Level restart, clear, or reset procedures as appropriate.

A failure at the Physical and/or Link Control Level is defined as a condition in which the DCE cannot transmit and receive any frames because of abnormal conditions caused by, for instance, a line fault between the DTE and the DCE. When a failure is detected, the DCE will transmit to the remote end a RESET INDICATION indicating Out of Order for a permanent virtual circuit and a CLEAR INDICATION indicating Out of Order for an existing VC. During the failure, the DCE will clear any incoming calls.

When the failure is recovered, the DCE will send a RESTART INDICATION packet indicating Network Operational to the local DTE; this will result in a RESET INDICATION indicating Remote DTE Operation being transmitted to the remote end of each permanent virtual circuit.

IV. Optional User Facilities

A. Introduction

CCITT Recommendation X.2 defines the availability of various optional user facilities as being universally available or only available in some countries. Recommendation X.25 defines the procedures associated with all optional user facilities, irrespective of their availability.

This section describes only those optional user facilities that are universally available.

B. Optional User Facilities

1. Closed User Group Facility

Closed User Group (CUG) is an optional user facility agreed to for a period of time between the Administration and a group of users. This facility permits the user in a CUG to communicate with each other, but precludes communication with all other users. A DTE may belong to more than one closed user group.

The calling DTE specifies the closed user group selected for a call using the optional user facility parameters in the CALL REQUEST packet. The closed user group selected for a call is indicated to a called DTE using the optional user facility parameters in the INCOMING CALL packet.

2. Flow Control Parameter Selection

Flow Control Parameter Selection is an optional user facility agreed to for a period of time that can be used by a DTE for its logical channels. The flow control parameters considered are the packet and window sizes for each logical channel for each direction of data transmission.

When the DTE has subscribed to the facility, it may, in a CALL REQUEST packet, separately request maximum packet sizes and window sizes for each direction of data transmission. The maximum packet sizes that may be supported on public data networks are 16, 32, 64, 128, 256, 512, 1024, 2048, and 4096 octets. If a particular packet or window size is not explicitly requested, the DCE assumes default requests of 128 octets and 2, respectively. The range of packet sizes supported by a given network is a subset of the above.

When the DCE transmits a CALL CONNECTED packet, it indicates in the facility field the flow control parameters to be used by the calling DTE. The only valid facility indications in the CALL CONNECTED packet as a function of the facility requests in the CALL REQUEST packet are specified by the following general negotiation rules:

1. Window sizes can be changed in the direction of $W = 2$.
2. Packet sizes can be changed in the direction of 128 octets.

When the called DTE subscribes to the facility, the DCE transmits flow control parameter facility indications to be used by the called DTE in selecting the flow control parameters for the call. The called DTE can change the indicated values using the above negotiation rules.

The flow control parameters for logical channels used for PVCs are established at subscription time.

The network may have to constrain the available parameter ranges in order to allow the call to be established. In this case, the network is involved in the negotiations discussed above. This would occur, for example, if a requested packet size, though available domestically, was not available on a particular international call.

3. Throughput Class Negotiation

Throughput Class Negotiation is an optional user facility agreed to for a period of time that can be used by a DTE for virtual circuits. This facility permits negotiation on a per call basis of the throughput classes. The throughput classes are considered independently for each direction of data transmission.

A Throughput Class for one direction of transmission is an inherent characteristic of a virtual circuit, related to the amount of network resources allocated to it. This characteristic is meaningful when the D bit is set to zero in DATA packets. It is a measure of the throughput that is not normally exceeded on the VC. However, owing to the statistical sharing of transmission and switching resources, it is not guaranteed that the throughput class can be reached 100% of the time.

Default values are agreed on between the DTE and the network. The default values correspond to the maximum throughput classes that may be associated with any virtual circuit.

4. Fast Select

The Fast Select facility is an optional user facility that may be requested on a per call basis. If requested, the CALL REQUEST, CALL CONNECTED, and CLEAR INDICATION packets can each contain user data fields of up to 128 octets. A suboption prohibits entry into the data transfer phase (i.e., the only valid response to a CALL REQUEST is a CLEAR INDICATION, both with up to 128 octets for user data).

The Fast Select option provides alignment with the OSI network service and is well suited to transaction applications requiring fast response times and exhibiting (short) inquiry/response operation.

5. Transit Delay Selection and Indication

The Transit Delay Selection and Indication facility allows the terminal to specify the transit delay applicable to the call. The network will route the call taking this information into account and indicate the estimated delay for the call in the CALL CONNECTED packet. The network could use this information for routing (e.g., satellite versus terrestrial trunks) and for selection of outgoing queues (e.g., multiple priority queues).

6. One-Way Outgoing Logical Channel

One-Way Outgoing Logical Channel is an optional user facility agreed to for a period of time. This user facility restricts the use of a range of logical channels to outgoing calls. One-way logical channels retain their duplex nature with respect to data transfer.

7. Incoming or Outgoing Calls Barred

Incoming or Outgoing Calls Barred are two optional user facilities agreed to for a period of time. These facilities apply to all logical channels used for switched virtual circuits.

Incoming Calls Barred prevents incoming calls from being presented to the DTE. The DTE may originate outgoing calls. Outgoing Calls Barred prevents the DCE from accepting outgoing calls from the DTE. The DTE may receive incoming calls.

V. Procedures for X.25 Circuit-Switched Access

A. Introduction

The general description of the X.25 interface given in Section I B applies to packet-mode terminals connected on a dedicated basis to a packet-switched network. In the early years of X.25 (i.e., the 1970s), X.25 terminal implementations were limited to minicomputers, host front-ends, and to terminal controllers and related technologies; X.25 is now being implemented at the single terminal level. A major obstacle to widespread terminal use of X.25 is high cost of dedicated access. This was recognized in the early 1980s, and has resulted in the introduction of Recommendation X.32 for packet-mode terminal access via the public switched telephone network (PSTN) via modems, or via digital circuit-switched connections, capitalizing on the digitization of the PSTN and on the introduction of ISDN B Channel services.

X.32 specifies physical level procedures for circuit-switched access, extensions to X.25 LAPB link control level procedures particularly with respect to link setup and terminal identification, and extensions to X.25 network level procedures with respect to terminal identification and customization. Given the importance of terminal identification and customization, this aspect of X.32 is further discussed in Section V B.

B. X.32 Terminal Identification and Customization

X.32 defines two sets of terminal services: one set for nonidentified terminals, and one set for identified terminals.

Characteristics of the service offered to nonidentified terminals include the following:

1. All charges to the remote (identified) terminal.
2. No packet network address assigned to the terminal (i.e., no incoming calls while connected to the PDN).
3. Predefined link level parameters.
4. No closed user group operation.

In addition, networks may offer a number of optional user facilities, the use of which may be requested through on-line facility registration.

Four distinct methods for terminal identification are specified:

1. Identification provided by the circuit-switched network (e.g., using the calling line identification functionality of common channel signaling).
2. Identification by means of a link level XID procedure.
3. Identification by means of network level registration procedure.
4. Identification by means of the network user identification (NUI) facility in call setup packets.

Identification procedures 2 and 3 above include the capabilities to verify and/or authenticate the terminal identification. Method 1 is inherently secure (being under network control); method 4 includes the conveyance of a password associated with each NUI.

Once terminal identification is completed, three separate services are defined as follows. The identified default service applies to those terminals that are not allocated a registered address by the network; these terminals cannot be called across the network. The X.25 subscription set is the same as for nonidentified terminals. Thus no terminal customization is provided.

The limited customized service provides for user selection (on a subscription basis) of default packet and window sizes, throughput classes, and closed user groups. Link control parameters and logical channel assignment are not user selectable.

Finally, in the fully customized service, full capabilities are provided for user selection including the ability of temporarily suspending the use of the registered address and substituting for it a different address (i.e., the temporary location option).

VI. Concluding Remarks

A. A Common X.25 DTE

From a DTE implementation point of view, a common X.25 interface can be defined [2, 3], which consists of the following universally available features:

1. An ISO-compatible frame level procedure (i.e., LAPB).
2. Use of Logical Channel Number one as the starting point for logical channel assignment.
3. Modulo 8 packet level numbering.
4. Dynamic $P(R)$ significance by use of the Delivery Confirmation bit.
5. A standard procedure for selecting packet and window sizes, with defaults of 128 octets and 2, respectively.
6. Two mechanisms for user control data transfer (i.e., qualified DATA and INTERRRUPT packets).
7. A standard way of specifying required call throughput.

B. X.25 and the Integrated Services Digital Network (ISDN)

X.25 continues to be a stable user–network interface to packet networks. Future enhancement will be introduced with full consideration of backward compatibility and will likely focus on new optional user facilities. Attention is now turning toward exploiting packet network access via ISDN interfaces.

Briefly, ISDN interfaces provide 64 kbps B channels, with signaling for these channels provided by D channels. Thus $2B + D$ "basic" and $23B + D$ "primary" interfaces are defined at 144 kbps and 1.544 Mbps, respectively. X.25 can be used on B channels, with these channels permanently established to the packet switch. In this case, no D channel signaling is required. B channels can also be used for digital circuit switched incoming and outgoing connections to the packet network.

ISDN also allows the sharing of the D channel by signaling and packet data. This multiplexing is provided by the use of level 2 logical links within the D channel Link Access Procedure (i.e., LAPD as opposed to LAPB, which provides a single logical link).

Applying OSI layers principles, the X.25 Packet Level procedures are specified to operate over LAPB. They can operate equally well over LAPD links. To meet signaling timing requirements, D channel DATA packets are constrained to a maximum length of 128 octets.

Finally, while technology and user needs will continue to evolve [4], X.25 virtual circuits will continue to be one of the basic services provided within telecommunications networks (both public and private). Access to X.25 VCs will be supported in diverse ways taking full advantage of the sound layered architecture adopted by X.25 over a decade ago.

References

[1] P. T. Kelly, "Public packet switched data networks, international plans and standards," *Proc. IEEE*, vol. 66, No. 11, November 1987, pp. 1539–1549.

[2] A. M. Rybczynski and J. D. Palframan, "A common X.25 interface to public data networks," *Comput. Networks J.*, vol. 4, No. 3, pp. 97–110, July 1980.

[3] Z. Drukarch *et al.*, "X.25: The universal packet network interface," Proc. Int. Conf. Comput. Commun., Atlanta, October 1980.

[4] A. M. Rybczynski, "Packet switching in the 1990s," Proc. of Globecom 87, Tokyo, October 1987, pp. 1406–1409.

Routing Protocols

Mischa Schwartz and Thomas E. Stern

I. Introduction

In this chapter, we provide an overview of routing techniques used in a variety of computer communication networks in current operation. These include the public data networks TYMNET and TRANSPAC (the former is a specialized common carrier network based in the United States, but with connections to Europe as well; the latter is the French government PTT data network), ARPANET, the U.S. Department of Defense Computer Network, and the commercial network architectures SNA (Systems Network Architecture) and DNA (Digital Network Architecture), developed by IBM and Digital Equipment Corporation, respectively. The networks are all examples of store-and-forward networks with data packets* moving from a source to a destination, buffered at intermediate nodes along a path. The path is defined simply as the collection of sequential communication links ultimately connecting source to destination.

The routing algorithms used in these networks all turn out to be variants, in one form or another, of shortest path algorithms that route packets from source to destination over a path of least cost. The specific

---

*We use the word "packet" here to represent a self-contained block of user data, of possibly varying size, that will traverse the network as one cohesive unit. In some networks, this is synonymous with a message. In others, a message may be broken at the source node into several smaller packets. For this reason, we make no real distinction between the two, and we shall, in fact, sometimes use the words interchangeably.

MISCHA SCHWARTZ AND THOMAS E. STERN • Center for Telecommunications Research, Columbia University, New York, New York 10027.

cost criterion used differs among the networks. As will become apparent in the discussion following, some networks use a fixed cost for each link in the network, the cost being roughly inversely proportional to the link transmission capacity in bits per second. For a network with equal capacity links, minimization of the path cost generates a minimum hop path. Links with measured congestion and/or high error rates may be assigned higher costs, steering traffic away from them. Costs may also vary with the type of traffic transmitted—whether interactive, asynchronous terminal type, synchronous traffic, or file transfers between computers. Other networks attempt to estimate average packet time delay on each link and use this to assign a link cost. The resultant source–destination path chosen tends to provide the path of minimum average time delay.

Since a least cost routing algorithm is used in all cases, we provide in the next section a unifying discussion of least cost routing to further demonstrate the similarities in the network algorithms.

Although the basic routing procedures are similar, differing primarily in the choice of a link cost function used to establish the minimum cost path, the routing techniques used tend to differ in implementation and the place at which the algorithms are run. The routing algorithm may be run in centralized fashion by a central supervisory program or Network Control Center, or may be carried out in a decentralized or distributed way with individual nodes in the network running the routing algorithm separately. In the former (centralized) case, global information about the network required to run the algorithm (current topology, line capacity, estimated link delays if required, condition of links and nodes, etc.) need only be kept by the central supervisor. Path setup is then accomplished through routing messages sent to each node along the path selected. In the latter (distributed) case, the required information must be exchanged among nodes in the network. This implies some means of disseminating changes in topology (nodes and links coming up or going down), congestion, and estimated time delay information if used in the algorithms. In the next section, we discuss two least cost routing algorithms, which are the basis for routing procedures in many networks.

The routing procedures adopted also differ in how dynamic they are—how rapidly and in what manner they adapt, if at all, to changes in network topology and/or traffic information. In some cases, routes are fixed during the time of a user session. (This is the length of a call from sign-on or connect time to sign-off or disconnect time.) A node or link failure during a session will then abort the call or may, in some cases, cause a new route to be selected, transparent to the user. In other cases, paths may be changed during a session (although unknown to the user and relatively slowly to avoid stability problems).

Once the path has been determined, routing tables, set at each node, are used to steer individual packets to the appropriate outgoing link. An

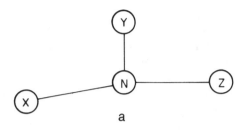

a

PACKET IDENTIFICATION	NEXT NODE ASSIGNMENT
(1,4)	X
(1,5)	Y
(2,4)	X
(2,5)	Z
(2,6)	Y
⋮	⋮

b

Fig. 1. Routing at a node in a network. (a) Current node and neighbors. (b) Routing table.

example appears in Fig. 1. A typical node N in a network is shown, with three neighboring nodes X, Y, Z to which it is connected. In part (b) of the figure, a partial routing table is shown, associating individual user packets with an appropriate outgoing line leading to one of the neighboring nodes. The packet identification requires two numbers. The two numbers could be source and destination address, or they could be a mapping of these two fields into a corresponding pair given by the incoming link number and a number associated with that link. (This is variously called the logical record number, the logical link number, next node indicator, etc.) The source-destination addresses could also be combined into one unique network-wide virtual circuit number, although this becomes difficult to monitor and assign with large networks. If all packets routed to a particular destination follow the same path, only a destination address is required to determine the proper outgoing link. (This would be the case, for example, if the paths chosen are independent of message class, type, etc.)

It is obvious that routing procedures play an important role in the design of data networks. Together with the techniques of flow and congestion control, they are implemented as part of the network level in layered protocols. This is the level just above the data link level that ensures correct transmission and reception of packets between any neighboring nodes in a network.

Because of their importance to the proper operation of data networks, routing techniques have received a great deal of attention in recent years.

They have been variously classified as deterministic, stochastic, fixed, adaptive, centrally controlled, or locally controlled [1].

The fixed versus adaptive classification is particularly vague, since all networks provide some type of adaptivity to accommodate topological changes (links and/or nodes coming up or going down, new topologies being established). In the past, the distinction had been made primarily on the basis of individual packet handling. In the original ARPA routing algorithm, routing tables could be updated at intervals as short as 2/3 s [2]. Routing changes were made by individual nodes in a decentralized manner. As a result, individual packets in a message could follow diverse routing paths. The ARPA adaptive routing algorithm was adopted by a number of other networks as well [3], [4]. The French Cigale network used a related decentralized algorithm [5].

The hope was that by adapting on a packet-by-packet basis, the network could be made more responsive to changes in traffic characteristics and to topology, enabling packets to arrive at their destinations more rapidly, as well as avoiding failed links and/or nodes and regions of congestion. This was the case to some extent, yet the ARPA experience indicated some fundamental problems arising—there were problems with message reassembly at the destination, packet looping, adaptation problems ("too rapid a response to the good news of added links and too sluggish a response to the bad news of deleted links") [2], and so forth. As a result, the ARPA algorithm has been changed, making it less dynamic.

Although these routing techniques will be called adaptive in the sense of responding to network changes, the time constants are considerably longer. In the case of the new ARPA algorithm, changes may take place about every 10 s. Details appear in Section III.

If the algorithms used in most of these networks are adaptive and of the shortest path type, how then are they to be distinguished? We have already indicated that they may differ in the cost criterion used, and as to whether the computations are done centrally or on a distributed basis. The rate of adaptation is another distinguishing characteristic. This has also been noted already—the ARPA network, as an example, will change routes, if necessary, every 10 s. TYMNET and SNA make changes from session to session only. DNA changes paths only when necessary.

Other differences arise due to the actual implementation: the size of routing tables, the routing overhead required, the time required to set up a path or change one if necessary; all of these will be found to differ in the networks to be described. Other differences will be noted during the discussion.

Interestingly, shortest path single routes turn out not to be optimum if the long-term average *network* time delay is to be minimized. In this case, multiple or "bifurcated" paths arise [1], [6]. Packets at a node are assigned to one of several outgoing links on a probabilistic basis. Bifurcated routing has not as yet been used in routing algorithms implemented in operating networks, although there are plans to incorporate this procedure in future routing mechanisms for the Canadian DATAPAC network [7].

In Section II, we provide a more detailed treatment of routing procedures in networks, focusing, as already noted, on shortest path (least cost) algorithms. In Section III, we then describe the routing implementations currently found in TYMNET, ARPANET, TRANSPAC, and the two commercial network architectures, IBM's SNA and Digital Equipment's DNA. In these last two cases, the routing procedures adopted are part of the overall protocol design and do not refer to a specific network implementation.

II. Structure of Routing Procedures in Packet-Switched Networks

Efficient utilization and sharing of the communications and nodal processing resources of a packet-switched communication network require various types of control, perhaps the most important of these being packet routing, that is, selecting paths along which packets are to be forwarded through the network. The objective of any routing procedure is to obtain good network performance while maintaining high throughput. "Good performance" usually means low average delay through the network, although many other performance criteria could be considered equally valid. Since poor routing algorithms often lead to congestion problems, and conversely, local congestion often requires at least temporary modification of routing rules, the routing problem cannot be completely divorced from that of congestion control, which is the subject of Chapter 10. Nevertheless, in this chapter, we restrict ourselves to routing under the assumption that the better the routing algorithm, the less congestion is likely to occur.

While routing procedures can be set up within a network more or less independently of the protocols seen by the users (i.e., the devices external to the network), the choice of an appropriate routing procedure is influenced to some extent by the network protocols operating at the network/user interface. It is convenient to classify these as either *virtual circuit-oriented* or *datagram-oriented*. In the former case, a device or a process within a device (e.g., an application program within a computer) prepares to communicate with another device by exchanging a number of control messages with the network. The purpose of these messages is to determine whether the destination device is connected and ready to receive messages, to agree on certain aspects of the transmission protocol, and to set up a virtual circuit (VC) from source to destination.* Figure 2 illustrates three VCs connecting terminals $T1$, $T2$, $T3$ to a host H. The VC appears to the external devices as

---

*VCs set up in this manner are termed "switched" VCs, in contrast to "permanent" VCs, which require no call setup procedure.

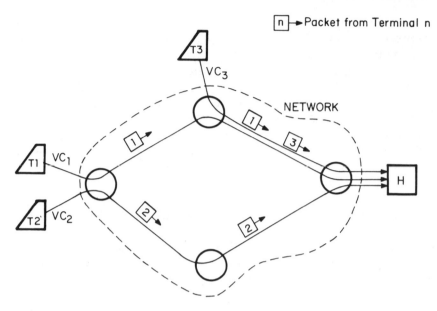

Fig. 2. Virtual circuits.

if it were a dedicated line; under normal operation, individual data packets arrive at the destination essentially without loss or error and in the proper sequence. It is important to note, however, that *within* the network, packets from many different virtual circuits are generally sharing the same communication lines; errors, losses, and changes of packet order may occur. However, it is the function of the internal network protocols to correct for all of these effects. In the datagram-oriented case, communication is on a datagram-by-datagram basis. Each message or packet (called a *datagram* in this case) must therefore contain its own destination address, but no preliminary control messages are required to set up a communication path.

A. Functions of Routing Procedures

In an idealized situation where all parameters of the network are assumed to be known and not changing, it is possible to determine a routing strategy which optimizes network performance for some class of users, e.g., minimizes average network delay for the interactive user or maximizes throughput for the batch user. The routing problem posed in this form is equivalent to the multicommodity flow problem well known in the operations research literature, and has been treated extensively in the communications network context [6], [8]–[11]. Changing situations in real networks such as a line failure or a change in the traffic distribution, necessitate some degree of adaptivity. Any adaptive routing procedure must perform a number of functions:

1. Measurement of the network parameters pertinent to the routing strategy.
2. Forwarding of the measured information to the point(s) [Network Control Center (NCC) or nodes] at which routing computation takes place.
3. Computation of routing tables.
4. Conversion of routing table information to packet routing decisions. (This may include dissemination of a centrally computed routing table to each switching node as well as the conversion of this information to a form suitable for "dispatching" packets from node to node.)

Typical information that is measured and used in routing computation includes states of communication lines, estimated traffic, link delays, and available resources (line capacity, nodal buffers). The pertinent information is forwarded to the NCC in a centralized system and to the various nodes in a distributed system. In the distributed case, two alternatives are possible: (1) forward only a limited amount of network information to each node (i.e., only that which is required for computing its local routing decisions), or (2) forward "global" network information to all nodes. (See Section III A 1 for a comparison of these two strategies in a specific network.) Based on the measured information, "costs" can be assigned to each possible source-destination path through the network. Routing assignments may be based on the principle of assigning a single path to all traffic between a given pair of source/destination nodes, or else traffic for a given source/destination pair might be distributed over several paths, resulting in the *multiple path*, or *bifurcated routing* procedure mentioned in Section I. In the latter case, single paths might still be maintained for each virtual circuit (if a VC-oriented protocol is used). This case is illustrated in Fig. 2, wherein VC_1 and VC_2 involve the same source/destination nodes, but take different paths. (Bifurcated routing on a packet basis is illustrated in Fig. 3.) While maintenance of single paths for each VC is not an optimal procedure, it has a number of practical advantages, an important one being the fact that packets always arrive at their destination in the proper order. It is therefore not surprising that most of the networks currently in operation use VC-oriented protocols with single-path routing per VC. These paths generally remain fixed for the duration of operation of the VC, unless a failure occurs.

Once one thinks in terms of single-path routing, it is natural to choose the "shortest" or, more generally, the *least cost* path whenever alternate paths exist. The path cost can, of course, be assigned using whatever cost

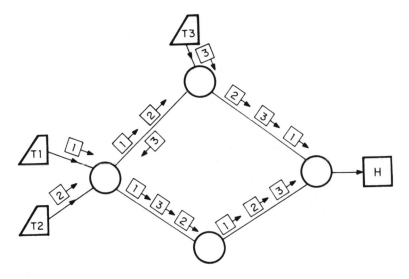

Fig. 3. Bifurcated routing.

functions seem appropriate (see above), the only essential property being that the *path* cost is computed as the sum of the costs of the *links* comprising that path. In such a case, the routing problem is equivalent to that of finding the shortest path through a graph, wherein link "length" is understood to have the more general meaning of link "cost." The set of shortest paths from all source nodes to a common destination node in a network forms a tree with the destination as root node. Thus, it is clear that if single-path routing on a source/destination basis is to be used, the path of a packet is uniquely determined by its destination alone. (Optimal bifurcated routing on a packet basis also only requires destination information.) On the other hand, single-path-per-VC routing requires either explicit or implicit VC identification for each packet; source/destination information alone is insufficient. This is because each time a new VC comes into operation, the costs determining the shortest path may be different since they generally change with time as network operating conditions change. Thus, VCs between the same source/destination pairs, established at different times, may take different paths as illustrated in Fig. 2.

B. Shortest Path Algorithms

The shortest path problem described above has received much attention in the literature. A variant of this problem, that of finding the k shortest paths between source and destination, is also applicable to the routing problem. (One is often interested in two or three alternate routes ranked in order of cost.) This too has been extensively treated. Since most operating networks use some version of shortest path routing, we discuss in this section the two algorithms most commonly used in communication network shortest path calculations. Algorithm A, due to Dijkstra [12], [13], is adapted to centralized computation, while B, a form of Ford and Fulkerson's algorithm [14], is particularly useful in distributed routing procedures. Since they are simple and intuitive, we present them informally, aided by an example.

Consider the network of Fig. 4(a) in which the numbers associated with the links are the link costs. (It is assumed for simplicity that each link is bidirectional with the same cost in each direction. However, both

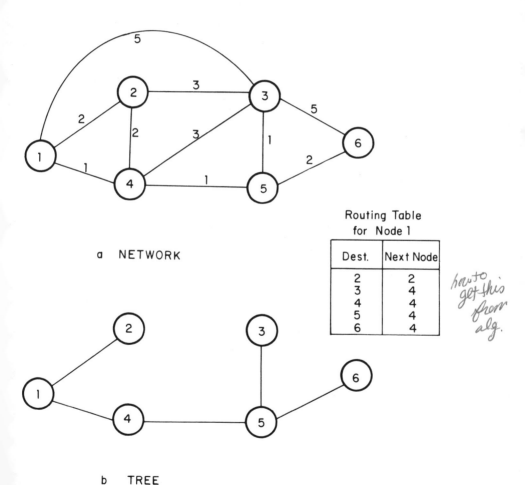

a   NETWORK

Routing Table
for Node 1

Dest.	Next Node
2	2
3	4
4	4
5	4
6	4

have to get this from alg.

b   TREE

Fig. 4. Example of shortest path routing.

Table I. Algorithm A

Step	N	$D(2)$	$D(3)$	$D(4)$	$D(5)$	$D(6)$
Initial	$\{1\}$	2	5	1	∞	∞
1	$\{1, 4\}$	2	4	1	2	∞
2	$\{1, 2, 4\}$	2	4	1	2	∞
3	$\{1, 2, 4, 5\}$	2	3	1	2	4
4	$\{1, 2, 3, 4, 5\}$	2	3	1	2	4
5	$\{1, 2, 3, 4, 5, 6\}$	2	3	1	2	4

algorithms are applicable to the case of links with different costs in each direction.) We first use algorithm A to find shortest paths from a single source node to all other nodes. The algorithm is a step-by-step procedure where, by the kth step, the shortest paths to the k nodes closest to the source have been calculated; these nodes are contained in a set N. At the $(k + 1)$th step, a new node is added to N, whose distance to the source is the shortest of the remaining nodes outside of N. More precisely, let $l(i, j)$ be the length of the link from node i to node j, with $l(i, j)$ taken to be $+ \infty$ when no link exists. Let $D(n)$ be the distance from the source to node n along the shortest path *restricted to nodes within* N. Let the nodes be indicated by positive integers with 1 representing the source.

(1) Initialization. Set $N = \{1\}$, and for each node v not in N, set $D(v) = l(1, v)$.

(2) At each subsequent step, find a node w not in N for which $D(w)$ is a minimum, and add w to N. Then update the distances $D(v)$ for the remaining nodes not in N by computing

$$D(v) \leftarrow \text{Min}[D(v), D(w) + l(w, v)]$$

Application of the algorithm to the network of Fig. 4(a) is shown in Table I, and the resultant tree of shortest paths appears in Fig. 4(b), together with a *routing table* for node 1, indicating which outbound link the traffic arriving at that node should take. (It should be clear that the same algorithm can be used to find shortest paths *from* all nodes to a common *destination*.)

Now consider algorithm B. This is an iterative procedure, which we will use in the same network to find shortest paths from all nodes to node 1, considered now as the common *destination*. To keep track of the shortest paths, we label each node v with a pair $(n, D(v))$, where $D(v)$ represents the current iteration for the shortest distance from the node to the destination and n is the number of the next node along the currently computed shortest path.

Table II. Algorithm B

Cycle	Node →	2	3	4	5	6
Initial		$(\cdot, \infty)$	$(\cdot, \infty)$	$(\cdot, \infty)$	$(\cdot, \infty)$	$(\cdot, \infty)$
1		$(1, 2)$	$(2, 5)$	$(1, 1)$	$(4, 2)$	$(5, 4)$
2		$(1, 2)$	$(5, 3)$	$(1, 1)$	$(4, 2)$	$(5, 4)$

(1) Initialization. Set $D(1) = 0$ and label all other nodes $(\cdot, +\infty)$.

(2) Update $D(v)$ for each nondestination node v by examining the current value $D(w)$ for each adjacent node w and performing the operation

$$D(v) \leftarrow \underset{w}{\text{Min}} \left[D(w) + l(v, w) \right]$$

Update of node v's label is completed by replacing the first argument n by the number of the adjacent node which minimizes the above expression. Step (2) is repeated at each node until no further changes occur, at which time the algorithm terminates.

Table II illustrates the procedure for the network of Fig. 4(a). Two complete cycles of updates are required, after which no further changes occur and the iteration is complete. The tree of shortest paths generated is, of course, the same as that of Fig. 4(b). In this case, the nodes were updated in numerical order; however, any arbitrary order, cyclic or acyclic, will work. For each nondestination node, the first argument of its final label indicates the next node on the shortest path to the destination, and thus supplies the necessary routing information (for this destination *only*).

A word of comparison is now in order. Construction of routing tables based on algorithm A requires a shortest path tree calculation for each node in the manner described above. The tree is constructed with the particular node chosen as source (root) node, and the routing information that is generated is used to construct the table for that node as illustrated in Fig. 4(b). The tree can then be discarded. It should be noted that tree construction for each node requires *global* information about the network. Construction of a routing table using algorithm B requires repeated application of the algorithm for each *destination* node, resulting in a *set* of labels for each node, each label giving the routing information (next node) and distance to a particular destination. Note that in this case, the algorithm can be conveniently implemented in a *distributed* fashion, in which case each node requires only information from its neighbors.

Evaluation of the comparative merits of the two algorithms depends on a number of factors, including amount of overhead required in passing measured information to the point(s) at which computation is performed, amount of data to be stored, complexity of the computation, and speed with which the algorithm can respond to changes in link costs. These comparisons can only be made meaningful in the context of a specific network. See Section III A 1 for an example, namely ARPANET.

Finally, it should be noted that the algorithms described here have been assumed to be operating under *static* conditions of topology and link costs. (Their convergence has been proved in the literature for this case only.) In some applications, the link costs are defined to depend in some fashion on link traffic, which in turn depends, through the routing algorithm, on link cost; the result is a feedback effect. By studying the dynamics of such situations, it has been shown [15] that poor choices of link cost functions can, in fact, produce instabilities in the resultant traffic patterns. Stability can, however, be ensured by making the link costs sufficiently insensitive to link flow.

C. Packet Routing Implementation

As indicated in Section II A, computation of the routing tables does not complete the routing procedure. These tables must be converted to a form appropriate for dispatching packets from node to node. In this section, we describe a method which underlies a number of schemes used for implementing routing on a single-path-per-VC basis in some existing or proposed systems [16], [17]. The essence of the procedure is that each VC has a *path number* (PN) associated with each link it traverses; if two VCs share a link, they obtain different path numbers on that link. Each packet carries the appropriate PN, which is updated or "swapped" as the packet traverses the network. The updating procedure is determined by, and replaces the routing table, at each node. The PN contains all information necessary for routing; thus, the packet need not carry a VC number. To illustrate, consider a set of four active virtual circuits traversing the network of Fig. 5. The second column of Table III indicates the node sequence for the paths chosen for these VCs. (Note that VC_1 and VC_2 have the same source/destination node pair, but different paths.) Let $PN(n)$ be a path number associated with a path on a link outbound from node n; each link will have as many path numbers as there are distinct active VCs sharing that link. The remaining columns of Table III show how a sequence of PNs is assigned at each node, serving to identify uniquely the path to be followed by a packet on each VC. When a packet is received on an inbound line at a node, its PN must be updated, and the packet must be placed on the proper outbound line or released to its destination. At each node, a simple table lookup procedure can perform this function. In Table IV we show the necessary table for node 4. Note that it is derived directly from the routing information in Table III.

The PN used in this section is roughly equivalent to the *logical record number* used by TYMNET [16] (see Section III A 2) and the *next node*

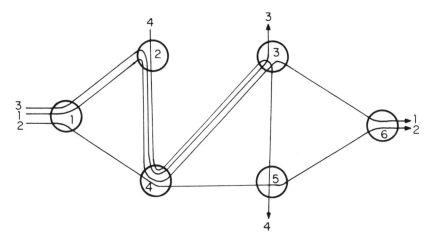

Fig. 5. Example of routing implementation.

indicator (NNI) once proposed for explicit routing [17]. It might be thought that it would be simpler to tag each packet with a unique VC number rather than using the procedure outlined here. However, there will generally be far more active VCs in the network (perhaps thousands) than there are distinct VCs sharing a link (up to 256 in the case of TYMNET, for example). Thus, the PN approach is generally more efficient in memory requirement and table lookup time than any method using VC numbers.

Table III. Path Numbers

VC#	Path	PN(1)	PN(2)	PN(3)	PN(4)	PN(5)
1	1-2-4-3-6	1	1	1	1	—
2	1-4-5-6	1	—	—	1	1
3	1-2-4-3	2	2	—	3	—
4	2-4-3-5	—	3	1	2	—

Table IV. Routing at Node 4

Arriving from node	Old PN	New PN	Next node
1	1	1	5
2	1	1	3
2	2	3	3
2	3	2	3

how to fill these tables in?

III. Examples of Routing Procedures Used in Practice

A. Computer and Data Networks

1. ARPANET, A Computer Network*

ARPANET [18] was created in 1969 as an experiment in computer resource sharing. Beginning with four nodes in 1969, it now runs as an operational system with hundreds of computers connected to nodes throughout the continental United States, Hawaii, and Europe. It is a distributed network with at least two paths between any pair of nodes. Most of its lines are 56-kbit/s synchronous links. It is a store-and-forward packet-switched network in which the routing protocol is datagram-oriented. Messages longer than the maximum packet length are segmented into up to 8 packets at the source node and are reassembled at the destination node. This requires special provision for buffer allocation at the nodes to prevent various types of deadlocks (a significant problem in the early stages of network development).

The network was originally operated with a distributed adaptive routing algorithm of the minimum cost, i.e., shortest path, type wherein link cost was evaluated in terms of measured link delay. Since the measured delays were determined by queue lengths encountered along a packet's transmission path, these quantities varied rapidly with time. Routing was on an individual packet basis where each packet was forwarded along the path that was perceived by the forwarding node to be the shortest in time to the packet's destination at the time of transmission. Since adaptivity was quite rapid, and different nodes could have different views of network conditions, perceptions of shortest paths could change during the period the packet traversed the network, typically ten to several hundred milliseconds. The shortest path algorithm used was essentially our algorithm B (Section II B), with information necessary for node updates passed among neighbors at 2/3-s intervals. Details of the algorithm can be found in [19]. A number of difficulties appeared in the algorithm, and it underwent several modifications [2] from the time it was implanted until May 1979 when it was replaced by a basically different procedure [20].

The new routing algorithm is distributed in the sense that each node independently computes its own routing tables using what is called a *shortest path first* algorithm (essentially our algorithm A with some modifications). That is, each node computes a shortest path tree with itself as the

*The authors are indebted to Dr. John McQuillan of Bolt Beranek and Newman for providing information on ARPANET.

root node. Since algorithm A requires availability of global network information at the node doing the routing computation, this procedure can also be viewed as a "partially" centralized method.

Link costs are evaluated in terms of time delays on the links. Each node calculates an estimate of the delay on each of its outbound links by averaging the total packet delay (processing, queueing, transmission, retransmission, propagation time) over 10-s intervals. (One of the problems with the first algorithm was that delay estimates were obtained too frequently to be accurate.) Since all nodes must be informed of any changes in link time delays, a "flooding" technique is used in the new method for forwarding the measured delays throughout the network. Each node transmits to all its neighbors delay information for all of its outgoing links. It also acts as a repeater, broadcasting to all of its neighbors the link delay information it has received from other nodes. (Transmitting delay information back to the adjacent node from which it was received provides an automatic positive acknowledgement mechanism.) Duplicate delay information packets are dropped, so that while the information propagates to all nodes in the network, it does not circulate indefinitely. To reduce the amount of communication overhead involved in this information exchange, the 10-s average link delay measurements are not always transmitted. Only when the *change* in link delay since the last transmission exceeds a certain threshold does a new transmission take place. The threshold is reduced as time increases since the previous transmission. (However, a change in the status of a line is reported immediately.) The total communication overhead involved in delay update exchanges is less than 1 percent.

Since a complete execution of algorithm A at each update requires considerable computation, the algorithm has been modified so that "incremental" computation can be performed. When a single link delay changes (or if a link or node is added or deleted from the network), each node does a *partial* computation to restructure its shortest path tree. (This, of course, implies that each node must store the most recently updated tree as a basis for future updates, imposing an additional memory requirement.) Also, to take care of the case where link or node failures cause a complete partition of the network, an indication of "age" is inserted in each delay update packet. In this way, "out of date" delay information can be recognized and discarded when lines are reconnected and routing tables are recomputed. Operational results indicate that complete processing of a routing update at a node requires several milliseconds on the average.

A series of tests were performed with the algorithm under actual operating conditions, revealing a number of its features: it responds fairly rapidly (100 ms) to topological changes (one of the problems with the earlier algorithm was that it responded too slowly to line failures); it usually does minimum hop routing, but under heavy load conditions it spreads traffic over lines with excess capacity; it can respond to congestion by choosing paths to avoid congested nodes; and it seems to be stable and free of sustained looping.

Based on the information available at this time, the new algorithm seems to show some advantages over the old in terms of speed of response to changing topology, stability, and suppression of looping. These advantages are apparently attained without undue overhead. It must be kept in mind that in going from algorithm B to A, many other aspects of the routing scheme were also changed, most importantly, the procedure for estimating and forwarding link delay information. Many of the problems encountered using algorithm B were due to the extremely rapid updating that was used based on information whose accuracy did not warrant such rapid adaptivity.

2. TYMNET Routing Algorithm

TYMNET is a computer-communication network developed in 1970 by Tymshare, Inc. of Cupertino, California. It has been in commercial operation since 1971 [1], [21], [22]. Originally developed for time-sharing purposes, it has more recently taken on a network function as well, and is classified by the Federal Communications Commission as a value-added specialized carrier. As of 1978, the network had 300 nodes in operation with lines ranging in speed from 2400 to 9600 bits/s, and was growing at the rate of 2 nodes/week [16]. Almost all nodes are connected to at least two other nodes in the network, giving rise to a distributed topology with alternate path capability. The network is designed primarily to handle interactive terminal users, although it does handle higher-speed synchronous traffic as well. The network covers the United States and Europe, with connections also made to the Canadian Datapac Network. Trans-Atlantic lines are cable with satellite backup. Satellites are avoided, where possible, for interactive users because of the substantial delay involved.

Individual user data packets or logical records, each preceded by a 16-bit header incorporating an 8-bit logical record number to be discussed below and an 8-bit packet character count, are concatenated to form a physical record of at most 66 8-bit characters, including 16 bits of header and 32 bits of checksum for error detection [1]. These data packets can range in length from a few characters to a maximum of 58 characters. (Physical records are transmitted as soon as available, without waiting for a specified size logical record to be assembled.)

TYMNET routing is set up centrally on a virtual circuit, fixed path, basis by a supervisory program running on one of four possible supervisory computers in the network.

A least cost algorithm [1], [16] is used to determine the appropriate path from source to destination node over which to route a given user's packets. The path is newly selected each time a user comes on the network, and is maintained unchanged during the period of the user connection or session. (In the event of an outage, the session is interrupted and a new routing path has to be computed. In TYMNET I, the first version of TYMNET, this could take up to 2.5 min as the supervisor learned of the incident and established the new topology. In the newer TYMNET II, which replaces the earlier version, rerouting in the event of an outage is carried out by the supervisor in a manner transparent to the user.) The algorithm used by the supervisor is a modification of Floyd's algorithm, a variation of our algorithm B.

Integer-valued costs are assigned to each link, and costs are then summed to find the path of least cost. The cost assignments depend on line speed and line utilization. Thus, the number 16 is assigned to a 2400-bit/s link, 12 to a 4800-bit/s link, and 10 to a 9600-bit/s link. A penalty of 16 is added to a satellite link for low-speed interactive users. This shifts such users to cable links, as noted above.

A penalty of 16 is added to a link if a node at one end complains of "overloading." The penalty is 32 if the nodes at both ends complain. Overload is experienced if the data for a specific virtual circuit have to wait more than 0.5 s before being serviced. This condition is then reported by the node to the supervisor. An overload condition may occur because of too many circuits requesting service over the same link, or it may be due to a noisy link with a high error rate, in which case the successive retransmissions which are necessary slow the effective service rate down as well. The penalty used in this case serves to steer additional circuits away from the link until the condition clears up.

Details of the specific algorithm used appear in [16]. In the absence of overloading, the algorithm tends to select the shortest path (least number of links) with highest transmission speed. As more users come on the network, the lower-speed links begin to be used as well. In lightly loaded situations, users tend to have relatively shorter time delays through the network. The minimum hop paths, favored in the lightly loaded case, also tend to be more reliable than ones with more links. Users coming on in a busy period may experience higher time delays due both to congestion and to the use of lower speed lines. The use of the overload penalties tends to spread traffic around the network, deviating from the shortest path case, but attempting to reduce the time delay. In practice, the average response time for interactive users is 0.75 s [16].

It takes 12 ms for the supervisor to find the least cost path using this algorithm [23]. Once the path has been selected, the supervisor notifies each of the nodes along the path, assigning an 8-bit *logical record number* to each link on that path. (This allows up to 256 users or channels to share any one link. In practice, the maximum number ranges from 48 for a 2400-bit/s line to 192 for a 9600-bit/s line. In addition, one number or channel is reserved for a node to communicate with the supervisor and one channel is reserved for communications with the neighboring node.) The supervisor also associates a logical record number on an incoming link to a node with a number on the appropriate outgoing link setting entries in routing tables, called permuter tables in TYMNET terminology. This process, described in more detail later, is basically the same as the method of *path number swapping* described in Section II C. In the TYMNET II version of the network, the nodal computers themselves establish the routing table sizes and entries, as well as the buffers associated with them, relieving the supervisor of this burden.

Routing information is sent to a particular node in a 48-bit supervisory record with the usual 16 bits of logical record overhead as part of a normal physical record. The data transmission overhead due to the dissemination of this routing information is calculated, on a worst case basis, to be 1.6 percent [23]. This assumes that the circuit to be set up is 4 links long, with 5 nodes to be notified (the average path in TYMNET is 3.1 links) during a busy period in which an average of 1 user/s requests entry to the network. The supervisory overhead is taken as distributed equally over a minimum of 8 outgoing 2400 bit/s links from the supervisor. This calculation does not assign any physical record overhead to the supervisory logical record. The assumption is made that there are always data waiting to be transmitted and that the supervisory record is piggybacked onto a normal data record, as noted earlier.

Each node acknowledges receipt of the routing information, again doing this as part of a physical record. (Nodes, in addition, report any link outages to the supervisor as part of a 48-bit record transmitted every 16 s.)

The procedure at a node for forwarding an incoming data packet (logical record) to the appropriate outgoing link, or to either a host computer or terminal if at the destination node, proceeds as follows. As noted earlier, there is a routing or permuter table associated with each link at a node. Each logical record number in either direction on the link is associated with an entry in the table. That entry, in turn, corresponds to the address of a pair of buffers at the node, one for each direction of data flow (inbound and outbound). For L links at a node, L permuter tables are needed, each receiving up to 256 buffer addresses. An error-free physical record arriving at a node is disassembled into its component data packets (logical records). Each data packet is steered by the permuter table entry to its appropriate buffer. Data in buffers destined for terminals and/or computers associated with this node are then transferred to the appropriate device. This node thus represents the destination node for these logical records. Logical records waiting in transit buffers are handled differently. A physical record for a given outgoing link is created, under program control, by scanning sequentially the entires in the permuter table for that link. As each buffer address is read, a determination is made as to whether its *pair*

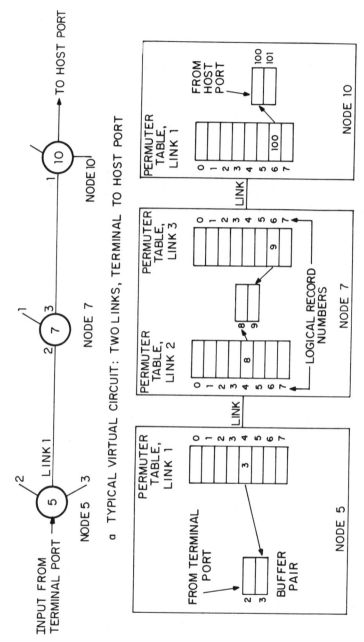

a TYPICAL VIRTUAL CIRCUIT: TWO LINKS, TERMINAL TO HOST PORT

b PERMUTER TABLES AND LOGICAL RECORD NUMBERS

Fig. 6. Routing example, TYMNET.

has had data entered. If so, the data are then formed into a logical record with their corresponding new logical record number. This logical record is incorporated in the physical record and is transmitted out over the link.

A specific example appears in Fig. 6 [23]. Figure 6(a) shows a typical two-link virtual circuit connecting nodes numbered 5, 7, and 10. In this example, terminal data enter the network via a terminal port at node 5, destined for a host computer connected to node 10. The link connecting nodes 5 and 7 is labeled 1, as seen at the node 5 side, and 2 as seen at the node 7 side. Similarly, the link connecting nodes 7 and 10 is labeled 3 at the node 7 side and 1 at the node 10 side.

Figure 6(b) portrays the logical record number assignments and permuter table entries in detail, node by node. (Eight possible logical record numbers only have been assumed for simplicity.) The logical record numbers 4 and 6 have been assigned to this virtual circuit over the two links shown, respectively. At node 5, the entry node, the number 3 in entry 4 in the permuter table for link 1 indicates that data with logical record number 4 are to be found in buffer 2, the mate of buffer 3.

At node 7, data coming from link 2 are stored in buffers designated by the contents of the permuter table for link 2 at that node. Continuing with this example, data arriving at that link with logical record 4 are to be further transmitted over outgoing link 3 to node 10. Their outgoing logical record number is to be changed to 6. To accomplish this, note that the contents of entry 4 of permuter table 2 and entry 6 of the permuter table 3 are paired together. Data arriving over incoming link 2 are stored in buffer 8. They are read out over link 3 when the entries for the permuter table for that link are scanned, entry 6 pointing to buffer 9, the mate of buffer 8. At node 10, the destination node for this virtual circuit, data arriving with logical record 6 are stored in buffer 100 of that node and are then transferred to the appropriate host.

3. Routing in TRANSPAC*

TRANSPAC, the French public packet-switching service [24], began operation in December 1978 with ten nodes (soon expanded to twelve) in a distributed network configuration. As is the case with most public packet-switching services, the network protocol for TRANSPAC follows the X.25 international standard protocol. Thus, this is a virtual-circuit-oriented system, and the routing procedures discussed below reflect this orientation. For purposes of reliability, there are at least two 72-kbit/s lines, following different physical paths, connecting each node to the remainder of the

*The authors are indebted to J. M. Simon of TRANSPAC for providing information used in preparing this section.

network. Each node consists of a control unit (CU) to which are attached a number of switching units (SU). Each incident link is controlled by an SU, which executes all data link procedures. The SUs also execute the access protocols for customers connected to the node. Routing is handled by the CU, using information from the Network Management Center (see below).

Network control is partially decentralized through six local control points which handle a certain amount of statistics gathering and perform test and reinitialization procedures in case of node or line failures. However, general network supervision, including the bulk of routing computation, is exercised through a single Network Management Center (NMC).

Routes in TRANSPAC are assigned on a single-path-per VC basis. The algorithm of interest to us here is that which governs the assignment of a route to a switched virtual circuit, i.e., a VC which is established temporarily in response to a "call request." The call request takes the form of a *Call Packet*, emitted by equipment connected to the originating network node, and requesting connection to a specified destination. The path that eventually will be retained by the switched VC is identical to that taken by the Call Packet as it is forwarded through the network. Routing of the Call Packet is effected through routing tables stored at each node; as indicated in Section II, the tables associate a unique outbound link with each destination node. The network as currently configured has two classes of nodes. One class is connected in a distributed fashion, with alternate route capability. The second class consists of nodes homing in via a single link to a node of the first class. Node 5 in Fig. 7 is an example of a node of the first type; node 6 is a node of the second type. Messages destined to nodes

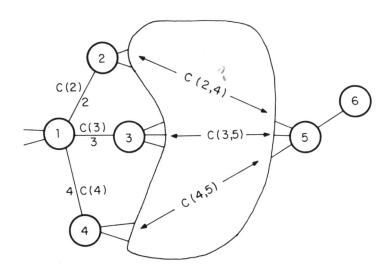

Fig. 7. Routing example, TRANSPAC.

of the second type are routed to the "target" node to which they are connected. In Fig. 7, messages destined for node 6 have node 5 as a target node.

The routing tables for the network are constructed in an essentially centralized fashion, using a minimum cost, i.e., shortest path criterion. Link costs are defined in terms of link resource utilization. Thus, the cost assigned to a link varies dynamically with network load. We shall first describe the method of evaluation of link cost and then the routing algorithm [25], [26]. Consider a full duplex link k connected between nodes m and n. Let $C_m(k)$, $C_n(k)$ be the cost assigned to link k as perceived by nodes m and n, respectively, and let $C(k) = \text{Max}[C_m(k), C_n(k)]$ be the "combined" estimate of link cost. The quantities $C_i(k)$ are the basic data on which routing computation is based; they are determined locally by each node's CU which gathers estimated and measured data from its associated SU's. Link cost is defined as a function of the level of utilization of two types of resources: line capacity and link buffers. The utilization of these quantities is evaluated both by estimation (based on the parameters of the active VCs using the link) and by measurement. The cost $C_i(k)$ is set to infinity if either the link is carrying its maximum permissible number of VCs or it has exceeded a preset threshold of buffer occupancy. Otherwise, $C_i(k)$ is defined as a piecewise constant increasing function of average link flow, quantized to a small number of levels and including a "hysteresis" effect. A typical function is shown in Fig. 8, with the arrows indicating the way link cost changes as a function of changing utilization. The nodes send updated values of their $C_i(k)$'s to the NMC whenever a change occurs; these events are infrequent owing to the combined effect of coarse quantization and hysteresis. At the NMC, the costs perceived by the nodes at both ends of each link are compared to form $C(k)$ as defined above.

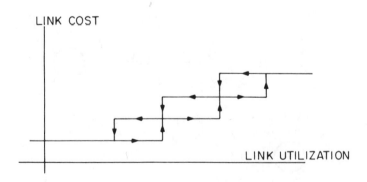

Fig. 8. A typical link cost relation, TRANSPAC.

The major part of the routing computation takes place at the NMC, but some local information is used at each node. The procedure is illustrated by an example in Fig. 7 in which a Call Packet arriving at node 1 (which may be either the originating node or an intermediate one) is to be forwarded through one of the adjacent nodes $2, 3, 4$ to the target node 5, and finally to the destination node 6. Let $C(k, n)$ (computed by the NMC) be the total cost associated with the minimum cost path between nodes k and n. Node 1 determines the "shortest" route to node 5 by choosing the value of k which minimizes $C(k, 5) + \text{Max}[C(k), C_1(k)]$, $k = 2, 3, 4$. In this way, node 1 chooses the intermediate node that would have been chosen by the NMC, unless the value of $C_1(k)$ has changed recently. Ties are resolved by giving priority to the shortest hop path. Because of the way in which link costs are defined, the routing procedure becomes a minimum hop method upon which is superimposed a bias derived from the level of link resource utilization.

Although the TRANSPAC routing algorithm has many of the features of a typical centralized routing procedure, its operation departs from being purely centralized by allowing the final routing decision to be made locally, based on a combination of centrally and locally determined information. This is similar to the concept of "delta routing" suggested by Rudin [27].

By examining the current topology of the network [24], one can deduce the order of magnitude of the computational load at the NMC. The $C(k, n)$'s must be determined for all k and n belonging to the subset of all possible target nodes. Only six out of twelve network nodes are in this category. Furthermore, rather than doing a complete shortest path computation to determine these quantities, the designers chose to limit the shortest path computation to a minimization over a prescribed subset of four or five paths joining each pair of nodes. Thus, the computation of all pertinent $C(k, n)$'s involves at most 75 path length evaluations.

B. Commercial Network Architectures

The examples discussed thus far have all been operational networks. Specific physical implementations exist, although the networks have been steadily growing and changing their topologies. In the two examples discussed in this section, we focus on another type of distributed network architecture for which routing procedures become important. These are the network protocols introduced by most large computer system manufacturers to enable private users to configure their own computer networks. Such networks are being increasingly developed to handle such diverse tasks as distributed processing, distributed data-base handling, and computer resource sharing. A computer manufacturer's protocol is designed to enable a user to interconnect a variety of computer systems and terminals in any desired configuration. All of these network protocols tend to follow a layered architecture, starting at the lowest level, that of setting up physical connections; continuing to the next, data link level, which controls the flow of data packets between neighboring nodes; then proceeding to the network and transport levels, involved with end-to-end (source to destination) control of packet flow, routing, and congestion control; and finally concluding, at the highest levels, with several levels of "handshaking" between users or programs at the two ends. Other chapters in this book discuss these network protocols in detail.

In this section, we describe the routing procedures defined for distributed versions of the IBM Systems Network Architecture (SNA) and the Digital Equipment Corporation's Digital Network Architecture (DNA). These are both relatively recent developments since earlier versions of both SNA and DNA were tailored primarily to star- or tree-type network configurations with no real need for routing. It will be noted that, unlike the network examples discussed previously, where networking is essentially transparent to the user, it is left to the user of either SNA or DNA to configure his own network. There is a certain flexibility in the routing procedures as well, with the user free to define his own link costs and paths to be taken. This is, of course, not the case in the earlier networks described.

1. IBM's Systems Network Architecture (SNA)*

The early versions of SNA, appearing in 1974, were designed for single-computer system tree-type networks [28], [29]. In these networks, it is apparent that routing was not really a significant problem. Later versions of SNA allow two or more such single-system networks to be interconnected, leading to the concept of cross-domain networking [28]. Here, too, routing requirements were quite simple. IBM's more recent SNA architecture, termed SNA 4.2 [30], envisions multiple computer systems interconnected to form a distributed network. Routing thus plays an important role in the architecture.

The routing procedure chosen for SNA incorporates predetermined fixed paths from source to destination. A multiplicity of possible routes is provided to increase the probability that a route will be available when needed to achieve load leveling, to provide alternate route capability in the event of node/link failures or congestion, and to provide different types of services for different classes of users [30], [31]. For example, batch traffic would normally be routed differently from interactive traffic. (Not only are the response time requirements different, calling usually for different capac-

*The authors are indebted to Dr. James P. Gray of IBM for help with this section.

ity links, as noted earlier in discussing the TYMNET routing procedure, but one would not normally want to have batch traffic interfering with, and hence slowing down, interactive traffic. In SNA 4.2, this can be done by assigning a lower transmission priority to batch traffic.) Some traffic may require high security handling and will therefore be routed differently.

Multiple routing is provided at two levels: when first initiating a session, the user specifies a name corresponding to a particular class of service. Examples of classes of service include low response time, high capacity lines, and more secure paths. Associated with each class of service name is a list of possible virtual routes for use by sessions specifying that name. This list provides load balancing and backup capability. A particular session uses only one of these virtual routes at a time. This corresponds to the first level of multiple routing. Each virtual route provides a full-duplex connection between source and destination nodes, and can support multiple users or sessions. Each virtual route in turn maps into a so-called explicit route, the actual physical path from source to destination. It is this path that has been precalculated to provide the desired performance. Multiple explicit routes will exist, on a unidirectional basis, between any source-destination nodal pair. The multiple explicit routes provide the second level of multiple route control noted earlier. In the current SNA 4.2 release, up to eight explicit routes can be made available between any source–destination nodal pairs. Several virtual routes may use the same explicit route.

Although explicit routes are established by the source node on a unidirectional basis, explicit routes are used in pairs that are physically reversible. This simplifies user notification of route failure.

Up to 24 virtual routes are currently available between any pair of nodes. These are grouped into three levels of transmission priority, with eight possible virtual route numbers associated with each level. The entire set of virtual routes, each identified by a virtual route number and transmission priority, is stored in a virtual route identifier list. Class of service names are then associated with subsets of this list, in some preassigned order. A user setting up a session specifies his class of service name. He is then assigned to the first virtual route in the virtual route list that is available or can be activated. Multiple sessions may be assigned to the same virtual route. The same virtual route is defined by four fields—the source and destination addresses, a virtual route number, and the transmission priority.

The explicit route corresponding to a specific virtual route is, in turn, defined by the source and destination addresses and an explicit route number. Each explicit route number represents one of the eight distinct routes possible between any source–destination nodal pair. A given explicit route is made up of a sequence of logical links connecting adjacent nodes along the path.

The term "transmission group" is used for logical link in the SNA terminology. Transmission groups may consist of multiple physical links. Thus, a set of parallel physical links between any two nodes can be divided into one or more transmission groups. This adds flexibility to the transmission function: physical links may be combined in parallel to provide higher capacity, links may be dynamically added or deleted without disruption, and scheduling of links is employed to optimize the composite bandwidth or capacity available. But the use of multiple-link transmission groups means that data packets or blocks may arrive out of sequence. Out-of-order blocks must thus be reordered at the receiving end of each transmission group along the composite path.

Routing of data packets is carried out by examining the destination address and explicit route number as a packet arrives at an intermediate node along the path. An explicit routing table at each node associates an appropriate outgoing transmission group with the destination address and explicit route number. An example of such a table at a particular node appears in Fig. 9. The letters represent the transmission groups to which packets with the corresponding address, route number pair are directed. By changing the explicit route number for a given destination, a new path will be followed. This introduces alternate route capability. If a link or node along the path becomes inoperative, any sessions using that path can be reestablished on an explicit route that bypasses the failed element. Explicit routes can also be assigned on the basis of type of traffic, types of physical media along the path (satellite or terrestrial, for example), or other criteria, as already noted. Routes could also be listed on the basis of cost, the smallest cost route being assigned first, then next smallest cost route, etc.

Note that the explicit routing concept is similar to that adopted by TYMNET in its virtual circuit approach. Here the path selected may be changed by the source node, however, by choosing a new explicit route number. In essence, a variety of alternate routes is laid out in advance. This introduces the alternate route capability noted above. In the current

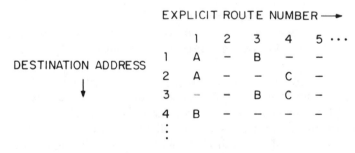

Fig. 9. Explicit routing table, SNA.

TYMNET approach, the central supervisor must set up a new path if one is desired.

The concept of explicit routing, as first enunciated and as noted in Section II earlier, is somewhat broader than the one described here [17]. There, rather than using a fixed explicit route number, a variable "next node index" (NNI) field was proposed for the packet header. The combination of the destination address and the NNI field then directs the packet to the appropriate outgoing transmission group. The NNI is changed at the same time as well. This allows more explicit routes to be defined than through the use of a fixed explicit route number. The idea is similar to the (variable) logical record number concept used by TYMNET. In addition, some form of intermediate or local node routing capability could be introduced through the use of the NNI. For by changing the NNI locally, a new path from that point on will be followed. This makes it possible to introduce alternate route capability along the initial path chosen in the event of localized congestion or some other delaying phenomenon. The British NPL, in a series of network simulation experiments, has shown the benefits of alternate route capability [32]. Rudin has proposed as well a routing strategy that combines centralized routing with a measure of local adaptability [27]. The general idea of explicit routing thus enables centralized, distributed, and local routing strategies (or some combination of them) to be introduced into the network. In the IBM implementation, however, only precalculated routes are used.

Three steps are required to activate a route. The individual links of the transmission groups forming the explicit route must be brought up. The explicit route is then activated. Finally, the virtual route to be used that maps into this explicit route must be activated. Special command packets are used for this purpose [31]. For example, an explicit route is activated by transmitting a specific activate command from node to node along the path. This packet verifies the routing tables of the nodes along the path. It ensures loop-free routes by checking the routing tables of nodes along the route. It verifies that there are no packets along the path with the same source–destination address pairs. It also measures the length of the explicit route in hops. Activation of the explicit route is considered completed when verified by a reply command from the destination. If the activation of the first-choice virtual route and its associated explicit route fails, the second-choice virtual route is tried, repeating with the third choice, and so on, if necessary.

The user is involved in setting the routing tables, and hence in route definition in the SNA architecture. Thus, the user can define the routes he desires, given his physical topology, by providing table entries at system definition time. The user can also ensure that a session is established on a desired route. For unique or specialized requirements, the user can write a

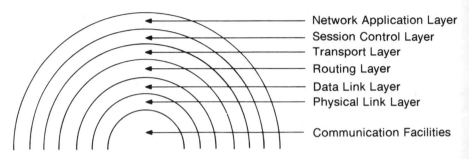

Network Application Layer
Session Control Layer
Transport Layer
Routing Layer
Data Link Layer
Physical Link Layer

Communication Facilities

Fig. 10. DNA layered architecture.

user-exit routine that is invoked during session establishment. This exit can assign a session to a specific route.

2. Routing in Digital's DNA *

Digital Equipment Corporation's network architecture is called Digital Network Architecture, or DNA for short. It provides the interfaces and protocols that enable users to create their own networks using Digital Equipment Corporation Systems. The family of network products supporting DNA is generally called DECNET. DNA and DECNET were first introduced in 1973.

As does the IBM SNA, DNA employs a layered architecture. Six levels have been defined, as shown in Fig. 10 [33]. Phase II provided for point-to-point connections with no routing capability required. The next phase, Phase III DECNET, had provision for store-and-forward distributed topologies requiring routing, flow and congestion control, and a network management capability. For this phase of DECNET, the Routing Layer was defined and provided the necessary routing and congestion control features. In this section, we focus on routing in Phase III of DECNET. Phase IV routing, described in [33], is of a hierarchical nature and incorporates Phase III routing at each of two levels.

The routing procedure adopted by Digital Equipment Corporation is based on a variation of the distributed shortest path algorithm (our algorithm B), with each node carrying out its own calculations. It is similar to the protocol analyzed by Tajibnapis [34], which has been implemented on the Michigan MERIT Computer Network. The routing algorithm adapts to changes in network topology (it does not use traffic flow information), and

---

*The authors are indebted to Anthony Lauck of the Digital Equipment Corporation for providing information used in preparing this portion of the chapter.

so needs to be invoked only when a link or node in the network comes up or fails. Unlike the IBM SNA approach, routing is done on a packet-by-packet or datagram basis, as contrasted to a virtual circuit service.

To carry out the least cost routing procedure, each link in the network is assigned a fixed cost. The specific cost is set by the user, but is approximately inversely proportional to link capacity. (Note that these assignments are then roughly similar to those used by TYMNET.) Paths with high capacity links are favored. These costs are used by each node to derive a routing data base (or routing table) which lists the cost to each destination using each of the node's outgoing lines. An example appears in Fig. 11(a). (Each node in the network is assigned a unique address. Naming and addressing are carried out at a level higher than the Routing level.)

Packets going to a particular destination are routed to the output link with the smallest cost. In the example of Fig. 11(a), packets going to

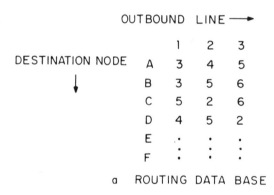

a   ROUTING DATA BASE

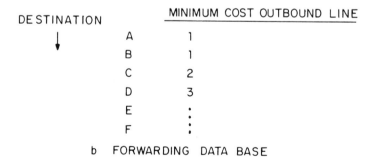

b   FORWARDING DATA BASE

Fig. 11. Typical nodal routing tables, DNA routing protocol. (Similar tables are kept at each node.)

destination C would take output line 2 with a cost of 2 units. Those going to destination B would take output line 1. The listing of minimum cost outgoing lines, one for each destination, that is used in routing the packets is kept in a second data base, called the forwarding data base. An example appears in Fig. 11(b) for the routing table of Fig. 11(a). (A third, Boolean, data base indicates whether each destination is reachable or not. This is discussed briefly later.)

As noted earlier, these tables are changed only on receipt of routing messages, triggered by a line (or node) coming up or going down. Specifically, a node, on learning that one of its links or a neighboring node has either been brought up or has failed, will update its tables. If the minimum cost to any destination has changed, the cost information is broadcast using a routing message to all the neighbors. These nodes, in turn, add to the cost forwarded for each destination the link cost for the link over which the message has arrived. The sum is then entered in the routing data base. Minimization is then carried out row by row (i.e., for each destination), and the forwarding data base is changed. If the resultant cost changes, this is, in turn, broadcast, using routing messages, to all neighbors. In this way, changes percolate throughout the network.

Routing messages contain 16 bits per destination, with a maximum of 128 nodes allowed at present. The routing message thus consists of a maximum of 256 bytes. Of the 16 bits, 11 are used to transmit total cost information and 5 bits represent a hop count that is transmitted as well. This hop count is incremented by 1 at each node, and is used for reachability analysis. To avoid indefinite ping-ponging, one node adding 1, its neighbor adding 1, back and forth, a destination is declared unreachable if the hop count reaches a specified maximum. This maximum could, for example, be one more than the maximum path length, or it could be the diameter of the network.

Packets going to an unreachable destination are discarded. However, there is an option of notifying the source that a particular destination is unreachable through the use of a "return to sender" packet. This would be used on setting up a connection or initializing the operation of the network.

If a link fails, packets queued on that link are discarded as well. To maintain end-to-end (source to destination) integrity, an acknowledgment and time-out procedure is carried out by the higher-level Transport Layer of DNA.

How does a node know if a link is down? This is based on the number of retransmissions of packets needed. If the number 7 is reached the link is declared down. In addition, provision is made for transmitting a low-priority "Hello" message to a neighboring node that has not been heard from for a while. If there is no acknowledgment, the node is declared down.

The actual software implementation of the routing procedure involves three processes: a *decision process*, which receives routing messages; an *update process*, which updates the routing tables; and a packet *forwarding process*, which uses the forwarding data base to route the packets. Normally, the third process only is used. The first two are run only when changes in the network topology dictate in the routing table. Provision is made to check the routing algorithm periodically, if desired, with the use of a timer. Such a check might be made once a minute, for example. Although the forwarding data base (minimum cost paths) is normally used for routing, the entire routing data base is retained as well at each node. This is required to run the distributed routing algorithm when needed. The routing data base can also be used to provide alternate path capability as well, if desired, or if necessary.

Some additional factors provided by the DNA Routing Layer in addition to routing include a packet lifetime control and a congestion control mechanism. The packet lifetime control is used to bound the time a packet spends in the network. A nodal visit count is kept in each data packet. If the number is too large, the packet is purged. The congestion control involved is the one analyzed by Irland [35]. The queues at each outbound link at a node are limited in size. Packets are discarded if the number queued will exceed this maximum value. Priority is, however, given to transit messages (those already in the network, as contrasted to packets originating at the node in question).

IV. Conclusions

After a brief discussion of routing in general, we have presented the basic features of routing procedures currently used in five representative packet-switched communication networks and network architectures. While the networks were chosen to represent a broad spectrum of operational characteristics, it is interesting to note that there are many similarities in their routing algorithms. At the same time, there is a great deal of diversity in the manner in which these algorithms are implemented. Most of the networks use some variation or approximation of a shortest path routing strategy. However, each network defines the "length" or "cost' of a communication link differently. Some use centralized computation, some decentralized, and some use a hybrid of the two. Adaptivity ranges from the bare minimum necessary to react to line failures to more sophisticated procedures sensing and responding to queueing delays, error rates, and line loading. Undoubtedly, a larger set of representative networks would have yielded a still richer set of alternative schemes for information gathering, routing computation, and packet forwarding. One can conclude from this survey that while the routing function is central to the smooth and efficient operation of packet-switched networks, no one scheme can be identified as "best." Many viable alternatives exist at all levels of the routing function.

References

[1] M. Schwartz, *Computer Communication Network Design and Analysis.* Englewood Cliffs, NJ: Prentice-Hall, 1977.
[2] J. M. McQuillan, G. Falk, and I. Richer, "A review of the development and performance of the ARPAnet routing algorithm," *IEEE Trans. Commun.*, vol. COM-26, pp. 1802–1811, Dec. 1978.
[3] T. Cegrell, "A routing procedure for the TIDAS message-switching network," *IEEE Trans. Commun.*, vol. COM-23, pp. 575–585, June 1975.
[4] F. Poncet and C. S. Repton, "The EIN communications sub-network: Principles and practice," in *Proc. 3rd ICCC*, Toronto, Ont., Canada, Aug. 1976, pp. 523–531.
[5] J. L. Grangé and M. I. Irland, "Thirty-nine steps to a computer network," in *Proc. 4th ICCC*, Kyoto, Japan, Sept. 1978, pp. 763–769.
[6] L. Fratta, M. Gerla, and L. Kleinrock, "The Flow Deviation Method: An Approach to Store-and-Forward Communication Network Design," *Networks*, vol. 3. New York: Wiley, 1973, pp. 97–133.
[7] W. Older and D. A. Twyver, personal communication.
[8] J. M. McQuillan, "Interactions between routing and congestion control in computer networks," in *Proc. Int. Symp. Flow Contr. in Comput. Networks*, Versailles, France, Feb. 1979, J. L. Grangé and M. Gien, Eds., Amsterdam: North-Holland, pp. 63–75.
[9] M. Schwartz and C. Cheung, "The gradient projection algorithm for multiple routing in message-switched networks," *IEEE Trans. Commun.*, vol. COM-24, pp. 449–456, Apr. 1976.
[10] R. Gallager, "An optimal routing algorithm using distributed computation," *IEEE Trans. Commun.*, vol. COM-25, pp. 73–85, Jan. 1977.
[11] T. E. Stern, "A class of decentralized routing algorithms using relaxation," *IEEE Trans. Commun.*, vol. COM-25, pp. 1092–1102, Oct. 1977.
[12] E. W. Dijkstra, "A note on two problems in connection with graphs," *Numer. Math.*, vol. 1, pp. 269–271, 1959.
[13] A. V. Aho, J. E. Hopcroft, and J. D. Ullman, *The Design and Analysis of Computer Algorithms.* Reading, MA: Addison-Wesley, 1974.
[14] L. R. Ford, Jr. and D. R. Fulkerson, *Flows in Networks.* Princeton, NJ: Princeton Univ. Press, 1962.
[15] D. P. Bertsekas, "Dynamic behavior of a shortest path routing algorithm of the ARPAnet type," presented at the Int. Symp. Inform. Theory, Grigano, Italy, June 1979.
[16] A. Rajaraman, "Routing in TYMNET," presented at the European Computing Conf., London, England, May 1978.
[17] R. R. Jueneman and G. S. Kerr, "Explicit routing in communications networks," in *Proc. 3rd ICCC*, Toronto, Ont., Canada, Aug. 1976, pp. 340–342.
[18] D. C. Walden, "Experiences in building, operating, and using the ARPA network," presented at the 2nd USA—Japan Comput. Conf., Tokyo, Japan, Aug. 1975.
[19] J. M. McQuillan, "Adaptive routing algorithms for distributed computer networks," BBN Rep. 2831, May 1974.
[20] J. M. McQuillan et al., "The new routing algorithm for the ARPAnet," *IEEE Trans. Commun.*, vol. COM-28, pp. 711–719, May 1980.
[21] L. Tymes, "TYMNET—A terminal oriented communication network," in *1971 Spring Joint Comput. Conf.*, AFIPS Conf. Proc., vol. 38, 1971, pp. 211–216.

[22] J. Rinde, "Routing and control in a centrally-directed network," in *1977 Nat. Comput. Conf., AFIPS Conf. Proc.*, vol. 46, 1977, pp. 603–608.

[23] J. Rinde, "TYMNET I: An alternative to packet technology," in *Proc. 3rd ICCC*, Toronto, Ont., Canada, Aug. 1976, pp. 268–273.

[24] A. Danet, R. Despres, A. LaRest, G. Pichon, and S. Ritzenthaler, "The French public packet switching service: The TRANSPAC network," in *Proc. 3rd ICCC*, Toronto, Ont., Canada, Aug. 1976, pp. 251–260.

[25] J. M. Simon and A. Danet, "Contrôle des resources et principes du routage dans le réseau TRANSPAC," in *Proc., Int. Symp. Flow Control in Comput. Networks*, Versailles, France, Feb. 1979, J. L. Grangé and M. Gien, Eds. Amsterdam: North-Holland, pp. 33–44.

[26] J. M. Simon, personal communication.

[27] H. Rudin, "On routing and 'Delta-routing': A taxonomy and performance comparison of techniques for packet-switched networks," *IEEE Trans. Commun.*, vol. COM-24, pp. 43–59, Jan. 1976.

[28] R. J. Cypser, *Communications Architecture for Distributed Systems*. Reading, MA: Addison-Wesley, 1978.

[29] P. E. Green, "An introduction to network architectures and protocols," *IBM Syst. J.*, vol. 18, no. 2, pp. 202–222, 1979.

[30] J. P. Gray and T. B. McNeill, "SNA multiple-system networking," *IBM Syst. J.*, vol. 18, no. 2, pp. 263–297, 1979.

[31] V. Ahuja, "Routing and flow control in systems network architecture," *IBM Syst. J.*, vol. 18, no. 2, pp. 298–314, 1979.

[32] W. L. Price, "Data network simulation experiments at the National Physical Laboratory, 1968–1976," *Comput. Networks*, vol. 1, no. 4, pp. 199–210, 1977.

[33] M. Schwartz, *Telecommunications Networks*. Reading, MA: Addison-Wesley, 1987.

[34] W. D. Tajibnapis, "A correctness proof of a topology information maintenance protocol for distributed computer networks," *Commun. Ass. Comput. Mach.*, vol. 20, pp. 477–485, July 1977.

[35] M. Irland, "Buffer management in a packet switch," *IEEE Trans. Commun.*, vol. COM-26, pp. 328–337, Mar. 1978.

Flow Control Protocols

Mario Gerla and Leonard Kleinrock

I. Introduction

A packet-switched network may be thought of as a distributed pool of productive resources (channels, buffers, and switching processors) whose capacity must be shared dynamically by a community of competing users (or, more generally, processes) wishing to communicate with each other. Dynamic resource sharing is what distinguishes packet switching from the more traditional circuit-switching approach, in which network resources are dedicated to each user for an entire session. The key advantages of dynamic sharing are greater speed and flexibility in setting up users' connections across the network and more efficient use of network resources after the connection is established.

These advantages of dynamic sharing do not come without a certain danger, however. Indeed, unless careful control is exercised on the user demands, the users may seriously abuse the network. In fact, if the demands are allowed to exceed the system capacity, highly unpleasant congestion effects occur which rapidly neutralize the delay and efficiency advantages of a packet network. The type of congestion that occurs in an overloaded packet network is not unlike that observed in a highway network. During peak hours, the demands often exceed the highway capacity, creating large backlogs. Furthermore, the interference between transit traffic on the highway and on-ramp and off-ramp traffic reduces the effective throughput of the highway, thus causing an even more rapid increase in the backlog. If

MARIO GERLA AND LEONARD KLEINROCK • Computer Science Department, UCLA, Los Angeles, California 90024.

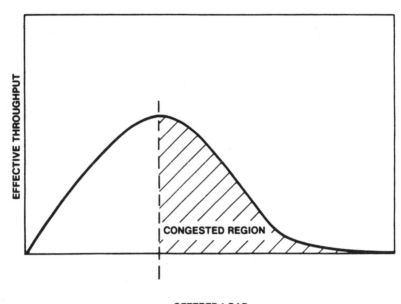

OFFERED LOAD

Fig. 1. Effective throughput versus offered load in an uncontrolled distributed dynamic sharing system.

this positive feedback situation persists, traffic on the highway may come to a standstill. The typical relationship between effective throughput and offered load in a highway system (and, more generally, in many uncontrolled, distributed dynamic sharing systems) is shown in Fig. 1.

By properly monitoring and controlling the offered load many of these congestion problems may be eliminated. In a highway system, it is common to control the input by using access ramp traffic lights. The objective is to keep the interference between transit traffic and incoming traffic within acceptable limits, and to prevent the incoming traffic rate from exceeding the highway capacity.

Similar types of controls are used in packet-switched networks, and are called *flow control* procedures. As in the highway system, the basic principle is to keep the excess load out of the network. The techniques, however, are much more sophisticated since the controlling elements of the network (i.e., the switching processors) are intelligent, can communicate with each other, and therefore can coordinate their actions in a distributed control strategy.

Internal network congestion may also be relieved by rerouting some of the traffic from heavily loaded paths to underutilized paths. It is important to understand, however, that *routing* can reduce and, perhaps, delay network congestion; it cannot prevent it. We will come back to the interaction between routing and flow control in Section VII. The interested reader is referred to the routing protocol survey by Schwartz and Stern in Chapter 9.

The main functions of flow control in a packet network are as follows:

1. Prevention of throughput and response time degradation and loss of efficiency due to network and user overload.
2. Deadlock avoidance.
3. Fair allocation of resources among competing users.
4. Speed matching between the network and its attached users.

Throughput degradation and deadlocks occur because the traffic that has already been accepted into the network (i.e., traffic that has already been allocated network resources) exceeds the nominal capacity of the network. To prevent overallocation of resources, the flow control procedure includes a set of constraints (on buffers that can be allocated, on outstanding packets, on transmission rates, etc.) which can effectively limit the access of traffic into the network or, more precisely, to selected sections of the network. These constraints may be fixed, or may be dynamically adjusted based on traffic conditions.

Apart from the requirement of throughput efficiency, network resources must be fairly distributed among users. Unfortunately, efficiency and fairness objectives do not always coincide. For example, referring back to our highway traffic situation, the effective throughput of the Long Island Expressway could be maximized by opening all the lanes to traffic from the Island to New York City during the morning rush hour, and in the opposite direction during the evening rush hour. This solution, however, would also maximize the discontent of the reverse commuters (and we all know how dangerous it is to anger a New Yorker)! In packet networks, unfairness conditions can also arise (as we will show in the following sections); but they tend to be more subtle and less obvious than in highway networks because of the complexity of the communications protocols. One of the functions of flow control, therefore, is to prevent unfairness by placing *selective* restrictions on the amount of resources that each user (or user group) may acquire, in spite of the negative effect that these restrictions may have on dynamic resource sharing, and, therefore, overall throughput efficiency.

Flow control can be exercised at various levels in a packet network. The following levels, shown in Fig. 2, are identified and discussed in this paper.

(1) Hop Level. This level of flow control attempts to maintain a smooth flow of traffic between two neighboring nodes in a computer network, avoiding local buffer congestion and deadlocks. (We shall devote Section III to the discussion of this form of flow control.)

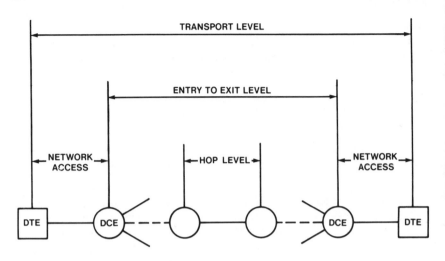

Fig. 2. Flow control levels.

(2) Entry-to-Exit Level. This level of flow control is generally implemented as a protocol between the source and the destination switch, and has the purpose of preventing buffer congestion at the exit switch (Section IV).

(3) Network Access Level. The objective of this level is to throttle external inputs based on measurements of internal (as opposed to destination) network congestion (Section V).

(4) Transport Level. This is the level of flow control associated with user sessions, i.e., the protocols which provide for the reliable delivery of packets on the "virtual" connection between two remote processes. Its main purpose is to prevent congestion of user buffers at the process level (i.e., outside of the network) (Section VI).

Some authors reserve the term *flow control* for the transport level, and refer to the other three levels of control as *congestion control* [34]. This terminology is used to emphasize the physical distinction between the first three levels, which are realized in the communication subnet (and therefore are the responsibility of the network implementer) and the fourth level, which is realized in the user devices (and therefore is the responsibility of the network customer). In this chapter, we have chosen to use the term "flow control" for all four levels.

The design of an efficient flow control strategy for a packet network is a complex task in many ways. The most critical issue is the fact that flow control is a multilayer distributed protocol involving several different levels. At each level, the flow control implementation must be consistent and compatible with other protocol functions existing at that level. Furthermore, the interactions between different levels must be carefully studied in order to avoid duplication of functions on one hand, and lack of coordination on the other.

The purpose of this chapter is to provide a taxonomy of flow control mechanisms based on the above-defined multilevel structure. First, we review problems, functions, and performance measures of flow control. Then, for each level we survey the most representative flow control techniques that have been proposed and/or implemented, providing a performance comparison among techniques at the same level, and discussing the interaction between techniques at different levels. Finally, we briefly mention some new flow control issues raised by novel computer network applications.

II. Flow Control: Problems, Functions, and Measures

Our overall problem is to identify mechanisms which permit efficient dynamic sharing of the pool of resources (channels, buffers, and switching processors) in a packet network. In this section, we first describe and use some toy examples to illustrate the congestion problems caused by *lack* of control. Then we define the functions of flow control and the different levels at which these functions are implemented. Finally, we introduce performance measures for the evaluation and comparison of different flow control schemes.

A. Loss of Efficiency

The main cause of throughput degradation in a packet network is the *wastage* of resources. This may happen either because conflicting demands by two or more users make the resource unstable (e.g., collisions on a random access channel); or because a user acquires more resources than strictly needed, thus starving other users (e.g., a slow sink fed by a fast source may create a backlog of packets within the network which prevents other traffic from getting through). The resources that are most commonly "wasted" in a packet network are *channel capacity, storage capacity*, and *processor capacity*.

Buffer wastage is an indirect consequence of limited nodal storage: a given end-to-end packet stream may be blocked at an intermediate node along the path because all of the buffers at that node have been "hogged" by other streams. This may happen even if channel bandwidth is plentiful along the path of our blocked stream, thus causing an unnecessary loss of throughput. The source of this throughput degradation is that some users unnecessarily monopolize (i.e., waste) the buffers at some congested node.

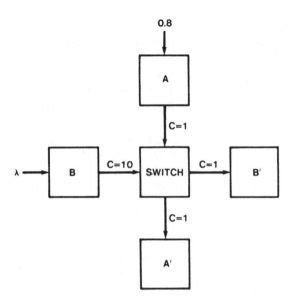

Fig. 3. Buffer interference example.

A simple example of throughput degradation caused by buffer interference is shown in Fig. 3. Two pairs of hosts, (A, A') and (B, B'), are engaged in data transmission through a single network node. Access line speeds (in arbitrary units) are given in the figure. The traffic requirement from A to A' is constant and is equal to 0.8 (measured in the same units as the line speed). The requirement from B to B' is variable, and is denoted by λ. When λ approaches 1, the output queue from the switch to host B' grows indefinitely large filling up all the buffers in the switch. Packets arriving when all the buffers are full are discarded, and are later retransmitted by the source host (we refer to this model as the *retransmit model*). If we plot the total throughput, i.e., the sum of (A, A') and (B, B') delivered traffic as a function of λ (as in the solid curve of Fig. 4), we note that for $\lambda = 1$, the throughput experiences a sharp drop from 1.8 to 1.1. The drop is due to the fact that the switch can handle the entire user demand $= \lambda + 0.8$ for $\lambda < 1$; while for $\lambda \geq 1$, the switch buffers become full, causing overflow. Consequently, large queues build up in both the A and B hosts. With a heavy load, the rate of packet transmissions (and retransmissions) from B is 10 times the rate from A because of the difference in line access speeds. Thus, packets from B have a 10 times better chance of being accepted when a buffer becomes free than packets from A, leading to a 10 to 1 imbalance in effective throughput. Since the (B, B') throughput is limited to 1, the (A, A') throughput is reduced to 0.1 (i.e., one tenth of the AA' throughput), yielding a total throughput $= 1.1$ for $\lambda \geq 1$.

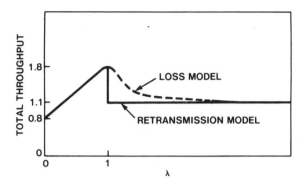

Fig. 4. Throughput degradation of system of Fig. 3 due to buffer interference.

In this example, we have observed a *decrease in useful throughput* caused by an *increase of offered load* beyond the critical system capacity. This throughput degradation is typical of congested systems, and is often taken as a definition of congestion as was mentioned in connection with Fig. 1 (i.e., a system is "congestion-prone" if an increment in offered load causes a reduction in throughput) [27].

In the previous example, we assumed that dropped packets would be retransmitted from the host. A similar analysis can be carried out assuming that dropped packets are lost (*loss model*). The throughput versus offered load performance is similar to that of the retransmit model, although the drop is somewhat smoother in this case (the dashed curve of Fig. 4).

Throughput degradation effects, caused by inefficient allocation (and therefore wastage) of buffers are found also in multinode networks as reported by several studies [27], [13], [18]. To prevent this type of degradation, proper buffer allocation rules are generally established at each node, as soon described.

Another cause of throughput degradation is channel wastage. This problem manifests itself very clearly in multiaccess channels (e.g., packet satellite, or packet radio channels), when users transmit packets at random times without prior coordination (random access). A well-known example is offered by the ALOHA channel [23]. Packets that collide are lost, thus causing channel wastage and consequently, throughput degradation. Congestion prevention in multiaccess channels is discussed in Chapter 6.

Finally, throughput degradation can be caused by the inefficient use (i.e., wastage) of packet processing capacity. Any switching processor has an upper limit on the rate of packets it can process. If the offered load exceeds such a limit, loss of performance can result as shown in the example of Figs. 5 and 6.

In this example, 80 packets/s must be transferred from A to A', while the requirement from B to B' is a variable denoted by λ. Packet length is

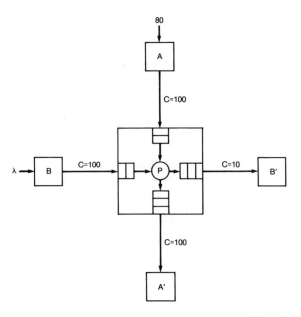

Fig. 5. Processor "wastage" example.

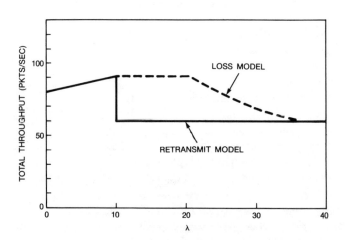

Fig. 6. Throughput degradation due to buffer wastage.

assumed to be 1000 bits. Channel speeds are expressed in Kbps. Switch processing capacity is 100 packets/s. Again, we note a mismatch between input and output channel speeds for the connection (B, B'). In order to avoid "buffer wastage" we allocate separate buffers to the output queues. We also allocate separate buffers to the input queues, i.e., the queues of packets that have just arrived to the switch and are awaiting processing. Packets arriving to a full queue, either input or output, may be lost (loss model), or may be retransmitted (retransmit model) as discussed before.

For fairness, we assume that the processor serves the input queues in a Round Robin fashion.

Total throughput as a function of λ is plotted in Fig. 6. In the retransmit model, throughput increases linearly with λ from 80 to 90 packets/s. Then, it suddenly drops to 60 packets/s. This is because the output queue to B' overflows (since $\lambda > 10$), thus causing packets to be dropped and subsequently retransmitted from B. The input queue from B then becomes full, and the switch processor then alternates between the two queues, serving 50 packets from each. All the packets from A are successfully forwarded to A', whereas only 10 packets/s can be forwarded from B to B'. Thus, the total throughput has dropped to 60 packets/s.

The loss model (see Fig. 6) exhibits a more gradual loss of efficiency, yet achieving the same asymptotic performance as the retransmit model. It is interesting to note the "plateau" between 10 and 20. In this region, the switch can still handle all the incoming packets, since the total arrival rate is < 100 packets/s. However, packets are rejected by the output queue to B'.

In the above example, the throughput degradation was caused by inefficient processor scheduling, i.e., the processor is dispatching packets to a queue that cannot take them. It is remarkable that even a careful buffer allocation does not prevent degradation. More sophisticated control schemes must be devised for this case. For example, the processor should not serve input queue B if output queue B' is full.

B. Unfairness

Unfairness is a natural byproduct of uncontrolled competition. Some users, because of their relative position in the network or the particular selection of network and traffic parameters, may succeed in capturing a larger share of resources than others, and thus enjoy preferential treatment.

One example of unfairness has already been given in Figs. 3 and 4 where the $(B \rightarrow B')$ flow is allowed to exceed the $(A \rightarrow A')$ flow by a factor of 10. Another obvious example of unfairness is offered by the single switch loss model in Fig. 7. The speed of the output trunk is 1. Hosts A and B are injecting data into the switch with rates 0.5 and λ, respectively. For fairness, the output trunk should be equally shared by the two hosts.

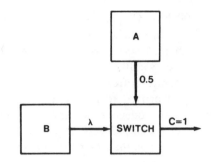

Fig. 7. Example of unfairness.

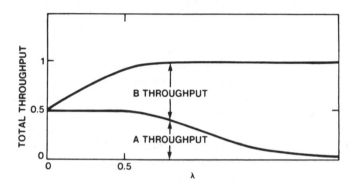

Fig. 8. Performance of system shown in Fig. 7.

However, the *loss model* performance results shown in Fig. 8 indicate that for large values of λ, host B captures the entire output trunk bandwidth, reducing the A throughput to zero. As previously observed, for $\lambda \gg 0.5$ host B has a far better chance to seize free buffers in the switch than host A. Specifically, the ratio of A packets to B packets in the switch at heavy load is roughly equal to $0.5/\lambda$. Thus, the ratio of A throughput to B throughput is also $0.5/\lambda$, explaining the behavior in Fig. 8.

Cases of unfairness have been reported in many multinode network studies, and several "fairness" techniques have been proposed. Unfortunately, the problem of fairness is considerably more difficult to deal with than the problem of total throughput degradation because a general, unambiguous definition of fairness is not always possible in a distributed resource sharing environment.

C. Deadlocks

A deadlock condition manifests itself by a (total or partial) network crash. Deadlocks often occur because of a cyclic wait of resources to become available. That is, one user is holding a portion of the resources

Fig. 9. Deadlock example.

that he currently needs and is waiting for another user to release the remaining resources necessary to complete his task and this user is waiting for yet a third user, etc., such that the sequence of "waiting" users closes into a cycle, and it is immediately seen that no user in the cycle can make any progress [3]. Thus, the throughput for this subset of users is reduced to zero.

Deadlocks are likely to occur in a network when the offered load exceeds network capacity. For a simple example of a deadlock, consider two switches, A and B, connected by a trunk carrying heavy traffic in both directions (see Fig. 9). Under the heavy traffic assumption, node A rapidly fills up with packets directed to B; and vice versa, B fills up with packets directed to A. If we assume that dropped packets are retransmitted, then each node must hold a copy of each packet (and therefore a buffer) until the packet is accepted by the other node. This may result in an endless wait in which a node holds all of its buffers to store packets being transmitted to the other node, and keeps retransmitting packets to the other node waiting for buffers to be freed there. Consequently, no useful data are transferred on the trunk. It turns out that this type of deadlock (known as direct store-and-forward deadlock [19]) is relatively easy to prevent by setting simple restrictions on buffer usage at each node. A more extensive discussion of deadlocks will be given in Section III. The reader is also referred to [63] for a thorough review of deadlock prevention schemes.

It is important to point out that buffer deadlocks are possible only in networks which retransmit dropped packets, i.e., which save a copy of a packet at each node while transmitting the packet to the next node on the path, and retransmit a copy of the packet in case of overflow (retransmit model). If dropped packets are not retransmitted (i.e., a loss model), the sending node is not required to save a copy of the packet until acceptance at the next node, thus removing a necessary condition for deadlocks. Thus, lossy networks are deadlock free; however, an additional recovery mechanism for lost packets must then be provided at the end-to-end level, as is done for example in DNA.

D. Flow Control Functions

Flow control may be defined as a protocol (or more generally, a set of protocols), designed to protect the network from problems related to overload and speed mismatches. Solutions to the three problems just discussed (maintaining efficiency, fairness, and freedom from deadlock) are accomplished by setting rules for the allocation of buffers at each node and by properly regulating and (if necessary) blocking the flow of packets internally in the network as well as at the network entry points. Actually, multiple levels of flow control are generally implemented in a real network, as we shall see.

Efficiency and congestion prevention benefits of flow control do not come for free. In fact, flow control (like any other form of control in a distributed network) may require some exchange of information between nodes to select the control strategy and possibly some exchange of commands and parameter information to implement that strategy. This exchange translates into channel, processor, and storage overhead. Furthermore, flow control may require the dedication of resources (e.g., buffers, bandwidth) to individual users, or classes of users, thus reducing the statistical benefits of complete resource sharing. Clearly, the tradeoff between gain in efficiency (due to controls) and loss in efficiency (due to limited sharing and overhead) must be carefully considered in designing flow control strategies. This tradeoff is illustrated by the curves in Fig. 10, showing the effective throughput as a function of offered load. The ideal throughput curve corresponds to perfect control as it could be implemented by an ideal observer, with complete and instantaneous network status information. Ideal throughput follows the input and increases linearly until it reaches a horizontal asymptote corresponding to the maximum theoretical network throughput. The controlled throughput curve is a typical curve that can be obtained with an actual control procedure. Throughput values

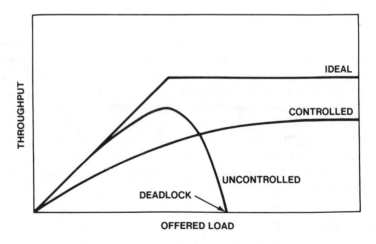

Fig. 10. Flow control performance tradeoffs.

are lower than with the ideal curve because of imperfect control and protocol overhead. The uncontrolled curve follows the ideal curve for low offered load; for higher load, it collapses to a very low value of throughput and, possibly, to a deadlock.

Clearly, controls buy safety at high offered loads at the expense of somewhat reduced efficiency. The reduction in efficiency is measured in terms of higher delays (for light load) and lower throughput (at saturation). Furthermore, experience shows that flow control procedures are quite difficult to design and, ironically, can themselves be the source of deadlocks and degradations. In particular, when one controls flow, one places *constraints* on the flow. If one cannot meet a constraint, then the result is a deadlock. Or, if one is slow in meeting the constraint, the result is a throughput degradation.

E. Levels of Flow Control

Flow control in a packet network can be best described as a multilayered structure consisting of several mechanisms operating independently at different levels. Since flow control levels are closely related to (and sometimes imbedded in) protocol levels, it is helpful for us to begin by briefly reviewing the network protocol structure, pointing to the flow control provisions existing at each level [17]. The flow control level structure will then be defined following the protocol structure model.

Figure 11 depicts the typical protocol layer architecture implemented in a packet network, using as a reference a network path connecting user devices called DTEs (data terminal equipment) through a number of intervening communications switches called DCEs (data circuit terminating equipment). For the user-to-network (i.e., DTE-to-DCE) interface, a standard set of protocol levels has been defined by ISO and ANSI [9]. For the internode protocols within the communications subnetwork, there is less emphasis on standardization since different network manufacturers tend to select different solutions to best exploit their equipment capabilities. In spite of these differences, it is still possible to define a set of reference levels for internal network protocols which closely parallel the DTE–DCE interface protocol levels.

Starting from the bottom of the protocol hierarchy, we have the *physical level* which has the function of activating and deactivating the electrical connection between the nodes. No flow control functions are assigned to this level.

Above the physical level, we have the *link level* which serves the purpose of transporting packets reliably across individual physical links. One of the functions of this protocol is related to flow control, and consists of retransmitting packets that are dropped because of congestion at the

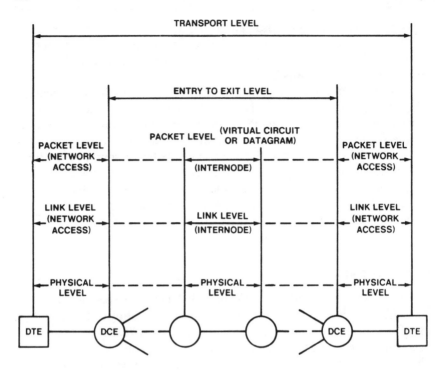

Fig. 11. Network protocol levels. (DTE: Data terminal equipment, e.g., host, terminal. DCE: Data circuit-terminating equipment, e.g., switching processor.)

receiving node. In some protocols, a congested receiver may stop the sender by using appropriate commands (e.g., RNR: receiver not ready, in HDLC and SDLC; or XOFF in asynchronous terminal connections). As mentioned before, we find two different types of links in the network: the internal (or node-to-node) link and the network access link. Correspondingly, we have (at the same level in the protocol hierarchy) two types of link protocol: the *network access protocol* and the *node-to-node protocol*. Typical examples of link protocol implementation are HDLC, SDLC, and LAPB (which is a subset of HDLC).

Above the link level, we have the *network or packet level protocol*, which defines the procedures for establishing end-to-end user connections through the network and specifies the format of the control information used to route packets to their destinations. Two different versions of packet protocol exist: the *virtual circuit* protocol and the *datagram* protocol.

When the virtual circuit (VC) implementation is used, a "virtual" circuit connection must be set up between a pair of users (or processes) wishing to communicate with each other *before* the data transfer can be started. The establishment of this circuit implies dedication of resources of one form or another along the network path. A typical virtual circuit implementation, used in TRANSPAC [7], assigns a fixed path to each connection at setup time. A virtual circuit ID number, stamped in the packet header, uniquely identifies the packets belonging to a connection, and is used to route packets to the destination using routing maps stored at each intermediate node at setup time. From the flow control point of view, the VC protocol has the distinguishing feature of permitting *selective* flow control on each individual user connection. This selective flow control can be applied at the internode level as well as at the network access level. Since a fixed path is maintained for the entire user session, the selective flow control can also be extended from entry to exit switch and if so desired, even from entry to exit DTE.

In contrast to the virtual circuit implementation, the datagram implementation does not require any circuit setup before transmission. Each packet is independently submitted to the network, and explicitly carries in its header all the information required for its delivery to destination [34]. Selective flow control, on a connection by connection basis, is not available in the datagram implementation since the packet header does not contain specific connection information (it merely posts source and destination DTE addresses).

Above the packet protocol, we find (within the subnet only) the *entry-to-exit* (ETE) protocol. The objective of this protocol is the reliable transport of single and multipacket messages from network entry to network exit node. Important functions of this protocol which are related to flow control are the reassembly of multipacket messages at the exit node and the regulation of input traffic using buffer allocation and windowing techniques. Some network implementations, e.g., DNA, do not have the ETE level of protocol. In this case, the ETE functions are relegated to higher-level protocols.

The highest level of network protocol which has impact on flow control is the *transport protocol*. This protocol provides for the reliable delivery of packets on the "virtual" connection between two remote processes. One of the flow control related functions of this protocol is the protection of destination buffers. The goal is to regulate the flow so as to make the most efficient use of network resources, while avoiding buffer overflow at the destination. "Window" and "credit" schemes are generally used for this purpose.

The above network protocol review has identified various flow control functions and capabilities built into different levels of protocols, and has brought to our attention the fact that each protocol level has its own distinct flow control responsibilities. It is now clear that the classification into the four types of flow control procedures mentioned earlier parallels

	ARPA			TRANSPAC	SNA			GMDNET	
	CQL	RFNM	NCP	X.25	SDLC	VR PACING	SESSION PACING	I-C	SBP
HOP LEVEL	●			●	●	●		●	●
NETW. ACC.				●			●	●	
ENTRY-EXIT		●		●	●		●		●
TRANSPORT			●	NOT DEFINED			●	NOT DEFINED	

Fig. 12. Classification of a sample of actual flow implementations.

the classification of network protocols. Recall that this is as follows:

1. Hop (or node-to-node) level (Section III).
2. Network access level (Section V).
3. Entry-to-exit level (Section IV).
4. Transport level (Section VI).

The diagram in Fig. 2 illustrates these levels of flow control for a typical network path. A comparison with Fig. 11 reveals the close relationship between flow control and protocol level structures.

Unfortunately, the true system behavior is far more complex than our models and classifications attempt (or can afford) to portray. Therefore, actual networks may not always mechanize all of the above four levels of flow control with distinct procedures. It is quite possible, for example, for a single flow control mechanism to combine two or more levels of flow control. On the other hand, it is possible that one or more levels of flow control may be missing in the network implementation. The matrix in Fig. 12 provides a synopsis of the main network implementations and flow control schemes that will be surveyed in this chapter. It is seen that some of the schemes cover more than one level.

F. Performance Measures

We wish to define a quantitative measure of flow control performance for various reasons. First, we wish to be able to "tune" the parameters of a given flow control scheme so as to optimize a well-defined performance criterion. Second, we wish to carefully weigh performance benefits against overhead introduced by flow control. Third, we are interested in comparing the performance of alternative flow control schemes in quantitative terms.

Throughput efficiency (where throughput is expressed in packets/s) is probably the most common measure of flow control performance. Total effective throughput (sum of all the individual contributions) is evaluated as a function of offered load. This representation is particularly useful to determine the critical load in an uncontrolled system and to assess the throughput efficiency of a controlled network at heavy load.

Another common measure is the *combined delay and throughput performance*. The delay-versus-throughput profile allows us to determine the delay overhead introduced by the controls (which the throughput versus offered load curve did not display). In general, it gives us a more complete picture of system performance than does throughput behavior alone. In fact, a system may be designed to deliver high throughput at heavy load, and yet it may experience intolerable delays at light load.

A more compact measure of combined throughput and delay performance is offered by the concept of *power* [13], [24], [58]. The simplest definition of power is the ratio of throughput over delay; it is, therefore, a function of the offered load. In fact, it defines the "knee" of the throughput–delay profile as that point where power is maximized, and as shown in Fig. 13 this knee occurs where a ray out of the origin is tangent to the performance profile [24]. A very nice characterization of this maximum power point is such that it occurs when the average buffer occupancy at each intermediate node on the path is unity. In [25], it was shown that blocking due to loss systems could easily be included in a more general definition of power (by multiplying the simple definition by one minus the blocking probability); this leads to system designs whose optimum operating point is easily found and which corresponds to the operating point one would intuitively choose.

In some important cases, power is maximized for a value of offered load which is approximately half of the saturation load [24]. The maximum power value reflects both delay performance (at light load) and throughput

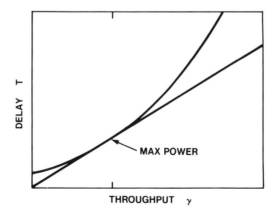

Fig. 13. Delay, throughput, and power.

performance (at heavy load) and therefore, represents a good figure of merit of the flow control implementation. Much more general definitions of power are also studied in [25].

III. Hop Level Flow Control

A. Objective

The objective of hop level flow control (HL) is to prevent store-and-forward buffer congestion and its consequences, namely, throughput degradation and deadlocks. Hop level flow control operates in a local, "myopic" way in that it monitors local queues and buffer occupancies at each node and rejects store-and-forward (S/F) traffic arriving at the node when some predefined thresholds (e.g., maximum queue limits) are exceeded. The function of checking buffer thresholds and discarding (and later retransmitting) packets on a network link is often carried out by the data link control protocol.

This locality of the control does not preclude, however, possible end-to-end repercussions of hop level flow control due to the "backpressure" effect [i.e., the propagation of buffer threshold conditions from the congested node upstream to the traffic source(s)]. In fact, the backpressure property is efficiently exploited in several network implementations (as soon described).

Store-and-forward congestion has two unpleasant consequences: throughput degradation and deadlocks. These conditions were described in Sections II A and II C, respectively. In the remainder of this section, we survey and compare a number of hop level flow control procedures, specifically designed to eliminate these problems.

B. Classification of Hop Level Control Schemes

The hop level flow control scheme can play the role of arbitrator between various *classes of traffic* competing for a common buffer pool in each node. A fundamental distinction between different flow control schemes is based on the way the traffic entering a node is subdivided into classes.

One family of hop flow control schemes distinguishes incoming packets based on the output queue they must be placed into. Thus, the number of classes is equal to the number of output queues; the flow control scheme supervises the allocation of store-and-forward buffers to the output queues. Some limit (fixed or dynamically adjustable) is defined for each queue; packets beyond this limit are discarded. Hence, the name *channel queue limit* schemes is generally given to such mechanisms (see Section III C).

Another important family of hop flow control schemes distinguishes incoming packets based on the "hop count" (i.e., the number of network links that they have so far traversed). This implies that each node keeps trace of $N - 1$ classes of traffic, where $N - 1$ is the number of different hop counts, and N is the number of nodes in the network (note that if loopless routing is assumed, no network path can exceed $N - 1$ hops in length), and allocates a (fixed or adjustable) number of buffers to each class. We will refer to this family of schemes as *buffer class* schemes (see Section III D).

A third family distinguishes packets based on the virtual circuit (i.e., end-to-end session) they belong to. This type of scheme requires, of course, a virtual circuit network architecture; it assumes that each node can distinguish incoming packets based on the virtual circuit they belong to and keep track of a number of classes equal to the number of virtual circuits that currently traverse it. Note that the number of classes varies here with time (since virtual circuits are dynamically created and released), as opposed to the previously mentioned schemes where the number of classes is merely a function of the topology. Upon creation, a virtual circuit is allocated a set of buffers (fixed or variable) at each node. When this set is used up, no further traffic is accepted from that virtual circuit. We will refer to this family of schemes as *virtual circuit hop level* schemes (see Section III E).

Many other traffic subdivisions are possible: for example, a traffic class may be associated with each traffic source; with each traffic destination; or with each source–destination node pair. Indeed, these are all legitimate and, in many respects, well justified choices for a link level flow control scheme. However, we will restrict our study to the three schemes just mentioned, since these are the only schemes which have been extensively analyzed in the published literature and implemented in real networks.

Apart from traffic class distinctions, another parameter that is often used to characterize and classify hop flow control schemes is the degree of dynamic sharing of the store-and-forward buffers. Here, several possibilities exist, namely:

1. Fixed, uniform partitioning of buffers among buffer classes (no sharing).
2. Buffer partitioning proportional to traffic in each class (no sharing).
3. Overselling (i.e., the sum of the buffer limits, one for each class, is larger than the total buffer pool).
4. Dynamic adjustment of buffer limits based on relative traffic fluctuations.

The following sections discuss each hop flow control class in more detail.

C. Channel Queue Limit Flow Control

In the channel queue limit (CQL) scheme, the traffic classes correspond to the channel output queues, and there are restrictions on the number of buffers each class can seize. We may define the following versions of the CQL scheme [20].

(1) Complete Partitioning (CP). Letting N be the number of output queues, and n_i be the number of packets on the ith queue and B the buffer size, we have the following constraint:

$$0 \le n_i \le B/N, \quad \forall i$$

(2) Sharing with Maximum Queues (SMXQ). Let b_{max} be the maximum queue size allowed (where, $b_{max} > B/N$); we have the following constraints:

$$0 \le n_i \le b_{max}, \quad \forall i$$
$$\sum_i n_i \le B$$

(3) Sharing with Minimum Allocation (SMA). Let b_{min} be the minimum buffer allocation which is guaranteed to each queue (typically, $b_{min} \le B/N$). The constraint then becomes

$$\sum_i \max(0, n_i - b_{min}) \le B - Nb_{min}$$

(4) Sharing with Minimum Allocation and Maximum Queue. This scheme combines (2) and (3) in that it provides for a minimum buffer guarantee and a maximum buffer allocation for each queue at the same time.

The above options assume that the buffer limit parameters are fixed in time and are the same for all queues. Additional flexibility may be introduced in these schemes by allowing the buffer parameters to change dynamically in time and from queue to queue based on traffic fluctuations.

Having defined a number of CQL flow control options, we now proceed to show that this form of flow control can eliminate some of the performance degradation and deadlock effects mentioned in Section II. Referring first to Fig. 2, we note that in the presence of CQL flow control, the traffic component (B, B') will no longer be permitted to seize all the buffers in the switch. Therefore, traffic can now flow freely from A to A', and the throughput degradation effect is removed. Similarly, the deadlock condition depicted in Figs. 8 and 9 cannot occur since the buffers in node A cannot be taken over completely by the channel (A, B) queue. Therefore, some buffers in A will always be available to receive packets from node B.

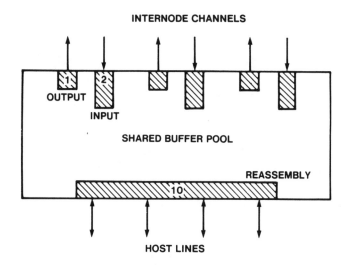

MINIMUM BUFFER ALLOCATION

Total buffer pool = 40 packet buffers

	minimum allocation	maximum allocation
Reassembly	10	20
Internode input queue	2	
Internode output queue	1	8
Total internode queues (i.e., total S/F buffers)		20

Fig. 14. Buffer allocation in ARPANET IMP (1972 version).

Some form or another of CQL flow control is found in every network implementation [30], [54]. The ARPANET IMP (Interface Message Processor) has a shared buffer pool with minimum allocation and maximum limit for each queue, as shown in Fig. 14 [30]. Of the total buffer pool (typically, 40 buffers of one packet), two buffers for input and one buffer for output are permanently allocated to each internode channel. Similarly, ten buffers are permanently dedicated to the reassembly of messages directed to the hosts. The remaining buffers are shared among output queues and the reassembly function, with the following restrictions: reassembly buffers ≤ 20, output queue ≤ 8, the total store-and-forward buffers ≤ 20.

Next we proceed to the evaluation and comparison of CQL implementations, and briefly review the main results available in the published literature [18], [20]. We first report on some throughput degradation conditions observed in absence of flow control. Figure 15 from [18] shows

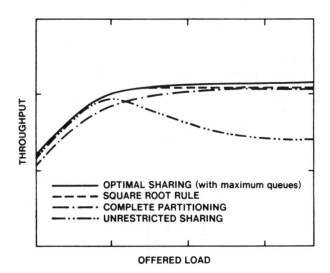

Fig. 15. Single switch buffer and allocation model. Throughput versus load behavior for various buffer management schemes (unbalanced load pattern).

throughput performance as a function of link load for a variety of buffer control policies. The curve labeled "unrestricted sharing" corresponds to a system without flow control. We notice that, for increasing input load, the throughput of the uncontrolled system reaches a peak and then degrades asymptotically to unity. This behavior confirms the throughput degradation predictions made in Section II.

Throughput degradation is easily corrected with the introduction of CQL flow control, as shown by the remaining curves in Fig. 15. The "no sharing" system (i.e., complete partitioning of the buffer pool among the outgoing queues) is, as expected, the most conservative scheme and least efficient with respect to throughput. The best scheme is the "optimal sharing" scheme, which corresponds to optimally reselecting a new buffer limit for each level of traffic (i.e., dynamic SMXQ). A heuristic approximation of the optimal scheme is offered by the "square root scheme," a load invariant scheme with fixed buffer limit $= \sqrt{B/N}$, where B is the total number of buffers and N is the number of output channels. The square root scheme is simpler to implement than the optimal scheme since it does not depend on traffic load and, therefore, does not require the reoptimization of the buffer limit values as a function of traffic pattern changes, and yet, it was shown to be practically as efficient as the optimal sharing for a number of cases [18].

Kamoun [20] used a similar switch model to investigate the sharing with minimum allocation (SMA) scheme. The results, obtained in a balanced load environment, show no substantial difference between SMXQ and SMA; in fact, neither scheme is consistently better over the entire range of offered loads. We conjecture, however, that with strongly unbalanced traffic SMA would exhibit better "fairness" since SMA guarantees minimum throughput (with low delay) for each output channel even when the shared portion of the buffer pool is captured by a few heavily loaded queues.

Summarizing various published results, we may state that CQL flow control is necessary to avoid throughput degradation, unfairness, and direct store-and-forward deadlocks. Furthermore, it seems that almost any form of CQL implementation will provide the minimum required protection. The safest scheme (for fairness reasons) seems to be the combination of SMXQ and SMA, which imposes a maximum and minimum limit on each queue (incidentally, this was the scheme used in ARPANET).

D. Structured Buffer Pool (SBP) Flow Control

We have shown in the previous section that CQL flow control eliminates direct store-and-forward deadlocks. However, there is another, more general form of deadlock which can arise in packet networks, namely, *indirect store-and-forward deadlocks* [19]. Figure 16 illustrates a typical indirect store-and-forward deadlock situation. Suppose that unfavorable traffic conditions in the ring topology shown in Fig. 16 cause each queue to be filled with Q_{max} packets, where Q_{max} is the limit imposed by the CQL strategy. Furthermore, assume that the packets at each node are directed to a node two or more hops away [e.g., all packets queued on link (A, B) are directed to C]. In these conditions, no traffic can move in the network since all the queues are full. Thus, we have a deadlock even if the network is equipped with CQL flow control (which is known to prevent direct store-and-forward deadlocks)!

Prevention of indirect store-and-forward deadlocks is obtained with the "structured buffer pool" strategy proposed by Raubold et al. [37]. In this strategy, packets arriving at each node are divided into classes according to the number of hops they have covered. For example, packets entering a node from the host belong to class 0 of that node, since they have not yet covered any hops. The highest class H_{max} corresponds to packets that have traversed H_{max} hops, where H_{max} is the maximum path length in the network (a function of the topology and the routing algorithm). The highest class H_{max} also includes all the packets that have reached their destinations and are therefore being reassembled into messages before delivery to the hosts. The nodal buffer organization reflects this class structure as shown in Fig. 17.

Each packet class has the right to use a well-defined set of buffers. Class 0 can access only the buffers available in set 0. Buffer set 0 is large enough to store the largest size message entering the network. Class $i + 1$

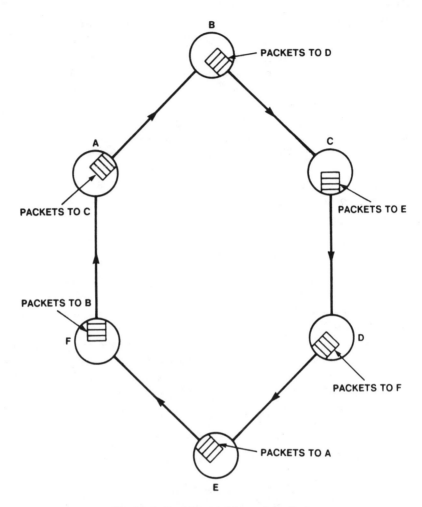

Fig. 16. Indirect store-and-forward deadlock.

can use all the buffers available to class i, plus one additional buffer. Finally, class H_{max} can access all the buffers available to class $H_{max} - 1$, plus a number of buffers sufficient to reassemble the largest message to be delivered to any destination (this provision is necessary, although not sufficient, to avoid "reassembly deadlocks," as will be shown in Section IV).

Under normal traffic conditions, only set 0 buffers are used. When the load increases beyond nominal levels, buffers fill up progressively from level 0 to level H_{max}. When at a given node the buffers at levels $\leq i$ are full, arriving packets which have covered $\leq i$ hops are discarded. Thus, in case

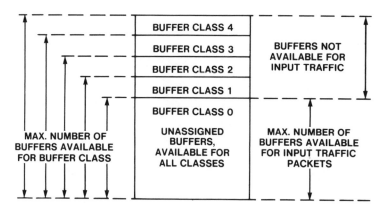

Fig. 17. Structured buffer pool.

of congestion, "junior" packets are dropped in the attempt to carry "senior" packets to their destination. This is a desirable property, since senior packets correspond to a higher network resource investment.

It can easily be shown that this strategy eliminates deadlocks of both the direct and indirect type [37]. To prove this, we consider the "resource graph" [3] associated with the packet-switched network. In this graph, there is an arc associated with each packet in the network. The arc originates from the buffer currently occupied by the packet and terminates in the (currently unavailble, but awaited) buffer in the next node on the path. A deadlock occurs if and only if there is a cycle in the graph, i.e., there is a chain of arcs which starts from one buffer, and terminates at the same buffer. The existence of cycles can easily be recognized in the deadlock situations depicted in Figs. 8 and 16.

With the structured buffer pool, however, no cycle can occur in the resource graph since each arc starts from a buffer of class i and points to a buffer of class $i + 1$ (recall that a packet gains seniority at each hop). An illustration of this property is shown in Fig. 18. Thus, both direct and indirect store-and-forward deadlocks are prevented.

The SBP method was developed by the GMD group in Darmstadt, Germany, for implementation in GMDNET, an experimental packet switch network [37], [59]. Before implementation, an extensive simulation effort was carried out to verify and evaluate the performance of all the network protocols, and of the SBP procedure in particular [13]. Early simulation results showed that the proposed flow control scheme was effective in eliminating deadlocks, but was not successful in preventing throughput degradation when the offered load exceeded the critical threshold (some SBP simulation experiments with typical packet switch network topologies showed that the throughput in heavy load conditions was four to five times lower than the maximum throughput).

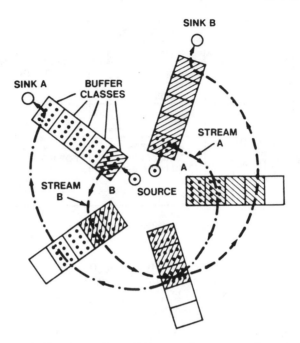

Fig. 18. Access to buffer classes. Example for two data streams. Dotted areas: buffers available for stream A. Hatched areas: buffers available for stream B.

To correct the loss of throughput efficiency under heavy loads, additional constraints were imposed on the number of buffers that each traffic class could seize. The most dramatic improvement was obtained by limiting the number of class 0 buffers that could be seized by input packets (i.e., packets entering the network from external sources). In the absence of this constraint, input packets had the tendency to monopolize all class 0 buffers, leaving only a "thin" buffer layer for the transit traffic to circulate. The control of input traffic, known as "input flow control" in GMDNET is a form of network access flow control and will be discussed more extensively in Section V.

Additional improvements in the SBP scheme were obtained in the case of datagram networks, by setting a specific buffer size constraint $L(i)$ on each class i[13]. (In other words, instead of having a nested buffer pool in which class i can access all buffers available to class $i - 1$, plus one buffer, a different constraint is set on each class.) The constraint $L(i)$ was dynamically adjusted to adapt to the relative demands of the various classes. It is interesting to note that the deadlock prevention property is *not* affected by dynamic changes in buffer class size (as long as at least one buffer is dedicated to each class at all times).

E. Virtual Circuit (Hop Level) Flow Control

We recall that packet switch networks can be subdivided into two broad classes: datagram (DG) networks and virtual circuit (VC) networks. In DG networks, each packet in a user session is carried through the network independently of the other packets in the same session; that is, packets in the same session may follow different routes, and may be delivered out of sequence to the destination. In VC networks, a physical network path is set up for each user session and is released when the session is terminated. Packets follow the preestablished path in sequence. Sequencing and error control are provided at each step along the path.

The previously mentioned flow control schemes, namely, CQL and SBP, are applicable to both DG and VC nets. In addition, VC nets permit the application of *selective flow control* to each individual VC stream (VC flow control). There are two forms of VC flow control: (1) hop level (or stepwise) VC flow control, which controls VC flow at each hop along the path, and is designed to avoid S/F buffer congestion; and (2) source-sink (or end-to-end) VC flow control, whose function it is to adjust source rate to sink rate so as to maximize VC throughput, yet avoiding sink buffer congestion.

In this section we will mainly deal with VC hop level (VC-HL) flow control; we discuss end-to-end VC flow control in more detail in Section IV.

The basic principle of operation of the VC-HL scheme consists of setting a limit M on the maximum number of packets for each VC stream that can be in transit at each intermediate node. The limit M may be fixed at VC setup time, or may be dynamically adjusted, based on load fluctuations. The buffer limit M is enforced at each hop by the VC-HL protocol, which regulates the issue of transmission "permits" and discards packets based on buffer occupancy.

The advantage of VC-HL (over CQL and SBP) is to provide a more efficient and prompt recovery from congestion by selectively slowing down the VCs directly feeding into the congested area. By virtue of backpressure, the control then propagates to all the sources that are contributing to the congestion, and reduces (or stops) their inputs, leaving the other traffic sources undisturbed. Without VC-HL flow control, the congestion would spread gradually to a larger portion of the network, blocking traffic sources that were not directly responsible for the original congestion, and causing unnecessary throughput degradation and unfairness.

As in the case of CQL and SBP schemes, various buffer sharing policies can be proposed. At one extreme, M buffers can be dedicated to each VC at setup time; at the other extreme, buffers may be allocated, on demand, from a common pool (complete sharing). It is easily seen that buffer dedication can lead to extraordinary storage overhead, since there is, generally, no practical upper bound on the number of VCs that can simultaneously exist in a network; furthermore, the traffic on each VC is generally bursty, leading to low utilization of the reserved buffers. For these reasons, most of the implementations employ dynamic buffer sharing.

The shared versus dedicated buffer policy also has an impact on the deadlock prevention properties of the VC-HL scheme. With buffer dedication, the VC-HL scheme becomes deadlock free. This can easily be deduced by considering the resource graph and recognizing that the graph cannot contain loops, since virtual circuits are loopless by construction. (For deadlock freedom, it actually suffices that at least one buffer be reserved for each virtual circuit.) If, on the other hand, no buffer reservations are made and buffers are allocated strictly on demand, deadlocks may occur unless additional protection (e.g., the SBP scheme) is implemented.

In the following, we briefly describe three different versions of VC-HL flow control implemented in existing networks and report on some performance results.

TYMNET is probably the earliest VC network developed [39]. As distinct from most VC networks, TYMNET uses a "composite" packet internode protocol. This means that data from different VCs traveling on the same trunk can be packed in the same envelope, for the purpose of link overhead reduction. TYMNET is a character-oriented network in the sense that data flows on a virtual circuit in the form of characters, rather than packets (i.e., characters are assembled into packets at the entry node, and are then disassembled at the exit node). The character-oriented nature of TYMNET implies that VC-HL buffer allocation is based on character (rather than packet) counts.

In TYMNET [39], a throughput limit is computed for each VC at setup time according to terminal speed, and is enforced all along the network path. Throughput control is obtained by assigning a maximum buffer limit (per VC) at each intermediate node and by controlling the issue of transmission permits from node to node based on the current buffer allocation. Periodically (every half second), each node sends a backpressure vector to its neighbors, containing one bit for each virtual circuit that traverses it. If the number of current buffered characters for a given VC exceeds the maximum allocation (e.g., for low-speed terminals—10 to 30 characters/s—the allocation is 32 characters), the backpressure bit is set to zero; otherwise the bit is set to one. On the transmitting side, each VC is associated with a counter which is initialized to the maximum buffer limit and is decremented by one for each character transmitted. Transmission stops on a particular VC when the corresponding counter is reduced to zero. Upon reception of a backpressure bit = 1, the counter is reset to its initial value and transmission can resume.

The effect of backpressure from an individual hop back along the VC in TYMNET constitutes a good example of the "hybrid" character of many practical flow control implementations, since we see here a mixture of hop level and transport level flow control. This was pointed out earlier in connection with Fig. 12, and we shall encounter other examples as we proceed.

TRANSPAC, the French public data network, is a VC network which uses X.25 as an internode protocol [42]. One of the distinguishing features of Transpac is the use of the throughput class concept in X.25 for internal flow and congestion control. Each VC call request carries a throughput class declaration which corresponds to the maximum (instantaneous) data rate that the user will ever attempt to present to that VC. Each node keeps track of the aggregate declared throughput (which represents the worst case situation), and at the same time, monitors actual throughput (typically, much lower than the declared throughput) and average buffer utilization. Based on the ratio of actual to declared throughput, the node may decide to *oversell* capacity, i.e., it will attempt to carry a declared throughput volume higher than trunk capacity. Clearly, overselling implies that input rates may temporarily exceed trunk capacities, so that the network must be prepared to exercise flow control. Packet buffers are dynamically allocated to VCs based on demand (complete sharing), but thresholds are set on individual VC allocations as well as on overall buffer pool utilization. Of particular interest is the impact of overall buffer pool thresholds on VC-HL. Three thresholds levels [S_0, S_1, and S_2 (where $S_0 < S_1 < S_2$)] are defined and are used in the following way:

1. S_0: do not accept new VC call requests.
2. S_1: slow down the flow on current VCs (by delaying the return of ACKs at the VC level).
3. S_2: selectively disconnect existing VCs.

The threshold levels S_0, S_1, and S_2 are dynamically evaluated as a function of declared throughput, measured throughput, and current buffer utilization.

Another example of a VC network is offered by GMDNET [13]. As we mentioned before, GMDNET applies SBP flow control. In addition, it applies I-control (individual control) on each virtual circuit. I-control consists of two components: end-to-end flow control and hop level flow control. End-to-end and hop level flow control are implemented using variable size windows PUL_E and PUL_L, respectively (PUL = packet underway limit). The window is defined as the maximum number of packets that a sender is allowed to transmit before receiving an ACK, or permit [5]. The windows PUL_E and PUL_L are dynamically adjusted based on sink congestion and intermediate node congestion, respectively; their values may vary within predefined ranges $(1 \leq PUL_E \leq W_E; \ 1 \leq PUL_L \leq W_L)$ [37], [13]. The buffer pool is completely shareable, without specific reservations for individual VCs.

Simulation results on the performance of the I-control scheme lead to the following important conclusions.

(1) I-control alone cannot prevent throughput degradation, unfairness, and deadlocks. Experimental results clearly show that an I-controlled network without SBP becomes deadlocked immediately after the applied load exceeds the critical value (this confirms our prediction that VC flow control without specific buffer reservations for individual VCs cannot prevent deadlocks).

(2) The end-to-end component of I-control is very effective in preventing network congestion in the case of source rates exceeding sink rates. Without I-control (i.e., the SBP control alone), a fivefold throughput degradation was observed in a typical network overload experiment.

IV. Entry-to-Exit Flow Control

The main objective of the entry-to-exit (ETE) flow control is to prevent buffer congestion at the exit node due to the fact that remote sources are sending traffic at a higher rate than can be accepted by the hosts (or terminals) attached to the exit node. The cause of the bottleneck could be either the overload of the local lines connecting the exit node to the hosts, or the slow acceptance rate of the hosts. The problem of congestion prevention at the exit node becomes more complex when this node must also reassemble packets into messages, and/or resequence messages before delivery to the host. If fact, reassembly and resequence deadlocks may occur, which require special prevention measures.

In order to understand how reassembly deadlocks can be generated, let us consider the network path shown in Fig. 19, where three store-and-forward nodes (node 1, node 2, and node 3, respectively) relay traffic directed to host 1. In the situation depicted in Fig. 19, three multipacket messages A, B, and C are in transit towards host 1. Without loss of generality we assume that the maximum message size is 4 packets and that 4 packet

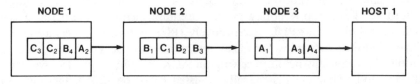

Fig. 19. Reassembly buffer deadlock.

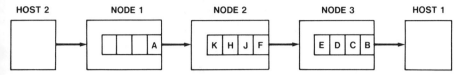

Fig. 20. Resequence deadlock.

buffers are dedicated to messages being assembled at a node; furthermore, a channel queue limit $Q_{max} = 4$ is set on each trunk queue, for hop level flow control. We note from Fig. 17 that message A (which has seized all four reassembly buffers at node 3) cannot be delivered to the host since packet A_2 is missing. Packet A_2, on the other hand, cannot be forwarded to node 2 since the queue at node 2 is full. The node 2 queue, in turn, cannot advance until reassembly space becomes available in node 3 for B or C messages. Deadlock!

A very similar order of events leads to resequence deadlocks as shown in Fig. 20. Assume that a sequence of single packet messages $A, B, \ldots, K$ originating from host 2 and directed to host 1 is traveling through a three-node network. If messages must be delivered in sequence, messages B, C, D, E in node 3 cannot be transmitted to host 1 until message A is received at node 3. However, due to store-and-forward buffer unavailability in node 2, message A cannot reach node 3. Deadlock!

Various schemes can be used to prevent these types of deadlocks. In the ARPANET, for example, reassembly deadlocks are avoided by requiring a reassembly buffer reservation for each multipacket message entering the network; resequence deadlocks are avoided by discarding out-of-sequence messages at the destination. Other networks (e.g., TELENET) have sufficient nodal storage to permit out-of-sequence messages to be accepted at a destination node with the understanding that these may be discarded later if storage congestion occurs; again, the existence of a source copy saves the day [54], [64]. These and other schemes are discussed in more detail in the following sections.

While the main objective of ETE controls is to protect the exit node from congestion, an important byproduct is the prevention of global (i.e., internal) congestion. Virtually all ETE controls are based on a window scheme that allows only up to W sequential messages to be outstanding in the network before an end-to-end ACK is received. If the network becomes congested (this may occur independently of destination node congestion), messages and ACKs incur high end-to-end delays. these delays, combined with the restriction on the total number of outstanding messages, effectively contribute to reduce the input rate of new packets into the network.

Several varieties of ETE flow control schemes have been proposed and implemented. We first describe four representative examples, and then briefly review some analytical and simulation models for the performance evaluation and comparison of such schemes.

A. ARPANET RFNM and Reassembly Scheme

ETE flow control in ARPAnet is exercised on a host-pair basis [30], [23]. Specifically, all messages traveling from the same source host to the same destination host are carried on the same logical "pipe." Each pipe is individually flow controlled by a window mechanism. An independent message number sequence is maintained for each pipe. Numbers are sequentially assigned to messages flowing on the pipe, and are checked at the destination for sequencing and duplicate detection purposes. Both the source and the destination keep a small window w (presently, $w = 8$) of currently valid message numbers. Messages arriving at the destination with out-of-range numbers are discarded. Messages arriving out of order are discarded since storing them (while waiting for the missing message) may lead to potential resequence deadlocks. Correctly received messages are acknowledged with short ETE control messages called RFNMs (ready for next message). Upon receipt of an RFNM, the sending end of the pipe advances its transmission window, accordingly.

RFNMs are also used for error control. If an RFNM is not received after a specified time out (presently about 30 s), the source IMP sends a control message to the destination inquiring about the possibility of an incomplete transmission. This technique is necessary to keep source and destination message numbers synchronized and also to request a retransmission from the host in the case of message loss.

The window and message numbering mechanisms described so far support ETE flow control, sequencing, and error control functions in the ARPANET. A separate mechanism, known as reassembly buffer allocation [30], is used to prevent reassembly deadlocks. Each multipacket message must secure a reassembly buffer allocation at the destination node before transmission. This is accomplished by sending a reservation message called a REQALL (request for allocation) to the destination and waiting for an ALL (allocation) message from the destination before attempting transmission. To reduce delay (and, therefore, increase throughput) of steady multipacket message flow between the same source–destination pair, ALL messages are automatically piggybacked on RFNMs, thus eliminating the reservation delay for all messages after the first one. If a pending allocation at the source node is not claimed within a given time out (250 ms), it is returned to the destination with a "giveback" message. Single-packet messages are transmitted to their destinations without buffer reservation. However, if upon arrival at the destination, all the reassembly buffers are full, the single-packet message is discarded and a copy is retransmitted from the source IMP after an explicit buffer reservation has been obtained. Some pitfalls inherent in such schemes are described in [23].

B. SNA Virtual Route Pacing Scheme

The IBM systems network architecture (SNA) is an architecture aimed at providing distributed communications and distributed processing capabilities between IBM systems [15], [16]. SNA was first announced in 1974. Since then, the original set of functions which supported single rooted networks (i.e., single host) have been enhanced to support multiple-domain (i.e., multiple host) networking. In this chapter, we refer to SNA release 4.2 [16].

SNA devices can be subdivided into four main categories: host computers (e.g., system/370), communications controllers (e.g., 3704 and 3705), terminal cluster controllers, and terminal devices (e.g., TTYs, CRTs, readers, and printers). Distributed communications with full routing, flow control, and global addressing capabilities are provided only on store-and-forward networks interconnecting host computers and communication controllers. These nodes are called *subarea* nodes in SNA. Terminals and terminal cluster controllers are considered *peripheral* nodes and are connected into the high level net at subarea nodes, which provide the necessary boundary functions (e.g., global/local address conversion). Thus, for purposes of this section, SNA can be viewed as the usual two-level network architecture, and terminals and terminal cluster controllers at the lower level, and hosts and communications controllers at the higher level.

SNA is essentially a virtual circuit network, in the sense that each user session is associated with a physical route at session setup time. The routing policy is a static, multipath policy which maintains up to eight distinct routes between each source–destination pair in the high-level network (i.e., between subarea nodes). These routes are called ERs (explicit routes), to distinguish them from VRs (virtual routes) defined below. ERs are defined as an ordered sequence of network trunks, and are uniquely identified by ER numbers. When a failure is detected on an ER currently being used, the next ER on the list is "switched in." One difficulty here is that the list of ERs must be updated by the network designer each time the network topology is changed.

Next, virtual routes (VRs) are defined between each source–destination node pair of the high-level network. A VR is essentially a virtual pipe which is constructed on top of an ER and is subject to flow control. Three sets of VRs, each with a different level of priority are maintained between each subarea node pair. Each set may consist of up to eight VRs, thus allowing for up to 24 VRs between each high-level network node pair.

Active VRs are identified by VR numbers and are stored in lists at each node.

At session setup time, the entry node scans the VR list and assigns the user session to the first available virtual route of desired priority. Several user sessions may be multiplexed on the same VR. In turn, several VRs may be multiplexed on the same ER. Finally, several ERs can be multiplexed on the same trunk.

The rationale for the distinction between virtual routes and explicit routes (a unique SNA feature among all VC networks, which typically associate a virtual route with a fixed path) is to "...insulate the virtual route layer from the physical configuration" [16]. As a consequence, user packets are driven through the network using the ER ID number, while the VR ID number needs to be checked only at the end points of the path. This feature considerably reduces storage and processing overhead with respect to conventional VC schemes, which typically require large maps at each intermediate node to store the information relative to all virtual circuits traversing that node.

In the high-level network, flow control is applied independently to each VR from entry to exit node. This scheme, known as *VR pacing*, is actually a combination of ETE and hop level flow control. It is based on a window mechanism, in which the entry node must request (and obtain) permission from the exit node before sending a new group of k packets, where k is the window size. The destination may grant (or delay) such permission depending on local buffer availability. The window size k varies from h to $3h$, where h is the path hop length. The value of k is dynamically adjusted not only by the exit node, but also by any intermediate node along the path on the basis of its buffer availability [1]. The fact that both the end node and the intermediate nodes can "modulate" the window size k makes VR pacing a hybrid ETE and hop flow control scheme. Details are given in Section XII of Chapter 11.

In addition to VR-pacing control, which operates between subarea nodes, the SNA architecture provides also for *session level pacing* which, for terminals and clusters, extends beyond subarea nodes and individually flow controls each user session between terminal and host computer. Session pacing is discussed in Section VI.

C. GMD Individual Flow Control

In GMDNET, entry-to-exit flow control is exercised individually on each virtual circuit, hence the name of individual flow control assigned to the scheme [37]. We recall that GMDNET is a VC network in which a fixed route is assigned to each user session at session setup time.

The main purpose of entry-to-exit flow control in GMDNET is to protect the exit node from overflow caused by low sink rates. When the source host rate exceeds the sink host rate, the flow control mechanism intervenes to slow down inputs from the source host into the entry node. This is achieved by maintaining a window of outstanding packets between entry and exit node for each virtual circuit. The window must be large enough to permit each virtual circuit to efficiently utilize the bandwidth available on the path. GMD simulation experiments have shown that $w = h + 1$ (where h is the hop length of the path) is a satisfactory choice under nominal load conditions. Window size can be reduced if the sink is slow in accepting packets. More precisely, when for a given VC the queue waiting to be transferred from exit node to sink reaches the value w, further arrivals to the exit node within that VC are discarded and a negative ACK is returned to the source node. Each negative ACK causes a window size reduction of 1 at the source node, until the minimum window size $w = 1$ is reached. Each positive ACK, on the other hand, increases window size by 1, until the maximum window size $w = h + 1$ is reached. In this way, window size is dynamically controlled in the range 1 to $h + 1$ by positive and negative acknowledgments [37].

In addition to the entry-to-exit flow control, each hop of the virtual circuit is also independently flow controlled (see Section III). The two layers of flow control, entry-to-exit and hop, are logically separated one from the other, in that the ETE window is controlled by exit buffer occupancy, while hop window is controlled by intermediate node congestion.

Packets within the same virtual circuit must be delivered to the host in sequence, and in case of multipacket messages, must be reassembled before delivery to the host. Fixed path routing and link level sequencing imply that packets arrive at their destination in sequence. This sequencing property, and the fact that a number of buffers sufficient to reassemble the largest size packet is permanently dedicated to traffic leaving the network, preclude the possibility of reassembly deadlocks and eliminate the need for reassembly buffer allocation schemes of the type implemented in ARPANET.

D. DATAPAC Virtual Circuit Flow Control

The Canadian public data network, DATAPAC, implemented with the Northern Telecom SL-10 Packet Switching System provides virtual circuit services using an internal transport protocol built on top of a datagram subnetwork [28], [54]. Flow control is exercised from entry to exit node on a virtual circuit basis, although no physical path is actually assigned to each virtual circuit, as was the case with SNA and GMDNET. The absence of a fixed path leads to some complications in the resequencing and loss recovery procedures, which will soon be discussed.

In DATAPAC, a virtual circuit is provided between the two end points of each user session. The virtual circuit is implemented at the concatenation of three protocol segments: a packet level X.25 protocol from the source device (i.e., data terminating equipment or DTE) to entry node (i.e., data communications equipment or DCE), an internal protocol from entry DCE to exit DCE, and a packet level X.25 protocol from exit node (DCE) to destination node (DTE). Each one of these protocol segments is flow controlled by a window mechanism. Of particular interest to us is the fact that window controls on these three segments are synchronized so as to provide a means of matching source DTE transmission rate with destination DTE acceptance rate. Window control synchronization is achieved by withholding the return of ACKs on a window if the downstream window is full.

As an example, let us assume that all windows are of size $w = 3$, and that the window between entry and exit DCE is full (i.e., there are three outstanding packets). The next packet arriving from the source DTE to the entry DCE will be accepted (assuming buffer space is available), but will not be immediately acknowledged; rather, the ACK will be withheld until an ACK from the exit DCE is received, thus opening up the downstream window [28].

Within the concatenated window mechanism the entry-to-exit flow control serves the function of promptly reflecting back to the source an exit segment congestion situation by withholding ACKs. Recall that in GMDNET the entry-to-exit flow control provided a similar service by dynamically adjusting the window with positive or negative ACKs. In DATAPAC, things are complicated, however, by the fact that the window mechanism is used not only for flow control, but also for sequencing, packet loss recovery, and duplicate detection. These latter functions are not required in the GMDNET, since sequencing is enforced there by the fixed path routing policy, and packet loss could occur only if a node along the path failed, in which case the virtual circuit would be automatically reinitialized.

The use of window ACKs for loss recovery in DATAPAC leads to the following problem. If the exit DCE does not return to the entry DCE an ACK for a correctly received packet (because the exit segment is congested), the entry DCE will retransmit the packet after a time out, under the assumption that the packet was lost (or was dropped by the exit DCE for lack of resequence space). If no ACK is received after a specified number of retransmissions, the entry DCE will clear the virtual circuit. In order to minimize the generation of duplicate packets, and avoid the unnecessary interruption of user sessions, the value of time out must be carefully adjusted as a function of window size and other network parameters.

E. Performance Models

The great majority of entry-to-exit flow control mechanisms are based on the window scheme. Critical parameters in the window implementation are the size of the window, and if error and loss recovery are to be provided, the retransmission time-out interval. Several analytic and simulation models have been developed recently to investigate the impact of these parameters on throughput and delay performance. This section briefly surveys some of the most significant contributions in this area.

We start with the Kleinrock and Kermani model of a single source-to-destination stream flow controlled by a window mechanism [26]. The network entry-to-exit delay is simplified as an $M/M/1$ queue delay, and the round trip delay therefore follows an Erlang-2 distribution. (This approximation is supported by simulation experiments showing that more accurate delay assumptions do not significantly change the nature of the results.) The exit node has finite storage and delivers packets to the destination host on a finite capacity channel. Consequently, the exit node may occasionally overflow and drop packets. To provide for transmission integrity, the entry node will retransmit an unacknowledged packet after a time-out interval. This simplified window model is solved analytically, yielding the optimal (i.e., minimum delay) window size and time-out interval for a given throughput requirement and destination buffer storage size.

In a subsequent paper [22], the same authors propose an adaptive policy (the "look-ahead" policy) for the dynamic adjustment of window size to time-varying traffic rate. In the proposed policy, the window size is dynamically controlled by the queue size at the exit node. Numerical results show that the delay versus throughput performance of the adaptively controlled scheme is somewhat superior to the performance of a scheme operated under static control, in which the window is adjusted in accordance with the traffic volume. These results are very encouraging, and are consistent with simulation experiments on dynamic window control carried out in multinode networks [1], [13].

The models in [26], [22] approximate the network as a single queue and therefore do not offer insight into the dependence of window size w on the window of intermediate hops. This issue is addressed by a simple multihop model developed by Kleinrock in [24]. In this model a packet stream from a single destination is transmitted across the network on a k-hop network path. Infinite buffer storage and negligible error rates are assumed on each hop. The stream is flow controlled by a window mechanism. In this model, as the window size w increases, the end-to-end delay grows without limit while the throughput asymptotically reaches the path capacity. In order to find a meaningful criterion for the optimization of w, the concept of "power" as defined in Section II F is used. We find that power is optimized by $w = k$. This implies that, at optimum, there should be on the average one packet in each intermediate queue. This result agrees with our intuition that the "entry-to-exit pipe should be kept full (in fact, *just* full)" for satisfactory performance. The general validity of this result is confirmed by actual window implementations. In fact, the SNA pacing scheme allows the window to dynamically vary from h to $3h$, where h is the number of intermediate hops. Similarly, the GMD individual flow control scheme uses a maximum window of $h + 1$. The performance of window flow control for different acknowledgment and credit schemes (sliding window, ACK-at-the-end, credit-at-the-beginning) was evaluated by M. Schwartz using the Norton equivalence theorem for networks of queues [65].

The main limitation of the previous models is the single-source, single-destination traffic assumption which excludes interference at a given node by other traffic traversing it. The model by Pennotti and Schwartz [32] includes the effect of interference in an approximate fashion in that it represents a virtual link situation in which end-to-end link traffic flowing on a multihop path must compete at each hop with external traffic. This is essentially a "one-hop" interference model in which some external traffic λ is injected into one node along the path and is transmitted to the next node on the path, where it then is removed from the network. The purpose of this study is to evaluate the possible path congestion caused by an increase in the virtual link rate λ_0, both with and without flow control. Congestion is defined as the relative average increase in time delay experienced by external users due to an increase in λ_0, taking $\lambda_0 = 0$ as a reference. Without flow control, congestion rapidly grows to infinity even for moderate values of λ_0. By introducing end-to-end window control which limits to w the number of packets outstanding on the virtual link at any one time, congestion can be bounded for any value of λ_0. The value of the upper bound varies with w, and decreases for decreasing w, as expected.

As an alternative to window flow control, hop flow control was also implemented in the Pennotti and Schwartz model by setting a limit on the number of link packets that would be stored at each intermediate node [32]. This scheme exhibited essentially the same performance as the window scheme. The above experiments show that flow control (either window or hop) can be used effectively to maintain fairness in a multiuser environment with conflicting requirements; that is, by adjusting the window parameter w, one can balance the relative user throughputs as desired.

The previously mentioned model offers some insight into multiuser flow control, but suffers from the limitation that only one virtual circuit can be flow controlled at a time, the remaining traffic components being kept constant. To remove this limitation, a number of multiple source, multiple destination models with selectively controlled user pairs have been developed. These models combine ETE flow control with network access flow control, and therefore may be regarded as hybrid models. Wong and Unsoy analyze a simple 5-node network to which individual entry-to-exit window control as well as isarithmic control are applied [41]. The isarithmic scheme is a network access flow control scheme which controls the total number of packets allowed in the entire network (see Section V for additional details). The major finding of this study is the fact that isarithmic control alone is not enough to guarantee efficient network operation. In fact, under some unfavorable traffic situations, one node pair may capture most of the permits, starving other pairs and leading to unfairness and to overall performance degradation. Similar results were found by Price in a series of simulation experiments [36]. The problem is corrected by introducing individual entry-to-exit flow controls in addition to isarithmic control.

The exact analysis of multinode networks with individually controlled node pairs becomes impractical for topologies with more than five or six nodes because of the rapidly increasing computational complexity of exact solution techniques [41]. To circumvent this problem, Reiser proposed an approximate solution technique based on a mean value analysis which is computationally affordable even for large networks, and which reaches a typical accuracy of 5 percent in throughput and 10 percent in delay [38]. With this technique it is now possible to analyze the interaction of various flow control schemes in a much more realistic environment (i.e., large networks, varied traffic patterns) than was possible with previous methods. Important design problems such as the optimization of window parameters for all source–destination pairs in order to maximize network throughput (within given fairness constraints), now become approachable. In particular, mean value analysis was used to study the interplay between routing and window flow control in [12b].

In spite of the previously mentioned advances in computational solution techniques, some window flow control issues are still too complex to be attacked analytically. For example, the dynamic control of window size in a multinode network is not amenable to a network-of-queues model even with the approximate solution methods. In these cases, simulation is still the leading performance evaluation tool [13], [1], [36].

V. Network Access Flow Control

A. Objective

The objective of network access (NA) flow controls is to throttle external inputs based on measurements of internal network congestion. Congestion measures may be local (e.g., buffer occupancy in the entry node), global (e.g., total number of buffers available in the entire network), or selective [e.g., congestion of the path(s) leading to a given destination]. The congestion condition is determined at (or is reported to) the network access points and is used to regulate the access of external traffic into the network.

NA flow control differs from HL and ETE flow control in that it throttles external traffic to prevent *overall internal buffer congestion*, while HL flow control limits access to a specific store-and-forward node to prevent *local congestion and store-and-forward deadlocks*, and ETE flow control limits the flow between a specific source–destination pair to prevent *congestion and reassembly buffer deadlocks at the destination*. The distinction, however, is not quite so clearcut, since as we mentioned earlier, both HL and ETE schemes indirectly provide some form of NA flow control by reporting an internal network congestion condition back to the access point either via backpressure (HL scheme), or via credit slowdown (ETE scheme).

Three NA flow control implementations will be discussed: the isarithmic scheme, a global congestion prevention scheme based on the circulation of a fixed number of permits [8]; the input buffer limit scheme, a local congestion scheme which sets a limit on the number of input packets stored at each node [27], [13]; and the choke packet scheme, a selective congestion scheme based on the delivery of special control packets of that name from the congested node back to the traffic sources [29].

B. The Isarithmic Scheme

Since the primary cause of network congestion is the excessive number of packets stored in the network, an intuitively sound congestion prevention principle consists of setting a limit on the total number of packets that can circulate in the network at any one time. An implementation of this principle is offered by the Isarithmic scheme proposed for the National Physical Laboratories network [8], [35].

The isarithmic scheme is based on the concept of a "permit," i.e., a ticket that permits a packet to travel from the entry point to the desired destination. Under this concept, the network is initially provided with a number of permits, several held in store at each node. As traffic is offered by a host to the network, each packet must secure a permit before admission to the high-level node is allowed. Each accepted packet causes a reduction of one in the store of permits available at the accepting node. The accepted data packet is able to traverse the network, under the control of node and link protocols, until its destination node is reached. When the packet is handed over to the destination subscriber, the permit which has accompanied it during its journey becomes free and an attempt is made to add it to the permit pool of the node in which it now finds itself.

In order to achieve a viable system in which permits do not accumulate in certain parts of the network at the expense of the other parts, it is necessary to place a limit on the number of permits that can be held in store by each node. If then, because of this limit, a newly freed permit cannot be accommodated at a node (overflow permit), it must be sent elsewhere. The normal method of carrying the permit in these circumstances is to "piggyback" it on other traffic, be this data or control. Only in the absence of other traffic need a special permit-carrying packet be generated.

A simulation program was developed by NPL to evaluate the performance of the isarithmic scheme in various network configurations and in the presence of different network protocols [35]. The main conclusion of these simulation studies was that the isarithmic scheme is a simple congestion prevention mechanism which performs well in uniform traffic pattern situations, but may lead to unnecessary throughput restrictions, and therefore, to poor performance in the case of nonuniform, time-varying traffic patterns. In particular, in the presence of high bandwidth data transfers, there is the possibility that permits are not returned to the traffic sources rapidly enough to fully utilize network capacity (the "permit starvation" problem). This would be the case when the destination node redistributes the overflow permits randomly in the network. If, on the other hand, the destination systematically returns all the permits to the source, the source-destination pair may end up capturing most of the network permits, thus causing unfairness. Tradeoffs between different permit distribution schemes are investigated with an analytical model in [41]. Finally, a delicate problem in isarithmic control is the bookkeeping of permits, to avoid unauthorized generation or disappearance of permits.

In spite of the above limitations, the isarithmic scheme proved to be very effective in *weakly controlled* networks (namely, networks without hop level flow control), eliminating congestion and deadlocks that had occurred without flow control. Some simulation experiments were also carried out on networks with hop level control (specifically CQL), and with a simple form of local access control (one buffer on each output queue was reserved for store-and-forward traffic). For this class of networks (called *strongly controlled* networks), it was found that the network performance did not show congestion tendencies even without isarithmic control in the case of a fixed routing discipline. When the fixed discipline was replaced with an adaptive routing discipline, it was found that the network would become easily congested since the simple form of network access control implemented would not prevent external traffic from flooding all the queues in the entry node. Again, the introduction of the isarithmic scheme was successful in eliminating the congestion problem for the adaptive routing case [36].

Critical parameters in the isarithmic scheme design are the total number of permits P in the network and the maximum number of permits L that can be accumulated at each node (permit queue). Experimental results show that optimal performance is achieved for $P = 3N$, where N is the total number of nodes, and $L = 3$. An excessive number of permits in the network would lead to congestion. An excessive value of L would lead to unfairness, accumulation of permits at a few nodes, and throughput starvation at the others.

C. Input Buffer Limit Scheme

The input buffer limit (IBL) scheme differentiates betwen input traffic (i.e., traffic from external sources) and transit traffic, and throttles the input traffic based on buffer occupancy at the entry node. IBL is a *local* network access method since it monitors local congestion at the entry node, rather than global congestion as the isarithmic scheme does. Entry node congestion, on the other hand, is often a good indicator of global congestion because of the well-known backpressure effect which propagates internal congestion conditions back to the traffic sources.

The function of IBL controls is to block input traffic when certain buffer utilization thresholds are reached in the entry node. This flow control approach clearly favors transit traffic over input traffic. Intuitively, this is a desirable property since a number of network resources have already been invested in transit traffic. This intuitive argument is supported by a number of analytical and simulation experiments proving the effectiveness of the IBL scheme.

Many versions of IBL control can be proposed. Here, we describe and compare four different implementations that have been experimentally evaluated.

The term input buffer limit scheme refers to a scheme restricting the number of buffers made available to input traffic and was first introduced by the GMD research group [37], [13]. The scheme proposed for GMDNET is a by-product of the nested buffer class structure used to allocate buffers to different classes of traffic. We recall from Section III D that the ith traffic class consists of all the packets that have already covered i hops. Input traffic is assigned to class zero (zero hops covered). Traffic class zero is entitled to use buffer class zero, which is a subset of the nodal buffer pool (in general, class i is entitled to use all buffer classes $\leq i$). Thus, input packets are discarded when class zero buffers are full. The size of buffer class zero (referred to as input buffer limit) was found to have a significant impact on throughput performance under heavy loads. Simulation experiments indicate that for a given topology and traffic pattern there is an optimal input buffer limit which maximizes throughput for heavy offered load. The use of lower or higher limits leads to a substantial drop in throughput [13].

A version of IBL control that is simpler than the GMD version was proposed by Lam [27] and analytically evaluated in an elegant model. Only two classes of traffic—input and transit—are considered in this proposal. Letting N_T be the total number of buffers in the node and N_I the input buffer limit (where $N_I \leq N_T$), the following constraints are imposed at each

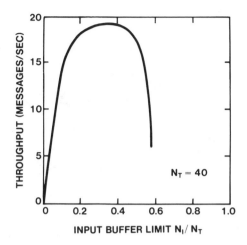

Fig. 21. Input buffer limit scheme: throughput versus buffer limitation for heavy offered load.

node:

1. Number of input packets $\leq N_I$.
2. Number of transit packets $\leq N_T$.

The analytical results confirm simulation results independently obtained by the GMD group. There is an optimal ratio N_I/N_T, which maximizes throughput for heavy offered load, as shown in Fig. 21. A good heuristic choice for N_I/N_T is the ratio between input message throughput and total message throughput at a node. As shown in the figure, throughput performance does not change significantly even for relatively large variations of the ratio N_I/N_T around the optimal value, thus implying that the IBL scheme is robust to external perturbations such as traffic fluctuations and topology changes. One shortcoming of this model is that all nodes in the net are assumed to have the same blocking probability, a somewhat unrealistic assumption.

A scheme similar to Lam's IBL scheme was proposed earlier by Price [35]. In order to prevent input traffic from monopolizing the entire buffer pool, one buffer in each output queue was reserved for transit traffic. This is essentially equivalent to setting an input buffer limit $N_I = N_T - C$, where C is the number of output channels. Simulation studies showed that this simple network access control based on source buffer utilization was quite successful in single level networks.

Kamoun [21], [60] proposes yet another version of IBL control, in which an input packet is discarded if the *total* number of packets in the

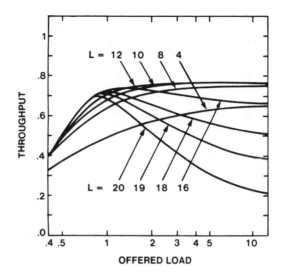

Fig. 22. Throughput versus load for a 121-node network for drop-and-throttle flow control.

entry node exceeds a given threshold (whereas in Lam's scheme an input packet is discarded when the number of *input* packets exceeds a given threshold). Transit packets, instead, can freely claim all the buffers. The scheme is called drop-and-throttle flow control (DTFC) policy since a transit packet arriving at a full node is dropped and lost (loss model); while all previous schemes assumed link level retransmission of overflow packets (retransmit model). The DTFC scheme was analyzed using a network of queues model [21]. The results, shown in Fig. 22, clearly indicate that there is an optimal threshold value L which maximizes throughput for each value of offered load. Below the threshold, the network is "starved"; above the threshold, the network is congested.

DATAPAC [54] has implemented a "subnet admittance control" which resembles DTFC in the fact that it limits the acceptance of external datagrams based on global buffer availability at the entry node. A similar scheme, referred to as the *free flow* scheme, is described and analyzed by Schwartz and Saad in [41]. Preliminary results indicate that, while free flow and IBL throughput performances are compatible, the free flow scheme offers substantial delay improvements.

We have pointed out that IBL control prevents congestion by favoring transit traffic over input traffic. In most cases (indeed, in all cases analyzed in the previously referenced studies), this favoritism leads to throughput improvements. In some cases, however, *unfairness* may result. Consider, for example, the 4-node network shown in Fig. 23. In this network, two file transfers, A to A' and B to B', respectively, are simultaneously competing

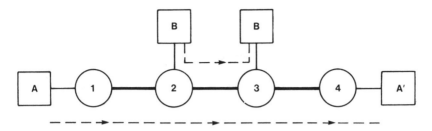

Fig. 23. Unfairness condition produced by input buffer limit and drop-and-throttle flow control schemes.

for trunk $(2, 3)$. Node 2 sees traffic A as transit traffic, so it gives it preferential treatment over traffic from B. Consequently, the $A \rightarrow A'$ packet stream can acquire more buffers in node 2, and thus achieve better throughput performance than the $B \rightarrow B'$ stream. The unfairness is particularly dramatic when DTFC is used. With the DTFC policy, if the A-packet queue in node 2 exceeds the buffer threshold (this could easily occur if $C_{23} < C_{12}$), B packets cannot be accepted by node 2. Consequently B traffic is completely shut off until the $A \rightarrow A'$ file transfer is completed.

D. Choke Packet Scheme

The choke packet (CP) scheme, proposed for the CYCLADES network [29], [57], is based on the notion of trunk and path congestion. A trunk (link) is defined to be congested if its utilization (measured over an appropriate history window with exponential averaging) exceeds a given threshold (e.g., 80 percent). A path is congested if any of its trunks are congested. Path congestion information is propagated in the network together with routing information, and thus each node knows hop distance and congestion status of the shortest path to each destination.

When a node receives a packet directed to a destination whose path is congested it takes the following actions:

(1) If the packet is an *input* packet (i.e., it comes directly from a host), then the packet is dropped.

(2) If the packet is a *transit* packet, it is forwarded on the path; but a "choke" packet (namely, a small control packet) is sent back to the source node informing it that the path to that destination is congested and instructing it to block any subsequent input packets to this destination. The path to the destination is gradually unblocked if no choke packets are received during a specified time interval.

This is a greatly simplified description of the CP scheme. Several other features (which are essential to make the scheme workable) are discussed in [29].

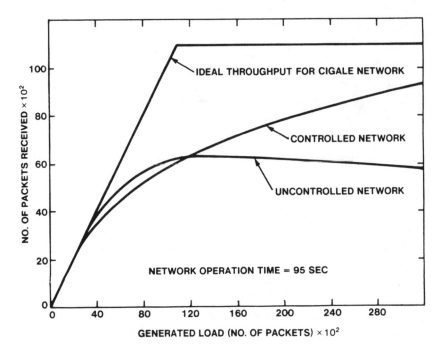

Fig. 24. Throughput performance in Cigale with and without flow control.

It is clear that the CP scheme attempts to favor transit traffic over input traffic, much in the same way as the IBL scheme did. The basic difference between the two schemes is the fact that IBL uses a local congestion measure, namely, the entry node buffer occupancy, to indiscriminately control all input traffic; whereas, CP uses a path congestion measure to exercise *selective* flow control on input traffic directed to different destinations.

Simulation experiments based on the Cigale network topology are given in Fig. 24 and show that the CP scheme can introduce substantial throughput improvements (with respect to the uncontrolled case) in sustained load conditions, asymptotically achieving the ideal performance for infinite load [29].

VI. Transport Level Flow Control

A. Objectives

A transport protocol is a set of rules that govern the transfer of control and data between user processes across the network. The main functions of this protocol are the efficient and reliable transmission of messages within each user session (including packetization, reassembly, resequencing, recovery from loss, elimination of duplicates) and the efficient sharing of common network resources by several user sessions (obtained by multiplexing many user connections on the same physical path and by maintaining priorities between different sessions to reflect the relative urgency).

For efficient and reliable reassembly of messages at the destination host (or more generally, the DTE), the transport protocol must ensure that messages arriving at the destination DTE are provided adequate buffering. The transport protocol function which prevents destination buffer congestion and overflow is known as *transport level flow control*. Generally, this level of flow control is based on a "credit" (or window) mechanism as discussed earlier. Specifically, the receiver grants transmission credits to the sender as soon as reassembly buffers become free. Upon receiving a credit, the sender is authorized to transmit a message of an agreed-upon length. When reassembly buffers become full, no credits are returned to the sender, thus temporarily stopping message transmissions [5].

The credit scheme described above is somewhat vulnerable to losses, since a lost credit may hang up a connection. In fact, a sender may wait indefinitely for a lost credit, while the receiver is waiting for a message. A more robust flow control scheme is obtained by numbering credits relative to the messages flowing in the opposite direction. In this case, each credit carries a message sequence number, say N, and a "window size" w. Upon receiving this credit, the sender is authorized to send all backlogged messages up to the $(N + w)$th message. With the numbered credit scheme, if a credit is lost then the subsequent credit will restore proper information to the sender [45].

Besides preventing destination buffer congestion, the credit scheme also indirectly provides global network congestion protection. In fact, store-and-forward buffer congestion at the intermediate nodes along the path may cause a large end-to-end credit delay, thus slowing down the return of credits to the sender, and consequently, reducing the rate of fresh message input into the network.

B. Implementations

Several versions of the transport protocol are in existence, each incorporating its own form of transport level flow control. Here, we briefly describe four representative implementations.

The earliest example of transport protocol implementation is the original version of the ARPANET network control program (NCP) [4]. NCP flow control is provided by unnumbered credits called "allocate" control messages (see Section IV D). Only one allocate could be outstanding at a time (i.e., window size $w = 1$).

The French research network CYCLADES provided the environment for
the development of the transport station (TS) protocol [50]. In the TS
protocol, the flow control mechanism is based on numbered credits, each
credit authorizing the transmission of a variable size message called a letter.
Flow control is actually combined with error control in that credits are
carried by acknowledgment messages.

The transmission control program (TCP) was a second-generation
transport protocol developed by the ARPANET research community in order
to overcome the deficiencies of the original NCP protocol [5]. As in the TS
protocol, flow and error control are combined in TCP. As a difference,
however, error and flow control are on a byte (rather than letter) basis. This
allows a more efficient utilization of reassembly buffers at the destination.

In SNA, the transport level flow control is provided by session pacing.
The purpose of session-level pacing is to prevent one session end from
sending data more quickly than the receiving session end can process the
data [16]. As in TCP and TS, session-level pacing is based on a window
concept, in which the receiving end grants "credits" to the sending end
based on its buffer availability and processing capability. As a difference,
however, subarea nodes in SNA can control the inbound flow from a
cluster controller into the network by intercepting and withholding the
credits (called pacing responses in SNA) for a given session, if the subarea
node buffers are congested or if the virtual route (VR) transmission queue
for that session is congested. Specifically, session-level pacing responses are
intercepted at the entry node to exercise *network access* flow control from
the terminal into the high-level network [16]. Thus, session pacing may be
viewed as a hybrid form of transport level flow control, which is obtained
by concatenating a network access level segment (from the terminal to the
high-level network node) and an entry-to-exit level segment (controlled by
virtual route pacing).

VII. Conclusions and Directions for Further Research

In this chapter we have proposed a taxonomy of flow control mechanisms based on a multilevel structure. We have defined *four levels of flow
control* and have shown how these levels are actually embedded into
corresponding levels of protocols. To the extent that these levels can be
independently defined, the analysis, design evaluation, and comparison of
flow control schemes is greatly simplified, since any complex control
structure can be decomposed into smaller modules, and each module
individually analyzed. The overall performance is then obtained by studying the interaction of the various modules.

Recent advances in queueing theory have led to reasonable success in
the *modeling and analysis of individual levels* of flow control. We have

reported on several performance results, and have used such results to compare different schemes.

In real life, however, some control structures defy the simple, hierarchical representation here proposed, and seem to combine two or more levels into *hybrid flow control solutions* (see Fig. 12). This is particularly common in homogeneous networks (e.g., SNA) in which a single manufacturer is responsible for the implementation of both DCE and DTE equipment and, therefore, has more freedom in the design of the various flow control levels.

The existence of multiple levels of flow control and the possible integration of some of these into hybrid arrangements immediately brings up a very critical issue in flow control which requires further study, namely the *interaction between levels.* Given that we understand the throughput and delay implications of each specific level of flow control, we still have to study the combined effect when these levels are operating simultaneously in the network. For instance, network experience seems to indicate that a network equipped with a very conservative hop level flow control, such as the SBP scheme in GMDNET or the VC-HL scheme in TYMNET, does not require strong network access or ETE flow control schemes since network congestion situations are immediately reported back to the entry node by back pressure through the hop level [36]. This type of issue can be fully investigated only by developing models which include multiple levels of flow control. An interesting example in this direction was the combined isarithmic and entry-to-exit flow control model presented in [47]. More research is required in this area.

Hybrid packet and circuit networks are now emerging as a solution to multimode (voice and data; batch and interactive) user requirements [11]. These networks must be equipped with novel flow control mechanisms. In fact, if the network were to apply conventional flow control schemes to the packet-switched (P/S) component only, leaving the circuit-switched (C/S) component uncontrolled, then the C/S component would very likely capture the entire network bandwidth during peak hours. If this does not cause congestion, since the C/S protocol is not as congestion prone as the P/S protocol, it certainly creates unfairness. Some form of flow control on C/S traffic which is sensitive to the relative P/S load is therefore required.

The *integration of voice and data* requirements in packet-switched networks has been vigorously advocated in recent years on grounds of improved efficiency and reduced cost [14]. Unfortunately, little attention has been given to the fact that integrated networks require a complete redesign of the conventional flow control schemes since voice traffic cannot be buffered and delayed in case of congestion. Priorities are of help only if the voice traffic is a small fraction of the total traffic. For the general case, new flow control techniques must be developed for voice. These techniques should be *preventive* in nature, i.e., they should block calls before congestion occurs, rather than *detecting* congestion and then attempting to *recover* from it, as is the case for most of the conventional flow control schemes for data [10], [31].

Progress in fiber optics and high-speed switching technology and increasing user demands have recently stimulated the research and development of *Broadband Integrated Services Digital Networks* (*B-ISDNs*). B-ISDNs will carry several services (data, voice, video, graphics, etc.) on very high speed fiber optics links. While several candidate B-ISDN architectures are currently being investigated, the prevailing thought is to use *Fast Packet Switching* to transport fixed size voice/data/video packets through the network [73]. In order to enable the Fast Packet Switch to process millions of packets per second, the internal network protocols have been drastically streamlined. In particular, almost all flow control features have been dropped from the link and packet level [73]. The lack of internal flow control mechanisms combined with the inadequacy of buffers to handle sustained overloads at such high speeds and the strict response time requirements of voice and video traffic make the problem of flow control and congestion protection of B-ISDNs a formidable one. This topic is the object of active investigation by potential B-ISDN service providers and users. It appears that congestion *prevention* mechanisms will play a major role also in this case [74].

Routing and flow control procedures have traditionally been developed independently in packet networks, under the assumption that flow control must keep excess traffic out of the network, and routing must struggle to efficiently transport to destination whatever traffic was permitted into the network by the flow control scheme. It seems, however, that routing and flow control can be brought together into useful cooperation in virtual circuit networks, where a path must be selected before data transfer on a user connection begins [12b], [12c]. In this case, the routing algorithm can be invoked first to determine whether a path of sufficient residual bandwidth is available. If no path is available, the virtual circuit connection is blocked immediately at the entry node by the network access flow control level, thus *preventing* congestion rather than allowing it to occur and then attempting to *recover* from it. A combined routing and flow control strategy is implemented in TYMNET [39], and is described in more detail in Section IV of Chapter 9.

Fair sharing of resources in a flow controlled environment is an important research topic which has received increasing attention in the past few years. By definition, flow control implies the selective reduction of user traffic rates in order to optimize global network performance (typically, the throughput). Unfortunately, one discovers that throughput optimization does not always correspond to uniform satisfaction of all users. In fact, some users may get penalized more than others (see, for example, the Drop-and-Throttle scheme in Section V C). Thus, one needs to define new optimization criteria which are intrinsically fair. A taxonomy of fairness measures was proposed in [68]. One of these measures was then used to optimize routes and user input rates in a packet-switched network [69]. The same measure was later used to optimize window parameters in a window controlled network [70]. Other recent contributions to the subject are reported in [71], [72]. While new fairness measures are being defined, and algorithms are being developed to optimize flow control parameters (input rates, windows, buffer allocation, etc.) according to such measures, the question still remains of how fair these measures really are. This is because there is still a great deal of subjectivity in defining fairness. The challenge for researchers in this area will be to develop a robust measure that includes all the desirable fairness properties and, at the same time, is simple enough to permit the design and optimization of network architectures and protocols.

Challenging flow control problems exist in *multiaccess broadcast networks*. In single hop multiaccess systems, congestion prevention and stability mechanisms are well understood, and are usually directly embedded in the channel access protocol [46]. In distributed multihop multiaccess systems (e.g., multihop ground radio networks), congestion prevention becomes a very hard problem because of the interaction between buffer and channel congestion. Conventional flow control schemes used in hardwired nets cannot be directly applied. In particular, the hop level flow control should be revised to combine the buffer allocation strategy with the retransmission control strategy. Some pioneering work in this direction is reported in [2], [48], [43].

Finally, growing user demands require the *interconnection* of networks which may implement different flow control policies and which may even be built on different media (e.g., satellite, radio, cable, or optical fiber). These networks are interconnected by gateways which provide for internet routing and flow control, as well as for protocol conversion between two adjacent networks [44], [6]. It appears that a new level of flow control must therefore be defined in our hierarchy, namely, the gateway-to-gateway level. This level should be designed to prevent the congestion of gateways along the path, and should be supported by explicit gateway-to-gateway protocols for the exchange of status information. The status information should include buffer occupancy at the gateway, and load conditions in the adjacent networks, and could probably be exploited also for gateway routing. Functionally, the gateway-to-gateway protocol is positioned between the entry-to-exit protocol and the transport protocol hierarchy in Fig. 2. All the other levels remain unchanged. The actual implementation of the gateway-to-gateway flow control will be dependent on the internet protocol used. If the CCITT X.75 Recommendation, which is an extension of the X.25 virtual circuit concept to internet connections [45], is adopted, the gateway-to-gateway flow control will be virtual-circuit oriented, and will be exercised on a connection-by-connection basis. Alternatively, datagram-oriented gateway level flow control schemes can also be implemented [54].

The design of efficient gateway flow control schemes is very challenging. It requires *vertical* consistency between the gateway level and all the other levels implemented in each individual network as well as *horizontal* consistency across the various networks on the internet path. Specifically, the gateway level flow control must be able to balance loads between extremely diverse network environments such as point-to-point, satellite, cable, and ground radio. These design requirements further emphasize the need for continuing research in *multilevel* flow control models in order to understand the vertical interactions between the various levels in the hierarchy, as well as the horizontal interactions between the various segments of a flow control chain along an internet path.

In summary, we have presented a framework for the study of flow control, showing that flow control mechanisms have advanced somewhat beyond simply being "a bag of tricks" [34], and indeed can be conceptually organized into a useful and well-structured system of controls. This structure is extremely helpful in the survey and comparison of existing flow control implementations, as well as in the development of flow control models. In particular, complex control systems can be (and should be) decomposed into smaller modules, thus simplifying the analysis of each module as well as the analysis of interactions between different modules. Furthermore, the proposed flow control structure is sufficiently flexible to permit extensions in response to new networking technologies and applications.

Although our focus has been on flow control models and performance criteria, we expect that the proposed structure will prove to be useful also for the actual implementation of flow control techniques. One must be aware, of course, of the fact that in actual networks, it is not always possible to develop and update flow controls in a well structured fashion. The designer, in fact, is usually confronted with a number of constraints imposed by the preexisting protocol structure (in which flow control mechanisms must be embedded) and by limited storage and processing resources. The designer must therefore avoid overburdening the switch with overly sophisticated flow control mechanisms, and creating inconsistencies and possibly deadlocks. These constraints, together with the fact that flow control is a distributed multilevel control function that cannot be confined to a well-defined modular "black box," make flow control design a very hard task. It is our strong opinion, however, that the only way to prevent flow control implementations from degrading to the state of an uncontrollable "bag of tricks" is to identify an underlying structure in the early stage of flow control design, and to continuously verify this structure during the various updates of protocols and flow control procedures.

Indeed, it is important that one be able to subject a proposed flow control algorithm to various tests of correctness, consistency, and proper termination [33], [49]. This is, in general, a very difficult task whose solution requires advances in the frontier of computer science. Unfortunately, since it is relatively difficult to create efficient, deadlock-free, flow control algorithms, we cannot totally ignore this need for verification. Moreover, many difficulties with flow control procedures often arise due to errors in the detailed implementation of otherwise correct algorithms. Consequently, it is important that a modular approach to flow control design be taken, that the code itself be confined to isolated portions of the network operating system (rather than sprinkled through thousands of lines of code), and that the mechanisms be simple enough to be understood and tested via simple procedures.

References

[1] V. Ahuja, "Routing and flow control in systems network architecture," *IBM Syst. J.*, vol. 18, no. 2, pp. 298–314, 1979.

[2] G. Akavia and L. Kleinrock, "Performance tradeoffs in distributed packet-switching communication networks," Dep. Comput. Sci., School of Eng. Appl. Sci., Univ. of California, Los Angeles, Tech. Rep. UCLA-ENG-7942, Sept. 1979.

[3] P. Brinch-Hansen, *Operating System Principles*. Englewood Cliffs, NJ: Prentice-Hall, 1973.

[4] S. Carr *et al.*, "Host/host protocol in the ARPA network," in *Proc. Spring Joint Comput. Conf.*, 1970, pp. 589–597.

[5] V. G. Cerf and R. Kahn, "A protocol for packet network intercommunication," *IEEE Trans. Commun.*, vol. COM-22, May 1974.

[6] V. G. Cerf, "DARPA activities in packet network interconnection," in *Interlinking of Computer Networks* (NATO Advanced Study Inst. Series). Dordrecht: Reidel.

[7] A. Danet *et al.*, "The French public packet switching service: The Transpac network," in *Proc. Int. Conf. Comput. Commun.*, Toronto, Ont., Canada, Aug. 1976.

[8] D. W. Davies, "The control of congestion in packet-switching networks," *IEEE Trans. Commun.*, vol. COM-20, June 1972.

[9] H. C. Folts, "International standards in computer communications," in *Proc. Nat. Telecommun. Conf.*, Nov. 1979, 59.5.1–59.5.5.

[10] J. Forgie and A. Nemeth, "An efficient packetized voice/data network using statistical flow control," in *Proc. Int. Conf. Commun.* Chicago, IL, June 1977.

[11] M. Gerla and D. DeStasio, "Integration of packet and circuit transport protocols in the TRAN data network," in *Proc. Comput. Network Symp.*, Liege, Belgium, Feb. 1978.

[12a] M. Gerla, "Routing and flow control in virtual circuit computer networks," in *Proc. INFO II Int. Conf.*, July 1979.

[12b] M. Gerla and P. O. Nielson, "Routing and flow control interplay in computer networks," *ICCC Proc.*, Atlanta, November 1980.

[12c] M. Gerla, "Bandwidth control in X.25 networks, *PTC Proc.*, Hawaii, Jan. 1981.

[13] A. Giessler *et al.*, "Free buffer allocation—An investigation by simulation," *Comput. Networks*, vol. 2, pp. 191–208, 1978.

[14]   I. Gitman and H. Frank, "Economic analysis of integrated voice and data networks," *Proc. IEEE*, pp. 1549–1570, Nov. 1978.

[15]   J. P. Gray, "Network services in systems network architecture," *IEEE Trans. Commun.*, vol. COM-25, pp. 104–116, Jan. 1977.

[16]   J. P. Gray and T. B. McNeill, "SNA multiple-system networking," *IBM Syst. J.*, vol. 18, no. 2, 1979.

[17]   P. E. Green, "The structure of computer networks," this book, Chap. 1; also in *IBM Syst. J.*, no. 2, 1979.

[18]   M. Irland, "Buffer management in a packet switch," *IEEE Trans. Commun.*, vol. COM-26, pp. 328–337, Mar. 1978.

[19]   R. E. Kahn and W. R. Crowther, "A study of the ARPA computer network design and performance," Bolt Beranek and Newman, Inc., Tech. Rep. 2161, Aug. 1971.

[20]   F. Kamoun, "Design considerations for large computer communications networks," Ph.D. dissertation, Univ. of California, Los Angeles, Eng. Rep. 7642, Apr. 1976.

[21]   F. Kamoun, "A drop and throttle flow control (DTFC) policy for computer networks," Proceedings of the 9th Int. Teletraffic Congr., Spain, Oct. 1979.

[22]   P. Kermani and L. Kleinrock, "Dynamic flow control in store and forward computer networks," *IEEE Trans. Commun.*, vol. COM-27, Feb. 1979.

[23]   L. Kleinrock, *Queueing Systems: Volume II. Computer Applications*. New York: Wiley-Interscience, 1976.

[24]   L. Kleinrock, "On flow control in computer networks," in *Proc. Int. Conf. Commun.*, June 1978.

[25]   L. Kleinrock, "Power and deterministic rules of thumb for probabilistic problems in computer communications," in *Proc. Int. Conf. Commun.*, June 1979.

[26]   L. Kleinrock and P. Kermani, "Static flow control in store and forward computer networks," *IEEE Trans. Commun.*, vol. COM-27, Feb. 1979.

[27]   S. Lam and M. Reiser, "Congestion control of store and forward networks by buffer input limits," in *Proc. Nat. Telecommun. Conf.*, Los Angeles, CA, Dec. 1977.

[28]   R. Magoon and D. Twyver, "Flow and congestion control in SL-10 networks," in *Proc. Int. Symp. Flow Control Comput. Networks*. Versailles, France, Feb. 1979.

[29]   J. C. Majithia *et al.*, "Experiments in congestion control techniques," in *Proc. Int. Symp. Flow Control Comput. Networks*, Versailles, France, Feb. 1979.

[30]   J. M. McQuillan *et al.*, "Improvements in the design and performance of the ARPA network," in *Proc. Fall Joint Comput. Conf.*, 1972.

[31]   W. E. Naylor, "Stream traffic communication in packet-switched networks," Ph.D. dissertation, Dep. Comput. Sci., School Eng. Appl. Sci., Univ. of California, Los Angeles, Sept. 1977.

[32]   M. Pennotti and M. Schwartz, "Congestion control in store and forward tandem links," *IEEE Trans. Commun.*, Dec. 1975.

[33]   J. Postel, "A graph model analysis of computer communications protocols," Ph.D. dissertation, Univ. of California, Los Angeles, Jan. 1974.

[34]   L. Pouzin, "Flow control in data networks—Methods and tools," in *Proc. Int. Conf. Comput. Commun.*, Toronto, Ont. Canada, Aug. 1976.

[35]   W. L. Price, "Data network simulation experiments as the National Physical Laboratory," *Comput. Networks*, vol. I, 1977.

[36]   W. L. Price, "A review of the flow control aspects of the network simulation studies at the National Physical Laboratory," in *Proc. Int. Symp. Flow Control in Comput. Networks*, Versailles, France, Feb. 1979.

[37]   E. Raubold and J. Haenle, "A method of deadlock-free resource allocation and flow control in packet networks," in *Proc. Int. Conf. Comput. Commun.*, Toronto Ont., Canada, Aug. 1976.

[38] M. Reiser, "A queueing network analysis of computer communication networks with window flow control," *IEEE Trans. Commun.*, pp. 1199–1209, Aug. 1979.

[39] J. Rinde, "Routing and control in a centrally directed network," in *Proc. Nat. Comput. Conf.*, Dallas, TX, June 1977.

[40] J. Rinde and A. Caisse, "Passive flow control techniques for distributed networks," in *Proc. Int. Symp. Comput Networks*, Versailles, France, Feb. 1979.

[41] M. Schwartz and S. Saad, "Analysis of congestion control techniques in computer communication networks," in *Proc. Int. Symp. Comput. Networks.* Versailles, France, Feb. 1979.

[42] J. M. Simon and A. Danet, "Controle des ressources et principes de routage dans le reseau TRANSPAC," in *Proc. Int. Symp. Comput. Networks.* Versailles, France, Feb. 1979.

[43] J. Silvester, "On spatial capacity of packet radio networks," Ph.D. dissertation, Dep. Comput. Sci., School Eng. Appl. Sci., Univ. of California, Los Angeles, Mar. 1980.

[44] C. A. Sunshine, "Interconnection of computer networks," *Comput. Networks*, vol. 1, 1977.

[45] C. A. Sunshine, "Transport protocols for computer networks," in *Protocols and Techniques for Data Communications Networks*, F. Kuo, Ed. Englewood Cliffs, NJ: Prentice-Hall, 1980.

[46] F. Tobagi, "Multiaccess link control," this book, Chap. 6.

[47] J. W. Wong and M. S. Unsoy, "Analysis of flow control in switched data networks,' in *Proc. Int. Fed. Inf. Processing Soc. Conf.*, Aug. 1977.

[48] Y. Yemini and L. Kleinrock, "On a general rule for access control or, silence is golden...," in *Proc. Int. Symp. Flow Control Comput. Networks*, Versailles, France, Feb. 1979.

[49] P. Zafiropulo, "A new approach to protocol validation," in *Proc. Int. Conf. Commun.*, June 1977.

[50] H. Zimmermann, "The Cyclades end-to-end protocol," in *Proc. 4th Data Commun. Symp.*, Quebec, P. Q., Canada, Oct. 1975, pp. 7 : 21–26.

[51] Schwartz and Stern, "Routing protocols," this book, Chap. 9.

[52] H. Rudin, "Congestion control: Preview and some comments," *IEEE Trans. Commun.*, vol. COM-29, no. 4, pp. 373–375, April 1981.

[53] D. G. Haenschke, D. A. Kettler, and E. Oberer, "Network management and congestion in the U.S. telecommunications network," *IEEE Trans. Commun.*, vol. COM-29, no. 4, pp. 376–385, April 1981.

[54] D. E. Sproule and F. Mellor, "Routing, flow, and congestion control in the Datapac network," *IEEE Trans. Commun.*, vol. COM-29, no. 4, pp. 386–391, April 1981.

[55] L. Tymes, "Routing and flow control in TYMNET," *IEEE Trans. Commun.*, vol. COM-29, no. 4, pp. 392–398, April 1981.

[56] C. Lemieux, "Theory of flow control in shared networks and its application in the Canadian telephone network," *IEEE Trans. Commun.*, vol. COM-29, no. 4, pp. 399–412, April 1981.

[57] L. Puzin, "Methods, tools, and observations on flow control in packet-switched data networks," *IEEE Trans. Commun.*, vol. COM-29, no. 4, pp. 413–426, April 1981.

[58] K. Bharath-Kumar and J. M. Jaffe, "A new approach to performance-oriented flow control," *IEEE Trans. Commun.*, vol. COM-29, no. 4, pp. 427–435, April 1981.

[59] A. Giessler, A. Jägemann, E. Mäser, and J. O. Hänle, "Flow control based on buffer classes," *IEEE Trans. Commun.*, vol. COM-29, no. 4, pp. 436–443, April 1981.

[60] F. Kamoun, "A drop and throttle flow control policy for computer networks," *IEEE Trans. Commun.*, vol. COM-29, no. 4, pp. 444–452, April 1981.

[61] L. Kaufman, B. Gopinath, and E. F. Wunderlich, "Analysis of packet network congestion control using sparse matrix algorithms," *IEEE Trans. Commun.*, vol. COM-29, no. 4, pp. 453–465, April 1981.

[62] J. Matsumoto and H. Mori, "Flow control in packet-switched networks by gradual restrictions of virtual calls," *IEEE Trans. Commun.*, vol. COM-29, no. 4, pp. 466–473, April 1981.

[63] K. D. Günther, "Prevention of deadlocks in packet-switched data transport systems," *IEEE Trans. Commun.*, vol. COM-29, no. 4, pp. 512–524, April 1981.

[64] D. F. Weir, J. B. Holmblad, and A. C. Rothberg, "An X.75 based network architecture," ICCC 80 Proceedings, Atlanta, GA, pp. 741–750.

[65] M. Schwartz, *Telecommunication Networks: Protocols, Modeling and Analysis*. Reading, MA: Addison Wesley, 1986.

[66] M. Gerla, "Routing and flow control in ISDN's," ICCC 86 Proceedings, Munich, pp. 643–647.

[67] J. Swiderski, "Examining packet-switching networks with local congestion control for exposure to store-and-forward deadlocks," ICCC 86 Proceedings, Munich, pp. 670–674.

[68] M. Gerla et al., "Fairness in computer networks," ICCC 85 Proceedings, Chicago, pp. 43.5.1–43.5.6.

[69] M. Gerla et al., "Routing, flow control, and fairness in computer networks," ICCC 84 Proceedings, Amstermdam, pp. 1272–1276.

[70] M. Gerla and H. W. Chan, "Window selection in flow controlled networks," 9th Data Communications Symposium Proceedings, Vancouver, Canada, Sept. 85, pp. 84–92.

[71] J. M. Selga, "Flow-control method for packet networks," ICCC 86 Proceedings, Munich, pp. 625–630.

[72] D. Bertsekas and R. Gallager, *Data Networks*. Englewood Cliffs, NJ: Prentice-Hall, 1987.

[73] J. S. Turner, "Design of an integrated services packet network," *IEEE JSAC*, vol. SAC-4, no. 8, pp. 1373–1380.

[74] M. Gerla and L. Fratta, "Design and control in processor limited packet networks," International Teletraffic Conference Proceedings, Torino, June 1–8, 1988, pp. 5.2B.5.1–5.2B.5.7.

Network Interconnection and Gateways

Carl A. Sunshine

I. Introduction

As computer networks proliferate, the importance of interconnecting networks increases. The recent explosion in the numbers of personal computers is leading to even greater growth in the local area network (LAN) area. Interconnecting these diverse networks presents many technical problems, and may be pursued in many ways [9], [16], [31], [32], [37].

The term "network interconnection" has been used broadly to mean any technique that enables systems on one network to communicate with or make use of services of systems on another network. In this chapter we explore this full range of meanings, but we focus on the problem of providing general purpose end-to-end communication at the network level in the protocol hierarchy, rather than the additional problems of integrating services at higher protocol levels.

Networks differ in geographic scope, type of using organization, types of services to be provided, and transmission technology. This leads to a variety of specific communication protocols and interfaces being used, at least at the lower levels, in different nets. There are good technical and marketing reasons for these different solutions, so diversity in network technologies is likely to persist. This suggests that for a network interconnection strategy to succeed, it must accommodate the autonomy and

CARL A. SUNSHINE • Unisys, West Coast Research Center, Santa Monica, California 90406.

differences of individual networks to the greatest extent possible. We shall see to what extent this can be accomplished in what follows.

We first consider the major technical problems of network interconnection, including stepwise versus end-point services, level of interconnection, addressing, routing, fragmentation, and congestion control, ending with a summary of functions performed by the "gateway" between networks. We next present several major current examples of internet systems, including the U.S. Department of Defense, Xerox Corporation, and public data networks. We finish with a review of standardization activity in ISO and likely future developments.

II. Major Technical Issues

The technical problems of network interconnection have much in common with problems of designing an individual network. Indeed, a common viewpoint sees the individual networks as links, and the *gateways* as switching nodes interconnecting these "links" to form a "supernetwork." This leads to consideration of issues common to any switching system, such as addressing, routing, congestion control, fragmentation, and multiplexing [16]. The following sections focus on the extra concerns that are important at the internet level in each of these areas, and on the extra issue of how to combine individual network services to provide end-to-end service.

A. Stepwise versus End-Point Services

A major question in designing network interconnection is whether services will be provided in a *stepwise* or *end-point* fashion. For simplicity, consider connecting two networks. One approach is to take the existing service (say virtual circuits) in each system, and concatenate them, hoping that they are close enough to each other to provide an essentially equivalent end-to-end service. The alternative is to use a more basic service in each network (e.g., "datagrams"), and to provide the bulk of the virtual circuit service in the two end points, one on each network.

The stepwise approach has the virtue of providing service via existing mechanisms, without requiring any new implementation. However, it can only provide services end-to-end that are common to all subsystems. There may also be some variation in the "flavor" of service available, requiring a translation at intermediate points [17], [48], [50]. Indeed, it has been argued that functionality mismatches are nearly inevitable in any attempt to create such *translating gateways*, such that their resulting service limitations and/or greater complexity make them undesirable [42]. Services not available on one side can also be made up by addition of a "convergence" sublayer protocol [48].

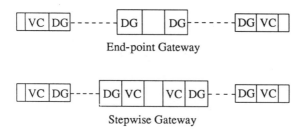

End-point Gateway

Stepwise Gateway

Fig. 1. Stepwise gateway is more complex than end-point.

The end-point approach guarantees a full service with common attributes at both ends by requiring implementation of a common protocol in the two end-point nodes. It makes use of simpler services on the individual networks along the way, and hence allows use of simpler gateways. There are fewer failure points since most errors along the path can be corrected by the end-point mechanisms.

One common example of this tradeoff occurs at the transport level where a choice must be made on how to provide reliable end-to-end "virtual circuit" (VC) service (Fig. 1) [47]. The end-point approach requires a common reliable transport protocol implemented in the end nodes, but makes use of simpler network service that need not be reliable or even connection oriented. The stepwise approach makes use of existing reliable VC service in each network, concatenating them to form an end-to-end VC. In this latter case, the gateways must function at the VC level, and any failure in the gateways would affect the end-to-end service.

B. Level of Interconnection

Another major question is at what level in the protocol hierarchy to interconnect networks. Alternatives exist all the way from the lowest (physical) level to the highest (application) level. In general, the lower the level of interconnection, the more similar the networks to be connected must be, while high-level interconnections support more specialized services.

When different networks and protocols are involved, the interconnection involves a conversion process between the services provided for comparable functions in each network [17]. The complexity of this process and the quality of end-to-end services resulting are largely determined by the level of interconnection chosen. The following sections summarize the key features of each major alternative.

1. Physical Level

Interconnection devices operating at the physical level are generally called *repeaters*. They forward individual bits of the packet as they arrive, perhaps translating from one medium to another (e.g., baseband coaxial cable to optical fiber). The resulting interconnected system functions essentially as a single network at the data link level, and hence all subnets to be so connected must have identical data rates and link protocols. This approach is typically used to interconnect several physically separate segments of a LAN system, perhaps separated by a point-to-point link.

2. Link Level

Interconnection devices operating at the link level receive entire frames from one link, examine the link level protocol header, and possibly forward the frame onto another link [2], [3]. They are typically called *bridges*, or more specifically *MAC bridges* (since they operate at the media access control level). As with repeaters, they may interconnect two or more local LAN segments, or may interconnect remote segments over a long distance link. The major motivation for their use is to interconnect LAN segments with different MAC protocols, or to increase network capacity by "filtering" incoming packets and forwarding only those whose link level destination is on another segment. MAC bridges transparently support systems with multiple network level protocols in use.

3. Network Level

Traditionally, interconnection at the network protocol level has been a Wide Area Network (WAN) problem, where different networks had independently developed different protocol mechanisms for a variety of network level functions such as routing, congestion control, error handling, and segmenting. If the networks are identical, then the problem becomes largely one of routing as with the X.25/X.75 approach in public data networks. When the networks differ, the complexity of protocols at the network level (e.g., X.25 versus ARPANET 1822) make a translation approach difficult. There has been some success in one vendor emulating another vendor's network behavior (e.g., "SNA gateways").

An alternative gaining wide acceptance places a common *internet protocol* (IP) sublayer on top of the different network protocols. As we shall see below, this has particular benefits for supporting the sophisticated routing procedures needed for large internet systems, and devices operating at this level are often called IP gateways. Choosing this level makes available the general purpose services of the network level, and allows the gateway implementor to take advantage of what is normally a well documented interface with many implementations. It allows each network to function autonomously with its own procedures internally, while requiring some standard "internet" procedures to be used on top of the normal network access for individual networks.

4. Transport Level

In the OSI architecture, the transport service is supposed to be an end-to-end service, so transport level gateways are strictly speaking a violation of the architecture. Nevertheless, they may be of practical benefit when common upper-level protocols are in use, but different transport protocols are available. References [18] and [49] describe how to convert TCP and ISO TP4 to maximize end-to-end features. Onions and Rose [40] discuss a transport level gateway providing ISO TP0 service on top of X.25 connections on one side and TCP on the other side. Another situation where a pragmatic solution may be needed is to support interoperation between ISO protocol users employing TP Class 2 and TP Class 4 (see Section IV D below) [25], [49].

5. Higher Levels

No gateways functioning at session or presentation level have been developed to date, but many application level gateways have been implemented to support specific services found at the application level. This type of gateway is essentially a "Janus host" [42] that implements two (or more) full protocol suites. Common examples are interconnecting terminal concentrators or PADs to provide an interactive terminal service [5], or electronic mail servers to form a mail-forwarding service [10]. Where only a specific application service is wanted and the desired application services on each net match closely, this type of gateway may be easy to set up with existing equipment. However, the service provided is clearly not general purpose, and the limitations imposed by providing only those service elements common to the interconnected systems are often more irksome than anticipated [42].

C. Naming, Addressing, and Routing

To understand the problem of delivering data to the correct destination in an internet, a clear distinction must be drawn between names, addresses, and routes [34]. Although these concepts are applicable at each protocol level, we shall be primarily concerned with the network level, where "hosts" or "end systems" and gateways are the relevant objects. A *name* serves to identify the host "logically," independently of its point(s) of attachment to the network(s). The same host may have several names to provide for convenient "nicknames" or aliases. An *address* identifies a point of attachment for purposes of delivering data to the host; since the same host may have multiple network interfaces, it may have multiple addresses. Finally, a *route* is the path taken from source to destination end system, and there are typically multiple routes available to the same destination.

The process of sending data to a destination generally involves first determining its address from its name using a directory service, and then determining the best route to that address. In large systems, this name lookup function is typically implemented in a distributed fashion, with a hierarchical name space where subdirectories are responsible for their portion of the name space [28], [44].

1. *Addressing*

A method must be found for uniquely identifying all network interfaces in an internet system. One straightforward method is to employ relatively large addresses from the start, making sure that each node receives a unique address (e.g., by incorporating the serial number as part of the address). This is the approach taken in the Xerox Network System using 48-bit addresses [13]. With an even larger address field, values can simply be assigned randomly, guaranteeing a sufficiently low probability of duplication (e.g., one billion nodes with 64-bit addresses implies one in ten billion chance of duplication).

If existing networks with possibly overlapping addresses in use must be combined, several alternatives are possible. One is to merge one network into another, moving all its addresses to an unused range of addresses in the other network (Fig. 2a). Of course, this is a major inconvenience for the moved addresses.

Another strategy is to map unused addresses in each net to desired addresses in other nets (Fig. 2b), a technique called "proxy mapping" in [39], which was chosen by IBM to interconnect SNA networks [41]. It has the advantage of avoiding any changes to existing network addresses, but it requires sufficient unused addresses in each net for all external destinations, and leads to different addresses for the same node depending on what net it is referenced from. Hence its major appeal is in situations where a limited number of networks and destinations is involved.

The most general strategy is to introduce a hierarchical address format where an explicit "network" prefix is added to existing "local" suffixes to form a complete address (Fig. 2c). This network prefix may have routing significance and correspond to a "physical" network, or it may have only administrative significance and correspond to an organizational domain.

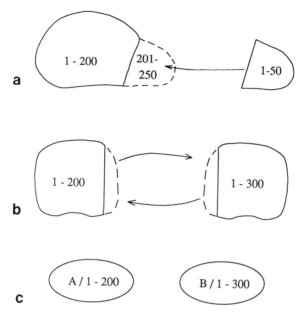

Fig. 2. Addressing alternatives: (a) moving; (b) mapping; (c) hierarchical addresses.

This is the approach used in the Federal Research Internet (see Section IV A below).

2. Routing

Once an addressing strategy is chosen, there is still the problem of routing, or how to best reach each possible destination given its address. In the case that unused local addresses are mapped into external destinations, all packets to those destinations must be routed to a suitable gateway, which then does the address mapping as part of the forwarding process (Fig. 3).

With a flat internet address space (or one with only administrative significance to its hierarchy), gateways must maintain path information for all destinations individually, and hence have large routing tables. This approach is used in MAC bridges, which must maintain a list of all addresses "local" to each interface so they can filter out (not forward) received packets destined for those local addresses. For nonlocal packets to be forwarded, they must also implement a routing procedure that prevents duplication or flooding if there are multiple paths through the interconnected networks, and yet efficiently finds all destinations.

Considerable ingenuity has been devoted to this problem, particularly in the LAN context. One approach being adopted by the IEEE 802.1

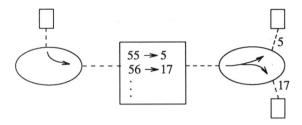

Fig. 3. Gateway intercepts and maps external addresses.

committee is based on formation of a minimal spanning tree through the internet [2]. An alternative favored within the token ring LAN community uses source routing [15]. A third approach discovers and establishes an optimal route the first time each destination is contacted, and then sends subsequent packets over that route without requiring either source routing or routing calculations in the bridges [36]. All of these approaches depend on the broadcast capabilities of LANs to send route finding packets efficiently, and hence the effective size of internets based on MAC level bridges is limited.

If a hierarchical address is used, as in an internet protocol, then routing can be done in steps, first to the final network (ignoring the local suffix), and then within the final network to the local address. This reduces the size of routing tables at the cost of some loss in optimality. There may also be rare cases when a network becomes "partitioned" (divided into two or more portions that cannot communicate with each other internally), such that destinations erroneously appear unreachable in a strictly hierarchical routing procedure [35] (see Fig. 4).

Hybrid approaches have been used in other communication areas (e.g., telephone switching) to obtain the benefits of both hierarchical and flat routing. The majority of destinations are handled with a hierarchical procedure to reduce the size of routing tables, while high traffic or error

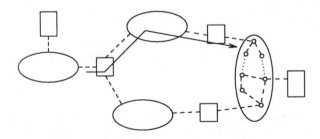

Fig. 4. Host unreachable due to partitioned net.

recovery conditions can trigger the insertion of additional routes to specific individual destinations.

In any approach based on the use of routing tables by gateways, the tables must be built and updated to reflect changing traffic conditions. Accomplishing this task in large networks or internets with high reliability, efficiency, and timeliness has been a very challenging problem that has led to the development of many routing information exchange algorithms. Some examples such as the External Gateway Protocol (EGP) and the ISO End System to Intermediate System protocol are presented below. Further discussion of this topic may be found in Chapter 9.

Another design choice concerns the frequency of routing decisions. For maximum robustness, each packet may cause a best route selection process to be carried out, as in the DARPA IP [29]. Other systems choose to perform the best route determination process only for the initial packet to a destination. This route is then remembered in the routers (or the source in [15]), and subsequent packets to the same destination follow the same route. Often this type of path setup is accompanied by an abbreviated addressing convention where only the first packet must carry full destination address, and subsequent packets carry only a shorter path identifier. Examples of this approach using hierarchical addressing include the Universe project [1] and CCITT X.75 (see Section IV C below), while examples using flat addressing include [2], [36] mentioned above. Some mechanisms for timing out such routes or recovering from breaks in the path must also be provided.

Yet another approach employs flooding to avoid the need for intelligence in packet forwarders. Since flooding is expensive of network resources, this is typically used only for control purposes, or for initially establishing a path that later packets to the same destination will follow [36]. Another way to avoid the need for intelligent routers is for the source to provide explicit path information in the packets it sends [15], [38].

A final complexity occurs when a node is *multihomed* (has multiple connections to the internet) and hence has multiple addresses. Normally the routing system is designed to find an optimal route to a single destination address carried in the packet. To take full advantage of multihoming, either the sources must select an optimal address in advance, or the routing system must be enhanced to deal with multiple addresses in the packet (select the best of best routes) or even to accept a name for the destination, and perform the name to (multiple) address lookup function in the gateways along a path to the destination [35], [39].

D. Congestion Control

The problems of congestion control in an internet system are much like those of individual networks. Speed mismatches are likely to be more severe between LANs and slower wide area networks (although recent advances in high-speed WAN service should reduce this). In some cases, the individual network procedures may be adequate (e.g., X.25 PDNs, where buffering resources are typically reserved at VC establishment). In others, some form of explicit internet level control may be needed.

Questions have been raised about the ability of connectionless systems to provide effective congestion control. This is a particular concern when connectionless or "datagram" internet service is used to support higher-level connection oriented services. Several techniques have been proposed in this area, including input buffer limits, buffer classes, fair queuing [45], slow start [46], and choke packets [29]. Once the sender has determined that congestion has occurred (by receiving an explicit signal from a host or gateway, or by timing out waiting for an acknowledgment), it must reduce its transmission rate for a while, and then try to increase it again. Various specific algorithms for this purpose have been proposed, and this is an active area of research [46]. Further information may be found in Chapter 10.

E. Fragmentation and Reassembly

When networks with differing maximum packet size limits are interconnected, the need to fragment large packets for traversal through networks with smaller size limits must be considered. These fragments can be reassembled at the exit from the individual small packet network, or allowed to propagate all the way to the final destination.

Mechanisms to support such fragmentation typically include some sort of additional sequencing information. The most general mechanisms allow further fragmentation of already created fragments, and proper reassembly of fragments from different (re)transmissions of the same data that may have overlapping boundaries (e.g., DARPA IP [29]).

Although fragmentation was intended to add robustness to the internet system, recent results suggest that it can also cause problems [43]. In congested systems, fragments may be repeatedly lost, drastically reducing performance. Fragmentation can also cause many more total packets to be used than if the source host limited itself to sending the maximum packet size allowed on the path. Kent and Mogul [43] suggest several ways to reduce or avoid the need for fragmentation and these negative consequences.

III. IP Gateway Functions

The primary function of an internet gateway is to implement the network access protocols of the two (or more) networks it interconnects.

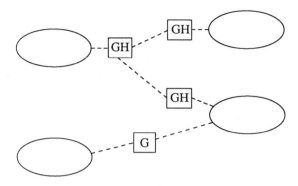

Fig. 5. Full gateways (G) and gateway halves (GH).

This allows the gateway to receive packets on one network, reformat them, and forward them into the appropriate next network.

To determine the proper direction to forward the packet, the gateway must perform a routing function based on addressing information contained in the packet and in internal routing tables. These routing tables must be updated according to the routing procedures used in the internet system.

If a stepwise approach to providing services is being used, the gateway must terminate the protocol in one net and translate the functions required into appropriate form on the next network. Congestion control and fragmentation may also be needed in the gateway as discussed above. If performance monitoring or accounting functions are required, these must also be implemented in the gateway.

All of these functions may be performed in a dedicated machine. Alternatively, they may be split into separate gateway "halves" associated with each network, connected to each other by a communications link (Fig. 5). This allows the gateway functions to be merged with other network functions in an existing switching node owned and operated by each network authority. However, it adds the extra complexity of defining a standard procedure for use across the link between gateway halves. The CCITT X.75 Recommendation (see below) is an example of this approach.

IV. Current Practice

A. Federal Research Internet System

One of the first major internet systems was developed by the Defense Advanced Research Projects Agency (DARPA) in the USA [9], [19], [30],

[31]. This system included the original ARPANET, packet radio nets, satellite networks, and various LANs. The system is now split into separate systems for research users and for operational military users, and has been enlarged by inclusion of IP networks run by other U.S. government agencies, most notably the National Science Foundation's NSFNET. Hence it is now coming to be called the Federal Research Internet (FRI).

FRI networks are interconnected by gateways that implement a connectionless or datagram Internet Protocol (IP) [14], [29] to provide maximum robustness and routing flexibility. Dedicated gateway machines of the 16-bit minicomputer class were originally employed, but new 32-bit microprocessor based products are now coming into use. Most of the individual networks provide connectionless service, although there is a provision for running the IP over connection oriented network services such as X.25 [11].

The major transport service is connection oriented, implemented by a common protocol called TCP that must be present at the end points (not in gateways). The IP also supports other types of transport protocols including datagram and "stream" mode (for packetized voice).

Addressing in the FRI is hierarchical, with gateways designed to route to the network portion of the address first, and then the local portion once the correct net is reached. The address length is fixed at 32 bits, and as the internet has expanded, this fixed hierarchical structure has caused significant difficulties and led to the invention of a "subnetting" approach for large campuses. Host name to address lookup was initially supported by a single flat directory, but as the number of hosts grew, a hierarchical distributed directory service was adopted [44].

The IP provides for fragmentation at gateways with reassembly at the final destination so that individual fragments may follow different routes. Only the header is checksummed to allow delivery of packets with some data errors (higher-level protocols may use their own checksums). A Time-to-Live field is also included to limit the maximum lifetime of packets in the system. Options are defined to allow inclusion of, for example, source routes, security markings, and timestamps.

There is a separate Internet Control Message Protocol (ICMP) used for signaling errors and diagnostic information. This includes destination unreachable, congestion control (choke packets), and redirect indications (giving a better route for a specific destination).

Internet routing information exchange was originally handled by a Gateway-to-Gateway protocol [19] that required interaction between all "neighboring" gateways. As the internet grew, a hierarchical scheme called the Exterior Gateway Protocol (EGP) [19] was developed to reduce the amount of routing traffic. In EGP, each "autonomous system" (typically a campus internet) elects one gateway to exchange routing data with a "neighbor" gateway in an adjacent autonomous system, and the systems then propagate the information to all their other gateways through an internal procedure. As currently defined, EGP requires all autonomous systems to use a single "core system" for interconnection with each other, rather than supporting a general mesh topology. Further extensions are under study.

B. Xerox

The original experimental Xerox internet system was developed about the same time as the DARPA system and shared many of its features, including hierarchical addressing and a connectionless internet layer [4], [9]. In addition to dedicated gateways, gateway halves were used to interconnect physically remote LAN segments via communication links. Fragmentation and reassembly were done on an individual network basis if needed. Because of the predominance of high-speed, low-error-rate LANs in the system, higher-level protocols were somewhat simpler than in the DARPA system.

Sophisticated directory servers were included to convert names to addresses [28]. This directory service made use of the broadcast capabilities of the individual LANs in the system to efficiently support distribution of its functions between user nodes, local servers, and centralized servers.

Several changes were then made to produce the current Xerox Network System (XNS) [12]. The major change was adoption of a large (48-bit) flat address to uniquely label each node ever manufactured, while an additional hierarchical network address was added to the internet packet format for routing purposes [13]. Further details may be found in Chapter 15.

C. Public Data Networks

In the late 1970s, the major public data networks developed the X.25 interface to provide connection oriented service. By modifying this procedure slightly to make it more symmetric and adding some new utility functions, they were able to create the X.75 interface for use on the links between PDNs [7], [9], [31], [33].

To provide internetwork service to X.25 users, Recommendation X.121 defined the address fields in the X.25 Call Request packet to include a hierarchical network portion of three digits followed by a local portion of up to ten digits [8]. Each PDN was responsible for interpreting these addresses and routing calls to an appropriate internetwork link according to its own internal procedures. In the 1984 revision of X.25, address extension facility fields were added to the Call Request packet, providing a limited form of source routing. The PDN internet routes the call to a "private gateway" based on the X.121 address, and then the gateway can forward the packet based on the address extension field.

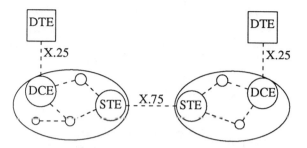

Fig. 6. Concatenated virtual circuits in PDNs.

In PDNs, the gateway half approach is used, with gateway functions typically performed by additional software in the existing network switching nodes. Virtual circuit network level service is provided by a stepwise concatenation of X.25-like links in each network (Fig. 6). This allows an easy (if conservative) method of resource allocation, and straightforward accounting and charging procedures. However, there is uncertainty about the end-to-end significance of some VC functions such as flow control and reset. End user service would be affected by the failure of any intermediate nodes performing VC functions.

Asynchronous terminal support is provided on top of these VCs in an end-point fashion between the packet mode subscriber on one end and the PAD on the other (Recommendation X.29).

Unlike the previous two examples, PDNs employ a connection oriented routing scheme whereby only the initial Call Request packet carries the full destination address and requires an optimal route decision in routing nodes. This first packet sets up a path entry in the routing nodes which is referenced by a short path number (see Chapter 9) in subsequent data packets, which are all sent over the same route. This reduces protocol overhead and processing time for data packets, at the expense of some robustness and optimality.

D. ISO

The International Standards Organization (ISO) has recently extended its OSI architecture to define three sublayers within the network layer [21]. The topmost layer corresponds to the internet protocol, and the middle layer is intended to adapt ("converge") specific network services to those required by the internet sublayer. One example would be use of a connectionless internet protocol over a connection oriented network, requiring a "connection management" intermediate layer protocol to set up and terminate connections as needed in order to send internet level datagrams [11], [24].

Various other combinations of link, network, and transport layer protocol types can be envisioned. The situation is complicated by the fact that most LANs provide connectionless service, while most PDNs provide connection oriented service. Interconnecting end users by means of both sorts of networks will require some "convergence" features, and the best strategy remains an open question.

One approach popular in Europe is to require all networks including LANs to provide a connection oriented service (e.g., see ISO standard 8881), and to use the TP Class 2 end-to-end. However, this may result in poor performance and inefficient use of the LANs [26]. The alternative of assuming only connectionless service on all networks requires all end systems to implement and employ TP Class 4. This is popular in North America and is mandated by the U.S. Government OSI Profile (GOSIP) [27], but may be inefficient over connection oriented networks. End systems taking the first approach will be unable to communicate with other systems taking the second approach unless some sort of transport level gateway as described in [25] is provided.

ISO has defined an internet sublayer protocol [6], [20] much like the DARPA IP [24]. Although the format of the ISO IP packet header is different, most fields have a one-to-one correspondence with the DARPA IP. However, the ISO IP does not include a field for "upper-layer protocol" (e.g., TCP or UDP) since this is viewed as part of the address information (see below). The ISO IP includes an error-reporting capability while the DARPA IP provides this through the separate ICMP protocol. The fragmentation (segmentation) fields are different, with the ISO IP including a field giving the total length of the original segment in each fragment to aid in assigning reassembly buffers.

The final major difference concerns the format of addresses at the network level, which is not part of the ISO IP itself, but covered in a separate document [22]. The ISO format is a variable length string that is intended to cover the requirements of both public and private, local and wide area networks for the foreseeable future. This involves a maximum of 16 octets of binary data, which could be alternately coded as 40 BCD digits. The first octet is an Authority/Format code meant to indicate what format the following data are in. Provision has been made to identify all the major current address formats as alternatives [X.121, F.69 (telex), E.163 (telephone), E.164 (ISDN), ISO 6523]. The address is assumed to be hierarchical, with each "domain" responsible for defining the meaning of the suffix portion of the address under its control (Fig. 7).

The initial routing information exchange protocol defined by ISO is intended for *end systems* (hosts and workstations) to interact with *intermediate systems* (gateways and bridges). This ES-IS protocol [23] provides functions for end systems to announce and learn correspondence of network (IP) level addresses to MAC level addresses for nodes on the same

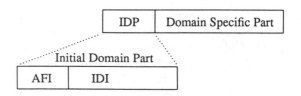

Fig. 7. ISO network service access point address. AFI, authority and format identifier: format and length of IDI; authority responsible for allocating IDI values; abstract syntax of the DSP. IDI, initial domain identifier: NSAP address subdomain; authority responsible for allocating DSP values.

network, and to learn the addresses of any gateways on their network that can serve as a path to other networks. The more difficult problem of IS-IS protocol is still under study.

V. Conclusions

The variety of individual network technologies is likely to continue increasing. Fortunately, by introducing standards at the internetwork level, it is possible to interconnect diverse networks while preserving their individual autonomy to a large degree. Truly universal addressing and routing schemes are now emerging from ISO work. The next few years should prove very interesting in determining the combinations of LAN and PDN protocols at the lower layers that are needed to provide good worldwide service.

References

[1] C. Adams et al., "Protocol architecture of the UNIVERSE project," Proc. 6th Int. Conf. Computer Commun., IEEE, 1982.
[2] F. Backes, "Transparent bridges for interconnection of IEEE 802 LANs," IEEE Network, vol. 2, pp. 5–9, Jan. 1988.
[3] E. Benhamou and J. Estrin, "Multilevel internetworking gateways: Architecture and applications," Computer, pp. 27–34, Sept. 1983.
[4] D. Boggs et al., "PUP, An internetwork architecture," IEEE Trans. Commun., vol. COM-28, April 1980.
[5] R. Braden et al., "A distributed approach to the interconnection of heterogeneous computer networks," Proc. ACM SIGCOMM Symp., March 1983.
[6] R. Callon, "Internetwork protocol," Proc. IEEE, vol. 71, pp. 1388–1393, Dec. 1983.
[7] CCITT, Recommendation X.75, "Terminal and transit call control procedures and data transfer system on international circuits between packet-switched data networks," Red Book, 1985.
[8] CCITT, Recommendation X.121, "International numbering plan for public data networks," Red Book, 1985.

[9]   V. Cerf and P. Kirstein, "Issues in packet network interconnection," *Proc. IEEE*, vol. 66, pp. 1386–1408, Nov. 1978.

[10] D. Cohen, "Internet mail forwarding," *Proc. IEEE COMPCON*, pp. 384–390, Feb. 1982.

[11] D. Comer and J. Korb, "CSNET protocol software: The IP-to-X.25 interface," *Proc. ACM SIGCOMM Symp.*, March 1983.

[12] Y. Dalal, "Use of multiple networks in Xerox' network system," *Proc. IEEE COMP-CON*, pp. 391–397, Feb. 1982.

[13] Y. Dalal and R. Printis, "48-bit internet and ethernet host numbers," *Proc. 7th Data Commun. Symp.*, pp. 240–245, Oct. 1981, ACM/IEEE.

[14] Department of Defense, "Internet protocol," MIL-STD-1777, 1983.

[15] R. Dixon and D. Pitt, "Addressing, bridging, and source routing," *IEEE Network*, vol. 2, pp. 25–32, Jan. 1988.

[16] M. Gien and H. Zimmermann, "Design principles for network interconnection," *Proc. 6th Data Commun. Symp.*, Nov. 1979, ACM/IEEE.

[17] P. Green, Jr., "Protocol conversion," *IEEE Trans. Commun.*, vol. COM-34, pp. 257–268, March 1986.

[18] I. Groenbaek, "Conversion between the TCP and ISO transport protocols as a method of achieving interoperability between data communications systems," *IEEE J. Selected Areas Commun.*, vol. SAC-4, pp. 288–296, March 1986.

[19] R. Hinden, J. Haverty, and A. Sheltzer, "The DARPA internet: Interconnection of heterogeneous computer networks with gateways," *IEEE Comput.*, vol. 16, pp. 38–48, Sept. 1983.

[20] International Standards Organization (ISO), "Protocol for providing the connectionless network service," IS 8473, March 1986.

[21] ISO, "Internal organization of the network layer," IS 8648, Feb. 1988.

[22] ISO, "Addendum to the network service definition covering network layer addressing," IS 8348 AD2, Nov. 1984.

[23] ISO, "End system to intermediate system routing information exchange protocol for use in conjunction with the protocol for the provision of the connectionless-mode network service," IS 9542.

[24] C. Kawa and G. Bochmann, "Hierarchical multinetwork interconnection using public data networks," *Proc. IEEE INFOCOM*, pp. 426–435, March 1987.

[25] MAP/TOP Users Group, "Position paper on a solution for CONS/CLNS interworking," Oct. 1987.

[26] B. Meister, P. Janson, and L. Svobodova, "Connection-oriented versus connectionless protocols: A performance study," *IEEE Trans. Comput.*, vol. C-34, pp. 1164–1173, Dec. 1985.

[27] National Bureau of Standards, "Government open systems interconnection profile (GOSIP)," Version 1.0, Oct. 1987.

[28] D. Oppen and Y. Dalal, "The clearinghouse: A decentralized agent for locating named objects in a distributed environment," *ACM Trans. Office Inf. Syst.*, vol. 1, pp. 230–253, 1983.

[29] J. Postel, C. Sunshine, and D. Cohen, "The ARPA internet protocol," *Comput. Networks*, vol. 5, pp. 261–271, July 1981.

[30] J. Postel, C. Sunshine, and D. Cohen, "Recent developments in the DARPA internet program," *Proc. 6th Int. Conf. on Comput. Commun.*, Sept. 1982.

[31] J. Postel, "Internetwork protocol approaches," *IEEE Trans. Commun.*, vol. COM-28, pp. 604–611, Apr. 1980.

[32] L. Pouzin, "A proposal for interconnecting packet switching networks," *Proc. Eurocomp*, 1974.

[33] A. Rybczynski, J. Palframan, and A. Thomas, "Design of the Datapac X.75 internet-working capability," *Proc. 5th Int. Conf. on Comput. Commun.*, Oct. 1980, pp. 735–740.

[34] J. Shoch, "Internetwork naming, addressing, and routing," *Proc. IEEE COMPCON*, pp. 72–79, Sept. 1978.

[35] C. Sunshine, "Addressing problems in multinetwork systems," *Proc. IEEE INFOCOM*, pp. 12–18, 1982.

[36] C. Sunshine *et al.*, Interconnection of broadband local area networks, *Proc. 8th Data Commun. Symp.*, 1983, ACM/IEEE.

[37] C. Sunshine, "Interconnection of computer networks," *Comput. Networks*, vol. 1, pp. 175–195, Jan. 1977.

[38] C. Sunshine, "Source routing in computer networks," *ACM Comput. Commun. Rev.*, vol. 7, pp. 29–33, Jan. 1977.

[39] S. Zatti and P. Janson, "Internetwork naming, addressing, and directory systems: Towards a global OSI context," *Comput. Networks ISDN Syst.*, vol. 15, pp. 269–283, 1988.

[40] J. Onion and M. Rose, "ISO-TP0 bridge between TCP and X.25," ARPANET Request for Comment 1086, Dec. 1988.

[41] J. Benjamin *et al.*, "Interconnecting SNA networks," *IBM Syst. J.*, vol. 22, pp. 344–366, 1983.

[42] M. Padlipsky, "Gateways, architectures, and heffalumps," in *The Elements of Networking Style*, Englewood Cliffs, NJ: Prentice-Hall, 1985.

[43] C. Kent and J. Mogul, "Fragmentation considered harmful," *Proc. ACM SIGCOMM Symp.*, pp. 390–401, Aug. 1987.

[44] P. Mockapetris and K. Dunlap, "Development of the domain name system," *Proc. ACM SIGCOMM Symp.*, pp. 123–133, Aug. 1988.

[45] J. Naigle, "On packet switches with infinite buffer storage," *IEEE Trans. Commun.*, vol. COM-35, pp. 435–438, Apr. 1987.

[46] V. Jacobson, "Congestion avoidance and control," *Proc. ACM SIGCOMM Symp.*, Aug. 1988, pp. 314–329.

[47] V. DiCiccio, C. Sunshine, J. Field, and E. Manning, "Alternatives for interconnection of public packet switching data networks," *Proc. 6th Data Commun. Symp.*, pp. 120–125, Nov. 1979, ACM/IEEE.

[48] G. Bochmann, "Principles of protocol conversion," Pub. #624, Dept. d'IRO, Univ. Montreal, May 1987.

[49] G. Bochmann and A. Jacques, "Gateways for the OSI transport service," *Proc. IEEE INFOCOM*, 1987.

[50] S. Lam, "Protocol conversion," *IEEE Trans. Software Eng.*, vol. 14, pp. 353–362, March 1988.

V

Higher-Layer Protocols

OSI Transport and Session Layers

Charles E. Young

I. Background

A. Purposes

The end-to-end protocol of the Transport Layer enhances the data transmission quality of the Network Service to a level satisfactory for use by the upper three OSI layers. Transport Services support the protocol of the Session Layer, which provides a wide variety of dialogue and synchronization control services for use by the upper two OSI layers. A basic knowledge of the Transport and Session Layers is fundamental to understanding OSI.

B. Origins

During the 1970s the importance of arranging data communications protocols into independent hierarchical layers was recognized by several different organizations at about the same time. During this period, CCITT introduced Recommendation X.25 [1], which specifies protocols for the lowest three layers (the ones now known in OSI as the Physical, Data Link, and Network Layers). Later, the Bell System introduced the BX.25 protocol set [2], which built upon X.25, adding a new layer to perform end-to-end functions now associated with the OSI Transport and Session Layers. IBM introduced the SNA protocol set [3], which performs functions in a layered

CHARLES E. YOUNG • AT&T Bell Laboratories, Holmdel, New Jersey 07733.

manner covering a similar range of the OSI layers. The U.S. Department of Defense (DoD) introduced the Transmission Control Protocol (TCP) [4], which performs Transport Layer functions, and the Internet Protocol (IP) [5], which supports TCP. The European Computer Manufacturers' Association (ECMA) introduced the ECMA Transport Protocol, ECMA-72 [6]. CCITT Study Group VIII introduced the Transport protocol for use in the CCITT Teletex Service, Recommendation S.70, now numbered T.70 [7]. Also, Xerox introduced its Pup Internetwork Architecture [8], and Digital Equipment Corporation introduced its Digital Network Architecture [9].

During the early 1980s layered protocols primarily intended to perform OSI Session Layer functions were introduced. The two most significant ones were CCITT Recommendation S.62, introduced by Study Group VIII and now numbered T.62 [10], and the ECMA-75 Session Protocol [11]. Thus, when joint CCITT/ISO work began in 1980 to develop specific OSI Transport and Session Layer protocols, the fundamental problem they faced was not the lack of such protocols—it was the overabundance of them.

C. CCITT / ISO Collaboration

In CCITT the responsibility for developing OSI Transport and Session Layer Recommendations for CCITT applications was assigned to Study Group VII. In ISO the OSI Transport and Session Layer responsibilities were assigned to Subcommittee 16 of Technical Committee 97 (ISO/TC 97/SC 16).

Collaborative work on the Transport Layer began in 1980, and the initial protocol and service definition standards [12]–[15] were completed in 1984. In ISO this effort was then transferred to ISO/TC 97/SC 6, which continues to work in close collaboration with CCITT Study Group VII to develop refinements and extensions for the Transport Protocol and Service Definition.

Collaborative work on the Session Layer began in 1982, following individual studies of the needs of Session Layer users in both CCITT and ISO. This effort was then assigned a high priority in both organizations, allowing the initial protocol and service definition standards [16]–[19] to be completed in 1984 in conjunction with completion of the companion Transport Layer standards. In 1985, as part of a reorganization in ISO, ISO/TC 97/SC 21 assumed responsibility in ISO for the OSI Session Layer. Thus, collaborative work on refinements and extensions for the Session Layer continues between CCITT Study Group VII and ISO/TC 97/SC 21.

The collaborative work by CCITT and ISO has resulted in Transport Layer standards that are almost, but not quite, alike in CCITT and ISO.

The texts of the Session Layer standards adopted by CCITT and ISO are identical, except that the CCITT service definition includes a tutorial appendix clarifying how the Session Services must be used to provide the CCITT Teletex Service. This tutorial information is not included in the corresponding ISO standard.

Concurrent with development of the Transport and Session Layer standards, CCITT Study Group VII and ISO/TC 97/SC 6 developed the OSI Network Service Definition [20], [21], which defines the lower layer services available for use by the Transport Protocol. Collaborative work by these two groups continues toward development of refinements and extensions, and to resolve detailed differences between the CCITT and ISO standards.

D. Current Work

Efforts have been underway since 1986 to improve the initial Transport and Session Layer standards in three general areas: improving the clarity and completeness of the current standards; planning for expanded functionality; and correcting minor discrepancies and ambiguities, which are inevitable in the initial version of any complex set of protocols.

Use of more formal definition techniques is being carefully studied, and trial specifications employing these new techniques are being tested. Some of these techniques show promise of being useful in automating parts of the implementation process.

Examples of expanded functionality currently being incorporated include: introduction of "connectionless" services and protocols [13], [15], [21]–[23]; addition of further OSI layer management protocols [15]; and addition of more efficient synchronization techniques in the Session Layer [17], [19]. Studies of the three upper layers of OSI are underway to assure that the Presentation and Application Protocols now being developed will be able to employ the Session and Transport Services in an optimal manner. Some adjustments are being made in the Session Protocol as a result of this work. Studies are also underway to determine the appropriate ways in which OSI systems can best employ the services provided by Integrated Digital Services Networks (ISDNs).

The purpose of this chapter is to help the reader establish a general understanding of the ideas behind the current Transport and Session Layer standards, in preparation for more detailed study of the standards and the proposed additions to them. Accordingly, the following sections provide an overview, which covers topics common to both the Transport and Session Layers, followed by specific discussions concerning each layer.

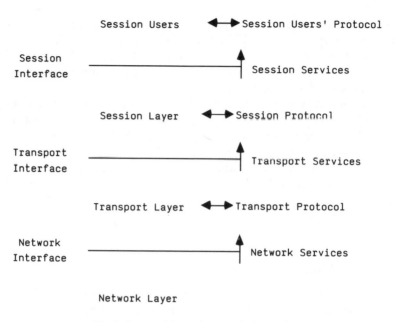

Fig. 1. Layer service and protocol relationships.

II. Overview

A. The Model

This section discusses topics that are common to both the Transport and the Session Layers. Subsequent sections deal with topics unique to each layer.

The OSI Reference Model [23], [24] prescribes the overall functions assigned to each layer and the services that each layer provides for use by the next higher layer. It also defines the terminology employed in OSI to describe the various layer functions and services. The reader is referred to Chapter 2 for an overall description of the OSI Reference Model. Figure 1 illustrates the layer service and protocol relationships relevant to the Transport and Session Layers, and Tables I–III list the associated service and protocol functions that are defined in the OSI Reference Model.

Within each OSI end system there are Transport and Session entities, which cooperate with peer entities in other OSI end systems to perform layer protocol functions necessary to provide services for the next higher layer. The layer services initially standardized are of the "connection oriented" type, wherein one of the service users must request establishment of an (n)-connection by the next lower layer before any other services of

**Table I. Session Services and Protocol Functions
Defined in the OSI Reference Model**

Session services
    Session connection establishment
    Session connection release
    Normal data exchange
    Quarantine service
    Expedited data exchange
    Interaction management
    Session connection synchronization
    Exception reporting
Session protocol functions
    Session connection to transport connection mapping
    Session connection flow control
    Expedited data transfer
    Session connection recovery
    Session connection release
    Session layer management

that layer can be employed. The three phases of an OSI (n)-connection are outlined below, where "n" indicates the layer providing the connection. This is followed by a brief description of connectionless-mode transmission, as it relates to the Transport and Session Layers. Work on connectionless-mode transmission for the Transport and Session Layers is being conducted within ISO, but not within CCITT.

**Table II. Transport Services and Protocol Functions
Defined in the OSI Reference Model**

Transport services
    Transport connection establishment
    Data transfer (normal and expedited)
    Transport connection release
Transport protocol functions
    Mapping transport address onto network address
    Multiplexing (end-to-end) transport connections onto network connections
    Establishment and release of transport connection
    End-to-end sequence control on individual connections
    End-to-end error detection and monitoring of the quality of service
    End-to-end error recovery
    End-to-end segmenting, blocking, and concatenation
    End-to-end flow control on individual connections
    Supervisory functions
    Expedited transport service data unit transfer

**Table III. Network Services Defined in the
OSI Reference Model**

Network addresses
Network connections
Network connection end-point identifiers
Network service data unit transfer
Quality of service parameters
Error notification
Sequencing
Flow control
Expedited network service data unit transfer
Reset
Release

B. (n)-Connection Establishment Phase

An OSI (n)-connection extends between two (n)-service access points called (n)-SAPs, as shown in Fig. 2.

An (n)-SAP is an abstract point at which an (n)-user and an (n)-provider interact, within an OSI system. All (n)-services are provided via (n)-service primitives that are transferred between an (n)-provider in the (n)-layer and an (n)-user in the (n + 1)-layer. Each (n)-SAP has associated with it a unique (n)-address, which is used in both the (n)- and the (n + 1)-layer. Each (n)-connection, as currently defined for the OSI Network, Transport, and Session Layers, has exactly two end points, which may be located at the same or two different (n)-SAPs, as illustrated in Fig. 2. There are six steps involved in establishing an (n)-connection:

1. The originating (n)-user issues an (n)-connect request service primitive.
2. The (n)-provider conveys this information, via the (n)-protocol, to the receiving (n)-entity that serves the addressed (n)-SAP in the destination OSI end system.

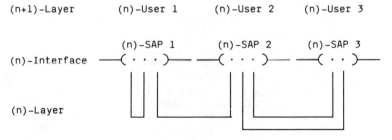

Fig. 2. (n)-Connections.

3. The receiving (n)-entity issues an (n)-connect indication service primitive, via the addressed (n)-SAP, to its associated (n)-user.
4. The receiving (n)-user then decides whether or not to accept the (n)-connection. An affirmative decision is indicated by means of an (n)-connect response service primitive indicating acceptance of the (n)-connection.
5. The (n)-provider conveys this information, via the (n)-protocol, to the originating (n)-entity serving the originating (n)-user.
6. The originating (n)-entity then issues an (n)-connect confirm service primitive to the originating (n)-user.

The (n)-connection establishment service is an example of a "confirmed" (n)-service, in that request, indication, response, and confirm primitives are all employed. Two other types of (n)-services may be provided during the data transfer and release phases. One is called a "nonconfirmed" service, in that only request and indication primitives are employed. The other is called a "provider-initiated" service, in that only indication primitives are employed.

After the (n)-connect confirm primitive is issued, the establishment phase is complete. Note that either of the (n)-entities or the called (n)-user may elect to prohibit the establishment of an (n)-connection. Also, the originating (n)-user may request that the establishment attempt be halted before it is completed. There are a number of reasons why these actions may be necessary, including lack of resources needed to support the (n)-connection.

In addition to the (n)-address, there are other parameters conveyed within the (n)-connect service primitives, each having a range of values. These serve to determine various characteristics of the (n)-connection. The values of many of these parameters may be negotiated among the (n)-users and the (n)-provider, according to specific rules stated in the applicable (n)-service definition. An example of such a parameter is the average throughput to be allowed during the data transfer phase of the (n)-connection.

C. (n)-Connection Data Transfer Phase

During the data transfer phase, there may be additional confirmed, nonconfirmed and provider-initiated services provided, depending on the particular (n)-service definition. It is during the data transfer phase that a sharp distinction may be noted between the Session and Transport Layers. The Transport Layer provides only two services during this phase (normal and expedited data transfer); however, the Transport Protocol, which is required to support these two services over Network Connections having

wide variations in quality, is complex. The Session Layer is quite different,
in that it provides a wide variety of different services to its users, each
requiring specific elements in the Session Protocol. Thus, the Session
Protocol is also complex, but for quite different reasons from those for the
Transport Layer.

The sharp division of responsibilities between the Session and Transport Layers provides the best example of interlayer protocol independence
in all of OSI. The Session Layer is almost entirely "upward" oriented, while
the Transport Layer is almost entirely "downward" oriented. This distinction extends throughout both the data transfer and release phases of
Session and Transport Connections.

D. (n)-Connection Release Phase

Either of the (n)-users or the (n)-provider may, at any time, elect to
enter the (n)-connection release phase. The (n)-provider only initiates this
action when it is necessary to abort the (n)-connection, due to some
difficulty that prevents it from continuing to provide the (n)-services
negotiated during the establishment phase. Either (n)-user may also initiate
entry into the release phase, in order to abort the (n)-connection. The
Transport Service also permits either user to initiate a normal disconnect at
any time. Upon receipt of such a request, the Transport Layer simply
discards all data that are currently enroute and notifies the other user (via
an indication service primitive) that the Transport Connection has been
released.

The services available during the release phase of a Session Connection
are much more complex, illustrating the sharp difference in the basic nature
of the two layers. The Session Services, for example, permit the two Session
Users to enter into the release phase, then to negotiate their way back into
the data transfer phase, with no loss of user data during the negotiations.
Complex Session Services, such as this negotiation of release, can only be
conducted based on complete confidence in the error-free data transfer
services of the Transport Layer. The Session Protocol provides no mechanisms to protect against errors in the Transport Services. For example,
whenever the Transport Provider initiates the abort of a Transport Connection, the Session Provider immediately aborts the corresponding Session
Connection. (Only one Session Connection is permitted to be assigned to a
given Transport Connection at a time.)

E. (n)-Connectionless-Mode Transmission

The Transport and Session Layer standards jointly adopted by ISO
and CCITT require the support of Network Connections, and only connection oriented Transport and Session Services are provided. ISO is currently

also standardizing connectionless services and protocols in the Network and Transport Layers, wherein no establishment or release phases occur. A nonconfirmed service employing (n)-UNITDATA request and indication service primitives, supported by an (n)-connectionless-mode protocol, is provided for transfer of (n)-Service Data Units [(n)-SDUs].

Either connection oriented or connectionless-mode Network Services [21] may be used by the Transport Layer in support of connection oriented Transport Services [13], [15] or connectionless-mode Transport Services [13], [22].

The existing Session Protocol, however, requires the connection oriented Transport Service, and no applications requiring connectionless-mode transmission have yet been proposed for standardization.

A variation of one class of the Transport Protocol (Class 4) is able to employ the connectionless-mode Network Service in support of the connection oriented Transport Service, owing to its error detection and correction capabilities. This is an example of crossover in the Transport Layer between different modes of Network and Transport Services. Such crossovers are permitted in the Transport Layer, but fully open real systems must be able to operate without requiring them.

It should be noted that end systems designed to employ only connectionless Network Services cannot interwork with end systems designed to require use of Network Connections. The type of Network Service employed for any given instance of communications must be the same in both end systems. It is anticipated that a directory attribute associated with each NSAP Address will indicate the mode(s) of Network Service able to be provided via the NSAP, thus facilitating selection of compatible Network Service modes.

F. Descriptive Techniques Used

CCITT Recommendation X.210 [25] and the ISO technical report 8509 [26] describe the service primitive naming conventions that are employed in the service definitions for the Network, Transport, and Session Layers. They also prescribe a uniform way to diagram the various types of services, such as confirmed and nonconfirmed services. The service primitive naming convention calls for each primitive to have three parts. The first part is a letter indicating the service interface across which the primitive is conveyed. The second part indicates the (n)-service of which the primitive is a part. The third part indicates the primitive type (request, indication, response, or confirm). Thus, for example, "N-CONNECT request" indicates the Network (N) Service primitive conveyed from a Transport entity to its Network Service provider requesting that a Network Connection be established. Chapter 17 describes formal definition techniques that are being developed by ISO and CCITT for use in future issues of standards.

Three descriptive techniques are used in the initial standards. First, written sentences and paragraphs are provided. Although great care has been taken to avoid ambiguities, this technique can only describe the simplest service and protocol interactions in a complete manner; it is difficult to use in describing the many second-order effects that can occur, owing to cross correlations between seemingly unrelated service primitives and protocol transfers. The second technique that is employed is the use of state transition diagrams (sometimes called bubble diagrams). These are instructive for indicating major service states and transitions, but are too cumbersome to describe more complex protocol interactions. This technique is used in the Network and Transport Service Definitions, but only for illustrative purposes. The third technique is a semiformal one used in defining the Session Services and in specifying the Transport and Session Protocols. It employs state tables, and is capable of handling more complex interactions than the other two techniques employed. State tables provide a shorthand notation for describing the behavior of an entity that has a finite number of states and acts in response to a well-defined set of events, one at a time, as they occur.

III. The Transport Layer

A. Transport Overview

The purpose of the Transport Layer is to enhance the quality of the underlying Network Service, so as to provide transparent transfer of data in a cost effective and reliable manner between Transport Users in OSI end systems. An integral part of the Transport Protocol is the manner in which it employs the underlying Network Service to attain this goal. Since Transport Layer relay entities are prohibited in OSI, the Transport Protocol operates only on an end-to-end basis between OSI end systems. Within this restriction a wide variety of techniques may be employed to attain an optimal balance between the cost and reliability of the Transport Service. These techniques are categorized in the Transport Protocol according to three types of situations. For a given set of cost and quality requirements requested for a Transport Connection, the quality of the available Network Service may be classified as being that of a Type A, B, or C network, as described below:

Type A: Entirely sufficient, requiring no quality enhancements by the Transport Protocol.

Type B: Insufficient only with respect to the rate at which the Network Service indicates abort or reset conditions, wherein data may be lost within the lower three layers.

Type C: Insufficient with respect to the rate at which errors may occur, which are not detected within the Network Service.

The cost of implementing and employing mechanism within the Transport Protocol to enhance the Network Services to acceptable levels increases with the categories of Network Service listed above. Accordingly, the mechanisms required in the Transport Protocol to enhance the quality of the Network Service have been categorized into Transport Protocol Classes. There are five such classes, numbered Class 0 through Class 4.

Classes 0 and 2 have been designed to operate most efficiently when the Network Service is of Type A quality, with respect to the needs of the Transport Users. Class 0 is the simplest class, while Class 2 allows options to permit independent Transport Layer flow control and multiplexing of multiple Transport Connections onto a single Network Connection, where those capabilities can be used to reduce costs.

Classes 1 and 3 have been designed to operate most efficiently when the Network Service is of Type B quality, with respect to the needs of the Transport Users. Both of these classes provide for recovery from Network Service abort and reset conditions, wherein data may be lost within the lower three layers. Class 3 also provides independent Transport Layer flow control and permits multiplexing of multiple Transport Connections onto a single Network Connection.

Class 4 is the only Transport Protocol class containing mechanisms intended to detect and correct errors in the Network Service. It also contains mechanism to permit multiplexing of multiple Transport Connections onto a single Network Connection. In addition, it is capable of performing splitting and recombining, wherein multiple Network Connections are used at the same time to share the load of transferring data for a single Transport Connection.

The Transport Service supports the establishment and release of a Transport Connection between two TSAPs and the transfer of normal and expedited data between Transport Users served via the TSAPs. Normal data are delivered in sequence, and no data length limits are placed on the data units transferred between the Transport Users. Expedited data in units of limited size are also able to be transferred between the Transport Users. This type of data is assured to be presented to the remote Transport User prior to any other data sent after it, and it may bypass and be delivered before normal data sent prior to it. It is subject to a separate flow control mechanism, and can be delivered even when normal data are blocked due to Transport Layer flow control. All five of the Transport Protocol classes provide the same Transport Service, with the exception that Class 0 cannot be used when the Transport Service of expedited data transfer is needed by the Transport Users. The basic functions specified in the Transport Protocol to support the Transport Service are described in the following sections, and are summarized in Table IV.

Table IV. Functions Performed by the Transport Protocol Classes

	Classes				
Functions	0	1	2	3	4
Functions not dependent on the connection phase					
TPDU transmission	×	×	×	×	×
Multiplexing and demultiplexing			×	×	×
Error detection					×
Error recovery		×		×	×
Connection establishment phase functions					
Network service negotiations	×	×	×	×	×
TSAP/NSAP address mapping	×	×	×	×	×
Transport connection negotiations	×	×	×	×	×
User data transfer	×	×	×	×	×
Data transfer phase functions					
Concatenation and separation		×	×	×	×
Segmenting and reassembly	×	×	×	×	×
Splitting and recombining					×
Flow control			×	×	×
Transport connection identification	×	×	×	×	×
Normal data transfer	×	×	×	×	×
Expedited data transfer		×	×	×	×
TSDU delimiting	×	×	×	×	×
Release phase functions					
Transport connection release and abort	×	×	×	×	×

It is important to note that much of the complexity of the Transport Protocol, as with any communication protocol, arises when multiple protocol mechanisms that are not entirely independent are active simultaneously. A basic understanding of the Transport Protocol requires an understanding of the individual mechanisms. An understanding of the interactions among the various mechanisms requires detailed individual study of the protocol state tables and the text of the Transport Protocol Specification.

B. Protocol Functions Not Dependent on the Connection Phase

1. TPDU Transmission

Transport Protocol Data Units (TPDUs) are able to be transferred during all phases of a Transport Connection. They are mapped onto a Network Connection at the sending end, transferred via the Network Service to the other end of the Network Connection, and then mapped to the appropriate Transport Connection at the receiving end.

2. Multiplexing and Demultiplexing

TPDUs for multiple Transport Connections may be mapped (multiplexed) onto the same Network Connection in Classes 2, 3, and 4. When this is done, a Transport Connection identifier is required to be included in every TPDU, so that the receiving Transport entity can perform the necessary demultiplexing function.

3. Error Detection

Only Class 4 contains mechanisms to detect errors in the Network Service that are not signaled. Such errors can include the loss, duplication, reordering, and misdelivery of Network Service Data Units (NSDUs), as well as the addition and deletion of individual bits, bytes, and TPDUs. These mechanisms include numbering of the TPDUs, retaining them at the sending end until they are acknowledged by the receiving Transport entity, insertion and verification of checksums in the TPDUs and the extensive use of 14 different timers.

4. Error Recovery

Classes 1, 3, and 4 contain error recovery mechanisms. Classes 1 and 3 contain mechanisms necessary to recover from errors signaled by the Network Service, such as a Network Connection abort or a reset signal indicating possible loss of data. These require the ability to reassign a Transport Connection to a different Network Connection, to resynchronize back to the earliest unacknowledged TPDU, and to repeat transmissions, as necessary, to avoid errors. They also require the ability to delay reusing a previous Transport Connection identifier while TPDUs associated with it may still exist.

In addition to the above, Class 4 requires the ability to retransmit TPDUs following an acknowledgment timeout, the ability to resequence TPDUs, the ability to deal with unsignaled terminations of a Network Connection, and the ability to distinguish between errors caused by the Network Service and Transport Protocol errors.

C. Connection Establishment Protocol Functions

1. Network Service Negotiations

The Transport Protocol requires that a Network Connection be established before it will transfer a connect TPDU to initiate the establishment of a Transport Connection. During the establishment of a Network Connection, through negotiations among the initiating Transport entity, the responding Transport entity and the Network Provider, the Network Service options and quality of service able to be provided are determined. Network Service options include the expedited data and receipt confirmation Network Services. These options may be used in Class 1 of the Transport Protocol, when they are available. The information negotiated during the establishment of a Network Connection is used in determining which Transport Protocol class is appropriate for a given Transport Connection.

The Network Service permits a limited amount of Network User data to be transferred in both directions during the establishment of a Network Connection. The Transport Protocol can utilize this capability to transfer protocol identification information to distinguish, for example, the currently defined OSI protocol from other protocols (for example, OSI system management protocols and private protocols that might also employ the same Network Service). Under study is a Network Connection Management Subprotocol (NCMS) which may be adopted in the future for negotiating additional information concerning use of Network Services (for example, negotiation of parameter values needed in performing Transport multiplexing).

2. TSAP / NSAP Address Mapping

When a Transport Connection is requested by a Transport User, a choice may need to be made by the initiating Transport entity between establishing a new Network Connection and using one that already exists. In either event, the initiating Transport entity must perform a mapping function between the TSAP addresses and the supporting NSAP addresses to be used at both ends. Where each TSAP has only one associated NSAP, this function is straightforward.

3. Transport Connection Negotiations

Several important negotiations take place during the transfer of a connection request TPDU sent by the initiating Transport entity and the return of a connection confirm TPDU sent by the responding Transport entity. These determine whether or not multiplexing will be employed, the maximum TPDU sizes, and the Transport options to be used. Numerous options and parameter values are negotiated, including use of normal or extended formats for certain numbering fields; selection of Transport Protocol class; selection of a Transport Connection identifier; use of the Network Service options of expedited and/or receipt confirmation, if Class 1 is to be used; use of flow control, if Class 2 is to be used; throughput for the Transport Connection; a value for the acknowledgment timer, if Class 4 is to be used; values of security parameters (determined by the Transport

Users); quality of service parameter values, such as the maximum allowed residual bit error rate; the Transport Protocol version number; use of checksums, if Class 4 is to be used; and whether or not the Transport expedited service is to be provided. The options and parameter values agreed during the Transport Connection establishment phase set the stage for their use during the data transfer phase that follows.

4. User Data Transfer

Although not currently used by the Session Protocol, the Transport Service allows a limited amount of Transport User data to be transferred transparently in both directions during the establishment of a Transport Connection. This capability could be employed in later versions of the Session Protocol for such purposes as passing system or Session Layer management information.

D. Data Transfer Protocol Functions

1. Concatenation and Separation

This function is performed by Classes 1–4. It allows several TPDUs to be mapped onto the same NSDU, then separated at the receiving end. It is useful in minimizing costs when short NSDUs are more expensive (per bit sent) than are longer NSDUs.

2. Segmenting and Reassembly

This function is performed in all classes. It permits a long TSDU to be mapped into several TPDUs for transfer, then reassembled at the receiving end. It is useful in minimizing costs, when long NSDUs, onto which TPDUs are mapped, are more expensive (per bit sent) than are shorter NSDUs, perhaps owing to average retransmission costs when bit error rates are high.

3. Splitting and Recombining

This function is performed only by Class 4. It permits simultaneous use of multiple Network Connections in support of a single Transport Connection. It is useful in certain cases to provide improved resilience of the Transport Connection, in the event of network failures, and to increase throughput. It can also be used in conjunction with diverse routing for security purposes.

4. Flow Control

This function regulates the flow of TPDUs. Classes 0 and 1 rely upon use of the Network Service to accomplish flow control, while Classes 3 and 4 employ explicit flow control in the Transport Protocol, thus facilitating multiplexing in the Transport Layer. Class 2 can negotiate use of either method.

5. Transport Connection Identification

This function is performed in all classes. A combination of the initiator's reference number and the responder's reference number (both assigned during the Transport Connection establishment phase) uniquely identifies a given Transport Connection during the data transfer phase.

6. Data Transfer

The Transport Services of normal and expedited data transfer are both able to be provided during the data transfer phase. Normal data transfer is always provided, and expedited data transfer is provided whenever the Transport Users request it during the Transport Connection establishment phase. Class 0 does not provide the expedited data transfer service.

7. TSDU Delimiting

The Session Protocol relies on the Transport Service to convey delimiting information for TSDUs. The Transport Protocol explicitly identifies TPDUs that contain the end of a TSDU. The Network Service is relied upon by all Transport Protocol classes to provide delimiting of NSDUs, onto which TPDUs are mapped. All of these functions are used together to provide TSDU delimiting.

E. Release Protocol Functions

Either Transport User may request that a Transport Connection be released at any time. The Transport Connection is then immediately terminated, with possible loss of any data in transit. In Classes 0 and 2 the Transport Connection is also terminated (as a provider initiated service) if a Network Service error or disconnect is signaled. Regardless of the Transport Protocol class, the Transport Service will (as a provider initiated service) abort a Transport Connection (and notify both users) if it is determined that the Transport Provider is no longer able to support the Transport Connection, as in the case of detected Transport Protocol errors.

IV. The Session Layer

A. Session Overview

The purpose of the Session Layer is to provide a common set of protocol mechanisms, whose services are able to be used by a wide variety of applications to govern their dialogue control and synchronization needs. Unlike the Transport Layer, where various complex protocol mechanisms are needed to support a few services that are easy to define, but complex to provide, sets of Session Layer protocol mechanisms are linked on a one-to-one basis with a like number of sets of services needed by Session Users. These sets of services and their corresponding sets of protocol mechanisms are referred to in both the Session Service Definition and the Session Protocol Specification as "functional units." There are 12 of them.

Much of the complexity of the Session Layer occurs as a result of the interactions among functional units, when more than one is being used at the same time. The following sections provide a basic description of the services and protocol mechanisms associated with the 12 individual functional units, leaving for deeper study on an individual basis the important, but complex, interactions that can occur among them.

Table V lists the functional units. It should be noted that the kernel functional unit and either the half-duplex or the duplex functional unit must be employed in every Session Connection. Use of half-duplex or duplex and the particular subset of the other nine functional units to be available for use during a Session Connection is negotiated among the Session Provider and the Session Users during the connection establishment phase. The selection of allowed subsets is subject to a few restrictions, mainly to avoid logical inconsistencies, and is not permitted to be renegotiated during the data transfer or release phases. Aside from these restric-

Table V. Session Functional Units

Kernel
Negotiated release
Half duplex
Duplex
Expedited data
Typed data
Capability data exchange
Minor synchronize
Major synchronize
Resynchronize
Exceptions
Activity management

Table VI. Basic Session Services

Kernel functional unit
    Session connection establishment
    Normal data transfer
    Orderly release
    U-Abort (user initiated)
    P-Abort (provider initiated)

tions, the sets of Session functional units needed for any given standard application, including which functional units may be optional, are specified as an integral part of the appropriate Session User's protocol specification. Thus, it is necessary for Session implementors to be aware of the applications intended to be served by their implementations, unless all functional units are to be implemented.

B. Basic Session Services

Table VI lists the services that are associated with the kernel functional unit. These are the basic Session Services that are always provided for every Session Connection. Provision of these services is not negotiable by the Session Provider.

1. Session Connection Establishment

Session Connection establishment is a confirmed service. Its major underlying purpose is to permit negotiations to be conducted among the two Session Users and the Session Provider concerning the particular combination of functional units to be employed during the Session Connection. This is necessary to assure that both end systems have implemented the necessary functional units and that the appropriate resources necessary to support the Session Connection can be allocated. Various other pieces of information may also be exchanged, and a unique Session Connection identifier can be established for use by the Session Users during the Session Connection, as well as after it has been released.

Session Users are able to transfer a limited amount of data in a parameter of the S-CONNECT request and response service primitives. This capability is very important to the designers of protocols for the upper two OSI layers. It is currently anticipated that the information necessary to establish a Presentation Connection and an Application Association will all be able to be sent within this Session User data parameter, whenever appropriate, thus saving additional round trip delays and associated processing costs. Also, discussions are currently underway in both ISO and

CCITT concerning the possibility of greatly enlarging the amount of such data permitted to be sent.

Another important function performed during the Session Connection establishment phase concerns the selection of tokens to be used and their initial allocation. Certain functional units require the use of tokens to control which Session User is currently authorized to request a particular Session Service, when only one of them is permitted to request the service at any given time. These functional units and their associated tokens are discussed in subsequent sections. During the data transfer phase, a Session User wishing to request such a service, when the other Session User has the token for it, may request that the other Session User relinquish the token. This request is made by means of the S-TOKENS-PLEASE (nonconfirmed) service. The Session User having the requested token may then decide whether or not to relinquish it to the requesting Session User. (Tokens may also be passed in groups, with or without any specific request being made.) Tokens are relinquished during the data transfer phase by use of the S-TOKENS-GIVE (nonconfirmed) service and the S-CONTROL-GIVE (nonconfirmed) service.

2. Normal Data Transfer

Normal data transfer is a nonconfirmed service that allows Session Users to request that data be transferred between the two SSAPs during the data transfer phase of a Session Connection. This phase begins immediately after the connection establishment phase is completed. The Session Service places no restrictions on the amount of Session User data that can be requested to be transferred as a single unit. The Session Protocol depends on the Transport Layer to perform any segmenting of data that may be needed to attain optimal transmission economy. The Session Protocol does provide the capability to negotiate (within the Session Layer) the ability to segment Session User data, where useful in accommodating efficient buffer usage and processing capabilities. The Session Protocol also provides the capability (within the Session Layer) to perform concatenation, which is the combining of multiple Session Protocol Data Units (SPDUs) within a single data transfer request to the Transport Service. This allows shorter SPDUs to be combined with each other and/or with longer SPDUs, thus increasing the average amount of information transferred per service request to the Transport Layer. For a given amount of data to be sent, the Transport Layer can often attain greater transmission cost efficiency if the data requested to be sent is provided in larger units.

Normal data transfer is the only data transfer service provided during the data transfer phase by the Kernel functional unit; however, there are other kinds of special data transfer services, which are provided when certain other functional units are used. These services are discussed later.

3. Orderly Release

Orderly release is a confirmed service permitted to be initiated by a Session User when that user (the requestor) has all of the tokens that are available for use with the Session Connection. All data in transit from the requestor, if any, are delivered to the other Session User, followed by an indication that the orderly release service is in progress. The requestor is prohibited from sending any more data, but the other Session User is permitted to continue sending data until it issues a response service primitive for the orderly release service. The response, when issued, must be positive (agreeing to release the Session Connection). Then any data in transit back to the requestor are delivered, followed by the confirm primitive. In this manner, the Session Connection is released with no loss of data. The negotiated release service, discussed later, permits a negative response primitive to be issued.

4. U-Abort and P-Abort

U-Abort is a nonconfirmed service able to be requested by either Session User at any time. The Session Connection is terminated immediately and the other Session User is informed. P-Abort is a provider-initiated service, which terminates the Session Connection, with notification to both Session Users, when the Session Provider recognizes that it cannot continue to support the Session Connection, usually due to recognition of Session Protocol errors or as a result of a failure of the underlying Transport Connection.

C. Negotiated Release Session Services

Table VII lists the services that are associated with the negotiated release functional unit. Negotiated release is similar to the orderly release service, except that it can only be requested by the Session User having the release token, and that the response primitive is permitted to be negative, indicating that the Session Connection should not be released. When a negative response primitive is issued, the responding Session User may continue to send data. The negative response information is conveyed in sequence with normal data to the requestor as a confirm (negative) primi-

Table VII. Negotiated Release Session Services

Negotiated release functional unit
Orderly release
Give tokens
Please tokens

Table VIII. Data Transfer Session Services

Half-duplex functional unit
  Normal data transfer
  Give tokens
  Please tokens
Duplex functional unit
  Normal data transfer
Expedited data functional unit
  Expedited data transfer
Typed data functional unit
  Typed data transfer
Capability data exchange functional unit
  Capability data exchange

tive. Then the requestor also reenters the data transfer phase. Use of the release token permits the Session Users to coordinate their negotiation of the release by avoiding race conditions that would be difficult to resolve. The release token may be passed between users by means of the give tokens and please tokens services.

D. Types of Data Transfer Services

Five of the 12 Session functional units involve various types of data transfers, each having its own distinctive features. The functional units that are associated with types of data transfers are listed in Table VIII.

1. Half-Duplex and Duplex Functional Units

During the Session Connection establishment phase, the Session Provider and both Session Users are required to negotiate and agree on the use of one of these two functional units. The duplex functional unit provides for duplex data transfers, wherein both Session Users may initiate data transfers at the same time. The half-duplex functional unit requires use of the data token, and only the Session User having the data token may initiate data transfers. The Session Protocol always operates in a duplex mode, with respect to transfer of its own protocol control information. It is only the Session Service that is governed by use of the data token.

2. Expedited Data Functional Unit

This functional unit permits either Session User to request the transfer of a limited amount of data, which is subject to flow control restraints that are independent from those for the other types of data transfers. Data sent in this manner are assured to be presented to the remote Session User prior to any other data sent after it, and it may bypass and be delivered before data of other types sent prior to it. The Session Protocol maps SPDUs carrying expedited data onto the Transport expedited data service.

3. Typed Data Functional Unit

The purpose of typed data is to provide a means by which either Session User may transfer control information, without being required to have the data token. This functional unit permits either Session User to request the transfer of typed data, without regard for the data token. Typed data are signaled as such to the accepting Session User, but are treated in every other way just the same as normal data. They are delivered in sequence with any normal data being transferred in the same direction, and are subject to the same flow control. The amount of data allowed in any given unit of typed data may be restricted within the Session Users' protocol, but the Session Service places no length constraints on it.

4. Capability Data Functional Unit

This functional unit is only permitted to be employed in conjunction with the activity management functional unit, which permits Session Users to establish periods within Session Connections called activities. A Session User having all of the tokens that are available for use in such a Session Connection is permitted to request transfer of a limited amount of capability data, but only when no activity is in progress. The purpose is to permit control information to be passed between the Session Users in a controlled manner, in preparation for the next activity. The activity management functional unit is described later. The capability data exchange service is a confirmed service, which, if used as prescribed in an annex to Recommendation X.215, supports compatibility with certain CCITT services, such as Teletex and the Message Handling Service.

E. Synchronization Session Services

Three functional units are dedicated to Session synchronization and resynchronization. Basically, these services permit Session Users to place marks in their data streams for synchronization purposes, and to resynchronize to such marks in a controlled and uniform manner. Table IX lists these functional units and their associated Session Services.

Table IX. Synchronization Session Services

Minor synchronize functional unit
　Minor synchronization point
　Give tokens
　Please tokens
Major synchronize functional unit
　Major synchronization point
　Give tokens
　Please tokens
Resynchronize functional unit
　Resynchronize

1. Minor Synchronize Functional Unit

The minor synchronization point service is a confirmed service, but the Session Layer does not police the return of a response primitive by the responding Session User. Thus the Session Users' protocol is free to specify the situations, if any, in which response primitives are to be issued. A synchronization point serial number and (optionally) a limited amount of user data are transferred as parameters within each primitive, and this information is delivered in sequence with normal data being transferred in the same direction. The serial number is incremented by one for each such mark, and a single serial number sequence is maintained for both directions of transfer. Thus, data are marked only in the direction in which the minor synchronization point information is conveyed, and a sync-minor token is required to avoid race conditions. The same serial number sequence is also employed in the major synchronization point service, which is described below.

2. Major Synchronize Functional Unit

The major synchronization point service is a confirmed service. It allows a Session User to request that a mark be placed in the data streams in both directions. To avoid race conditions, a sync-major token is employed. While the requestor dictates the exact position of the mark in the direction of the data it sends (by issuing the request in sequence with the data) the responder dictates the location of the mark in the data it sends (by issuing the response in sequence with the data it sends). Since the same serial number sequence is used for both major and minor marks, the requestor must have both the sync-minor and the sync-major tokens, if both functional units are in use.

3. Resynchronize Functional Unit

The resynchronize session service permits either Session User to request that the Session Connection be set to an agreed defined state, including the positions of all tokens and the synchronization point serial number. It is a confirmed service with three major options: restart, set, and abandon. With the restart option, the requestor may ask for the new synchronization point serial number to be any one previously assigned, back as far as the most recent major synchronization point. The Session Users are permitted to negotiate to a different value within this range. With the set option, the Session Users are allowed to negotiate to any value for the new synchronization point serial number. With the abandon option, the Session Provider is requested to assign a new synchronization point serial number, which is larger than any one previously assigned for the current Session Connection. Thus, a wide variety of possibilities exist for use in the Session Users' protocol.

F. Exception Reporting Session Services

The exceptions functional unit (see Table X) includes the provider exception reporting service and the user exception reporting service. Provider exception reporting is a provider-initiated service. It notifies both Session Users of unanticipated situations not covered by other services, including such things as user errors or inconsistencies in the use of Session Services. It is provided in cases where it is important for the Session Users to be informed, but not necessary that the Session Provider abort the Session Connection.

User exception reporting is a nonconfirmed service, which permits one Session User to notify the other one of an exception condition that requires action by the second Session User to clear the trouble, but which does not necessarily require aborting the entire Session Connection. The Session Provider prevents the second Session User from initiating any new requests for Session Services, until that user requests one of a set of Session Services necessary to clear the difficulty.

Used as prescribed in an annex to Recommendation X.215, these services support provision of certain CCITT services, such as Teletex and the Message Handling Service.

Table X. Exception Reporting Session Services

Exceptions functional unit
Provider exception reporting
User exception reporting

G. Activity Management Session Services

There are eight Session Services in the activity management functional unit (see Table XI) including the give tokens and please tokens Session Services discussed previously. They are all directed toward providing support for Session Users, with respect to units of work called activities, which may be conducted during Session Connections. Used as prescribed in an annex to Recommendation X.215, they provide support for certain CCITT services, such as Teletex and the Message Handling Service.

The support provided by the Session Services, with respect to activities, is largely a delimiting function, providing uniform ways in which activities may be started, interrupted, resumed after an interruption, discarded, and ended, as the names of the services in Table XI imply. The activity start and activity resume services are nonconfirmed services, while the activity interrupt, activity discard, and activity end services are confirmed services. An activity may be interrupted during one Session Connection and resumed later during the same or a different Session Connection. Thus, an activity identifier is required to be assigned by the Session User that requests the start of an activity. While only one activity can be delimited by the Session Service at a time, Session Users may have multiple simultaneous activities in progress, some nested within others, and some overlapping others. Proposals are currently under study within ISO and CCITT to expand the activity services to provide better support for Session Users in such cases.

In order to avoid race conditions the activity token is required to request any of the activity services. This is the same token as is required to request the major synchronization point service and, when used for both services, is called the major/activity token. During the periods between activities within a Session Connection, all of the various kinds of data are permitted to be transferred, and/or the Session Connection can be released or aborted. Capability Data is limited to being sent only outside of activities, as its purpose has to do with preparing for the start of a new activity or the resumption of an activity previously interrupted. Also, the give control service may only be used outside of activities. It results in

Table XI. Activity Management Session Services

Activity management functional unit
Activity start
Activity resume
Activity interrupt
Activity discard
Activity end
Give tokens
Please tokens
Give control

the transfer of all tokens in use for the Session Connection to the other Session User.

V. Summary

The Transport and Session Layers differ markedly in the most fundamental ways. While the Transport Protocol employs the small number of Network Services, and enhances their transmission quality without changing their basic nature, the Session Protocol employs the small number of Transport Services and provides a wide variety of dialogue control and synchronization services, without enhancing the transmission quality of the Transport Services at all.

Current standardization efforts for the Transport Layer include work concerning its formal description and use in various new environments, such as interworking between end systems served by local and wide area networks. Work is continuing in ISO concerning use of connectionless Network Services and provision of connectionless Transport Services.

Current standards work concerning the Session Layer is centered on refining the details of certain of its services, so as to satisfy better the detailed needs of the emerging Presentation and Application Layer Protocols for OSI. Work is also progressing on its formal description. A new synchronization technique aimed at improving the efficiency of the synchronization services, particularly for Session Users employing the duplex functional unit, is also being studied. This technique is called symmetric synchronization.

Standards work is also being conducted concerning the needs for additional protocol mechanisms at both of these layers, with respect to layer and system management functions. For example, a Network Connection Management Subprotocol (NCMS) is being standardized for use in the Transport Layer for the purpose of enhancing the management of Network Connections. Security considerations, common (n)-SAP addressing conventions, and a wide variety of other topics common to multiple layers are also being standardized.

Early signs of vendor acceptance of the OSI Transport and Session Layer standards are encouraging. CCITT Administrations are already employing compatible subsets of them in support of Telematic Services, such as Teletex and the Message Handling Service, and many computer vendors have announced plans to support them. In the United States over 40 of the major computer communication companies are cooperating through the Corporation for Open Systems (COS) to support and stabilize implementation of OSI standards. This will be done largely by selecting appropriate subsets of the standards to ensure compatible implementations, and by establishing a test facility to verify uniform conformance. In Europe the Standards Promotion and Application Group (SPAG), a group of 12 leading information technology companies, has been formed with similar goals for usage of OSI standards. Users' groups, such as those defining the Manufacturing and Automation Protocol (MAP) and the Technical and Office Protocols (TOP), supported by major users, such as General Motors and Boeing, are currently basing their plans on use of the OSI protocols.

References

[1]   CCITT Recommendation X.25, "Interface between data terminal equipment (DTE) and data circuit terminating equipment (DCE) for terminals operating in the packet mode and connected to public data networks by dedicated circuit," 1984.

[2]   Bell System Technical Reference, "Operations systems network communications protocol specification BX.25," Publication No. 54001, August, 1983.

[3]   J. H. McFadyen, "Systems network architecture: An overview," *IBM Syst. J.*, vol. 15, no. 1, pp. 4–23, 1976.

[4]   MIL-STD-1778, "Transmission control protocol," August 12, 1983.

[5]   MIL-STD-1777, "Internet protocol," August 12, 1983.

[6]   ECMA-72, "Transport protocol," 3rd Ed., March, 1985.

[7]   CCITT Recommendation T.70, "Network-independent basic transport service for the Telematic services," 1984.

[8]   D. R. Boggs, J. F. Shoch, E. A. Taft, and R. M. Metcalfe, "Pup: An internetwork architecture," *IEEE Trans. Commun.*, vol. COM-28, pp. 612–624, April 1980.

[9]   S. Wecker, "DNA: The digital network architecture," *IEEE Trans. Commun.*, vol. COM-28, pp. 510–526, April 1980.

[10] CCITT Recommendation T.62, "Control procedures for Teletex and Group 4 facsimile services," 1984.

[11] ECMA-75, "Session protocol," January, 1982.

[12] CCITT Recommendation X.214, "Transport service definition for open systems interconnection (OSI) for CCITT applications," 1984.

[13] ISO 8072-1985, "Information processing systems—Open systems interconnection— Connection-oriented transport service definition"; Addendum 1, "Connectionless-mode transmission."

[14] CCITT Recommendation X.224, "Transport protocol specification for open systems interconnection (OSI) for CCITT applications," 1984.

[15] ISO 8073-1985, "Information processing systems—Open systems interconnection— Connection-oriented transport protocol specification"; Addendum 1, "Network connection management subprotocol"; Draft addendum, "Operation over connectionless-mode network service."

[16] CCITT Recommendation X.215, "Session service definition for open systems interconnection (OSI) for CCITT applications," 1984.

[17] ISO 8326-1985, "Information processing systems—Open systems interconnection—Session service definition"; Draft addendum, "Symmetric synchronization."

[18] CCITT Recommendation X.225, "Session protocol specification for open systems interconnection (OSI) for CCITT applications," 1984.

[19] ISO 8327-1985, "Information processing systems—Open systems interconnection—Session protocol specification"; Draft addendum, "Symmetric synchronization."

[20] CCITT Recommendation X.213, "Network service definition for open systems interconnection (OSI) for CCITT applications," 1984.

[21] ISO 8348-1985, "Information processing systems—Open systems interconnection—Connection-oriented network service definition"; Addendum 1, "Connectionless-mode network service."

[22] Draft ISO 8602, "Information processing systems—Open systems interconnection—Protocol for providing the connectionless-mode transport service utilizing the connectionless-mode network service or the connection oriented network service."

[23] ISO 7498-1985, "Information processing systems—Open systems interconnection—Basic reference model"; Addendum 1, "Connectionless-mode transmission."

[24] CCITT Recommendation X.200, "Reference model of open systems interconnection (OSI) for CCITT applications," 1984.

[25] CCITT Recommendation X.210, "Open system interconnection (OSI) layer service definition conventions," 1984.

[26] ISO TR 8509-1986, technical report, "Information processing systems—Open systems interconnection—Service conventions."

13

OSI Presentation and Application Layers

Paul D. Bartoli

I. Introduction

This chapter discusses the Application and Presentation Layers of the Reference Model of Open Systems Interconnection (OSI) [1]. The Application and Presentation Layers perform functions necessary to exchange information between application processes; the Application Layer is concerned with the semantic aspects of the information exchange, while the Presentation Layer is concerned with the syntactic aspects. The ability to manage the semantic and syntactic elements of the information to be exchanged is key to ensuring that the information can be interpreted by the communicants.

II. The Application Layer

This section discusses the underlying assumptions and viewpoint that led to the development of the OSI Application Layer. The general structure of the Application Layer is presented [2]. This model describes how the Application and Presentation Layers cooperate to support the exchange of meaningful information between applications. The specific functions of each layer are then discussed.

PAUL D. BARTOLI • AT&T Bell Laboratories, Holmdel, New Jersey 07733.

A. The Model

Applications perform information processing to support the activities of an enterprise. An application may be expressed as a set of information processing functions.

The work on the upper two layers of the Reference Model was based on the premise that there are one or more applications that need to be performed on behalf of one or more users, and that the users and systems supporting these applications are distributed. Under these assumptions, an enterprise may be configured such that portions of the applications are implemented in different systems. We call each implementation of a distributed portion of the application on Application Process (AP). When the application is distributed the Application Processes must communicate with each other in order to cooperate to perform the application at hand.

B. Information Processing

Application Processes perform information processing in order to operate on (e.g., create, update, modify) information related to the overall purpose of the application at hand. The collection of information relevant to the overall objectives of the information processing task is called the Information Base. The result of information processing performed by APs is represented by the state of the Information Base.

For example, suppose that the enterprise of concern is an airline and that the application at hand is the provision and maintenance of airline reservations. Further, suppose that the airline reservations application is distributed so that there is an AP that manages the reservations terminal, and an AP that interacts with a flight reservation data base. In order to make a reservation, the AP managing the terminal must communicate with the AP managing the airline data base. The information processing objectives are to allow the APs to cooperate in order to modify the state of the airline data base to reflect that a particular reservation has been made.

Each AP has a Universe of Discourse (UOD), which is that part of the "world" relevant to its information processing activities. The UOD contains the objects that can be dealt with by the AP as well as the relationships among these objects and the rules for manipulating them. A description of the UOD is called a Conceptual Schema. In order for two APs to communicate, there must be an agreement on the semantics of the information that they will deal with in their dialog. This intersection (agreement) must encompass all matters about which exchange of information is intended [3].

This intersection is the shared UOD; it is the "shared world" that the communicating APs hold in common and is the context within which communication can occur. The description of the shared UOD is called the shared Conceptual Schema. For two Application Processes to communicate there must exist a shared UOD and a corresponding shared Conceptual Schema to provide the context for communication. We call the shared Conceptual Schema the Application Context.

C. Application Entities

An Application Entity (AE) is the part of an Application Process that is involved in OSI communications. An AE exists in the Application Layer within an Open System.

At this stage it is necessary to become more rigorous in talking about objects within the Application Layer of OSI. Application Processes reside in specific Open Systems. A particular instance of an Application Process in an Open System is called an Application Process Invocation. Each Application Process Invocation is independent of other invocations of the same Application Process. Application Process Invocations perform the information processing defined by the Application Process. This is analogous to invoking an instance of a program each time the capabilities of that program are required for use (i.e., one can think of an Application Process Invocation as an instance of the type "Application Process").

Similarly, Application Entity Invocations exist in the Application Layer of OSI. Each AE Invocation is a specific instance of a particular Application Entity, and is the part of an Application Process Invocation that is involved in performing communications functions. Thus, Application Process Invocations communicate with each other by using the capabilities of their Application Entity Invocations. These capabilities are provided by Application Service Elements within the individual AE Invocations. AE Invocations communicate using OSI Application Layer protocols. In the OSI environment, communication between AP Invocations is modeled as communication between AE Invocations.

In order for two AE Invocations to communicate, they must be logically "connected" (related) to each other. The term "Application Association" (AA) is used to describe this logical relationship between AE Invocations. Thus, in order for two AE Invocations to communicate with each other, an Application Association must exist between them. The establishment, termination, and maintenance of an AA is an Application Layer matter and is achieved through use of the Association Control Application Layer protocol.

An AE contains one or more application service elements (ASEs). An ASE is a set of functions that provide an OSI capability that supports communication between AEs. ASEs are defined by the specification of a set of application-protocol-data units and the procedures governing their use.

The Application Context describes the specific combination of ASEs comprising an AE and the rules describing the relationships among them.

1. Application Entity Structure

The following sections discuss the structure of an Application Entity. This structure is illustrated in Fig. 1.

Single Association Object. An Application Entity Invocation may support a number of Application Associations either consecutively or concurrently. A single association object (SAO) models the functions within an AE Invocation that are related to the operation of an individual application association. An SAO consists of one or more ASEs, one of which is always the Association Control Service Element. When an AE Invocation contains two or more SAOs, the combinations of ASEs within them do not have to be the same.

Single Association Control Function. The Single Association Control Function (SACF) coordinates the interaction among ASEs, including their use of the Presentation Service, within an SAO. The rules governing this coordination are defined by the application context used within a particular application association. The specifications of ASEs and application contexts contain the information required for the SACF to perform the coordination functions.

Multiple Association Control Function. In order to support certain information processing functions it may be necessary to coordinate the activities of several application associations supported by an AE Invocation. The set of functions in an AE Invocation that controls related activities across several associations is provided, within an AE Invocation, by the Multiple Association Control Function (MACF). The application

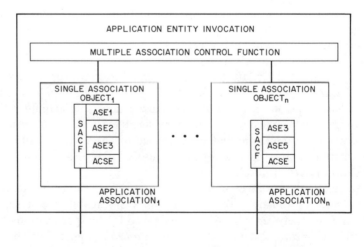

Fig. 1. Application entity structure. ASE, application service element; SACF, single association control function.

context may provide the information necessary for the MACF to perform its coordination functions.

2. Application Service Elements

Several application service elements have been defined thus far for OSI. Among them are the following:

- Association Control Service Element [4].
- File Transfer, Access and Management Service Element [5].
- Commitment, Concurrency, and Recovery (CCR) service element [6].*
- Virtual Terminal Service [7].
- Reliable Transfer Service Element [8].
- Remote Operations Service Element [9].
- Directory Service [10].
- Message Handling Systems [11].
- Office Document Architecture [12].

The above list, while not exhaustive, is intended to give the reader a sense of the types of service elements that are being developed in CCITT and ISO. Work is progressing in ISO on the definition of an ASE for transaction processing capabilities, and there is work in CCITT and ISO on developing authentication capabilities, a necessary building block for providing security in OSI.

The paragraphs that follow discuss the Association Control Service Element and the Commitment, Concurrency, and Recovery Service Element and their relation to the Application Layer structure. These service elements are chosen for discussion because they provide capabilities that form the basis of many OSI applications.

3. Association Control Service Element

The Association Control Service Element (ACSE) provides the capabilities to establish, maintain, and terminate application associations between application entities. These capabilities are required by applications that interoperate using OSI. The development of ACSE allowed the service and protocols for association control (capabilities that are required by applications to interoperate) to be standardized *once*. The result is that current and future application protocols can use the ACSE capabilities for performing association control, rather than reinventing these capabilities (and protocols) each time a new application service is developed. This approach has

*Subsequent to submission of this chapter ISO/IEC JTC1/SC 21 made extensive modifications to the CCR service and protocol. Those modifications are not reflected herein.

avoided duplication of effort and has established a single association control standard for use by OSI applications.

The ACSE provides the following services:

- A-Associate
- A-Release
- A-Abort
- A-P-Abort

A-Associate. The A-Associate service is used to initiate the beginning of use of an association by the application service elements identified by the application context parameters in the A-Associate request primitive. A-Associate is a confirmed service (see Chapter 2). The Association Control Protocol is the set of procedures and protocol data units that provide the Association Control capabilities. The Application Protocol Data Units that support the A-Associate service are carried in the User Data fields of the P-Connect primitives. Thus, the assoication is created simultaneously with the underlying presentation connection.

To establish an association, the requestor issues an A-Associate request primitive. The called Application Entity (AE) is identified by parameters on the request primitive (AE Title or Presentation Address). The ACSE provider issues an A-Associate Indication primitive to the responder. The responder may accept or reject the association request by sending an A-Associate response primitive with the appropriate result parameter. The ACSE provider issues an A-Associate confirm primitive conveying the result parameter to the requestor.

If the responder accepts the association request, the association is established and available for use. Service elements in the application context, other than A-Associate, may now be invoked by the AEs. If the responder rejects the association request, the association does not exist.

If the ACSE provider cannot support the requested association, it issues an A-Associate confirm primitive to the requestor with a negative Result parameter and the association does not exist.

A-Release. The A-Release service is used by either AE to indicate the completion of use of an association by the Application Service Elements identified by the application context that is in effect. A-Release is a confirmed service. The Application Protocol Data Units that support the A-Release service are carried in the User Data fields of the P-Release primitives. Thus, the association is terminated simultaneously with the underlying presentation connection.

A-Abort Procedure. The A-Abort is used by either AE to abnormally terminate the association. The peer AE is informed of the abort. Use of this service may result in loss of information in transit. The A-Abort service is

Table I. Association Context Service Primitives and Parameters

Service primitive name	Parameters
A-ASSOCIATE request A-ASSOCIATE indication	Calling AP title[a] Calling AE qualifier[a] Calling AP invocation-identifier[a] Calling AE invocation-identifier[a] Called AP title[a] Called AE qualifier[a] Called AP invocation-identifier[a] Called AE invocation-identifier[a] Responding AP title[a] Responding AE qualifier[a] Responding AP invocation-identifier[a] Responding AE invocation-identifier[a] Application context name[a] User information Result Result source Diagnostic[a] Calling presentation address Called presentation address Responding presentation address Presentation context definition list[a] Presentation context definition result[a, b] Default presentation context name[a] Default presentation context result[a, b] Quality of service Presentation requirements[a] Session requirements Initial synchronization point serial number Initial assignment of tokens Session-connection identifier
A-ASSOCIATE response A-ASSOCIATE confirmation	Responding AP title[a] Responding AE qualifier[a] Responding AP invocation-identifier[a] Responding AE invocation-identifier[a] Application context name[a] User information Result Diagnostic[a] Responding presentation address Presentation context definition result list[a] Default presentation context result[a] Quality of service Presentation requirements Session requirements Initial synchronization point serial number Initial assignment of tokens Session-connection identifier

(*Continued*)

Table I. (*Continued*)

Service primitive name	Parameters
A-RELEASE request	Reason for release[a]
A-RELEASE indication	User information[a]
A-RELEASE response	Reason[a]
A-RELEASE confirmation	User information[a]
	Result[e]
A-ABORT request	Abort source[a, b]
A-ABORT indication	User information
A-P-ABORT indication	Provider reason

[a] Note used in X.410 mode.
[b] Present in indication only.
[c] Present in response and confirmation only.

unconfirmed. The A-Abort procedure is supported by the P-U-Abort service of the Presentation Layer; when the A-Abort service is used, the association and the underlying Presentation Connection are simultaneously aborted.

A-P-Abort Procedure. The A-P-Abort is used to abnormally terminate an association because of problems in layers below the Application Layer. Occurrence of the A-P-Abort may result in loss of information in transit. The A-P-Abort service is provider initiated.

Table I provides a list of the ACSE primitives and their parameters.

4. Use of the Association Control Service Element

This section illustrates an example of the use of the Association Control Service Element in establishing an association between AE Invocations. The example illustrates the role of the SACF as described in Section 1 on Application Entity structure.

Consider an AE composed of the ACSE, the FTAM service element, and the Virtual Terminal (VT) application service element. It is the single association coordinating function (SACF) that is responsible for coordinating the use of the association control service element, the FTAM application service element, and the VT application service element over a particular association. The SACF is responsible for determining when it is appropriate to establish an association, to invoke the FTAM or VT service elements (or switch between the FTAM and VT service elements), and to terminate an association. In essence, the SACF is the collective intelligence of the application entity, which, on the basis of the overall activities of the application entity, can invoke the appropriate service elements to accomplish the communications objectives of the application entity.

In OSI, each ASE has been designed to operate without having to be aware of the details of other ASEs that may be included in the same application entity. Thus the conditions under which it is appropriate to establish an association, invoke an ASE, or switch from one ASE to another are not known to the individual ASEs. It is only the SACF that has the collective information of the application entity to perform such coordination. This information is obtained from the particular application context selected for use within the association.

5. The CCR Application Service Element

The CCR ASE provides the capabilities for AEs to coordinate their processing activities. A distributed application is performed through the cooperation of distributed application processes. The APs communicate with one another to cooperate to perform the application at hand. Pairwise communication between APs is supported by application associations between AEs. Application associations are provided by the ACSE as discussed above. In order to perform a distributed application it may be necessary to coordinate the processing activities occurring within multiple application associations. The CCR ASE provides the capabilities to coordinate the activities within multiple associations related to a particular application. The CCR ASE provides service primitives and protocol elements to coordinate activities across associations as an atomic action. The model used in the CCR standard describes an atomic action as a tree structure of AEs. The model consists of two or more communicating AEs; additional AEs may be added to the tree as a result of protocol interactions on existing associations.

The CCR ASE defines master/subordinate relationships. The master of an atomic action can determine whether its subordinates are prepared to perform a particular action. If all subordinates indicate that they are ready to carry out the action, all subordinates are ordered to commit (perform the action). If any one of the subordinates indicates that it is not prepared to carry out the requested action, the master orders all subordinates to rollback and the atomic action is terminated. This mechanism is called two-phase commitment. In Phase I the master determines whether all subordinates are ready to commit; in Phase II the master decides whether to order all subordinates to commit or rollback depending on the result of Phase I.

The CCR commitment capabilities allow applications to coordinate the modification of protected resources. Once commitment is performed, modifications to protected data are final in the sense that no CCR recovery capability exists to undo the effect of the committed actions.

The CCR model reflects the tree structure of atomic actions by allowing CCR users to serve as:

- A master in relation to one or more subordinates; or
- A subordinate in relation to one master; or
- A subordinate to one master and as a master to one or more subordinates.

It should be noted that the nature of an atomic action is determined by the particular Application Layer protocol used within the atomic action. Similarly, the action taken by an application when some or all subordinates have refused to offer commitment depends on the nature of the application at hand.

Table II presents the CCR primitives and their parameters.

Table II. Commitment, Concurrency, and Recovery Service Primitives and Parameters

Service primitive name	Parameters
C-BEGIN request C-BEGIN indication	Atomic action identifier Branch identifier Atomic action timer User data
C-PREPARE request C-PREPARE indication	User data
C-READY request C-READY indication	User data
C-COMMIT request C-COMMIT indication C-COMMIT response C-COMMIT confirmation	No parameters
C-REFUSE request C-REFUSE indication	User data
C-RESTART request C-RESTART indication	Resumption point Atomic action identifier Branch identifier Restart timer User data
C-RESTART response C-RESTART confirmation	Resumption point User data
C-ROLLBACK request C-ROLLBACK indication C-ROLLBACK response C-ROLLBACK confirmation	No parameters

III. The Presentation Layer

This section discusses the purpose of the Presentation Layer and its relation to the Application Layer. The discussions presented above on the Application Layer centered on the capabilities provided by various Application Layer components. In OSI Application Layer standards these capabilities are described in terms of their semantics and the required sequences of information that must be exchanged between Application Entities in order to carry out particular capabilities.

Thus, communication between application processes supporting a given distributed application is described in terms of:

1. The set of things that can be done as part of the application (i.e., the shared conceptual schema and application service elements).
2. The set of application protocol data unit (APDU) types associated with the Application Layer protocol to be used to support the application.
3. The required sequences of information (instances of APDU types) that must be exchanged to accomplish each of the actions described in (1) above.

In OSI, an Application Layer protocol is regarded as a *language* used between Application Entities to communicate in order to cooperate to perform aspects of the application at hand. Communication is the exchange of sentences, relevant to some UOD, between AEs; sentences represent semantics with respect to the relevant Universe of Discourse. Particular APDUs are sentences in a particular language (particular Application Layer protocol).

The term *abstract syntax* is used to refer to the categories of constructs in a language and the relationships among the constructs (e.g., terms, sentences); this includes the definition of what constitutes well-formed (grammatically correct) constructs in the language and is particularly concerned with what constitutes a well-formed sentence.

If expressions in a language are to be unambiguously manipulated by digital information processing machinery, the language must be formal. A language is formal if there exists an algorithm that can, by examination of the elements of the alphabet in an expression, determine whether the expression is grammatically correct in the language and unambiguously distinguish among the grammatical constructs in the expression (e.g., distinguish between names and sentences).

ISO and CCITT have defined a language for describing abstract syntaxes. This language is called Abstract Syntax Notation One (ASN.1) [13]. ASN.1 is a language that allows datatypes to be defined and values of these types to be specified. ASN.1 datatypes are defined independently of the way types and values are to be represented during information transfer between open systems. Complex datatypes may be defined as combinations of simpler datatypes.

All of these aspects are described as part of the Application Layer protocols without regard to the concrete representation used in the transfer of information between application entities in order to accomplish the application at hand. Selecting the concrete representation(s) for use in an application association and employing them is the purpose of the Presentation Layer [14].

The Presentation Layer is responsible for the representation of information, exchanged between application entities, in an agreed concrete *transfer syntax*. A concrete transfer syntax is defined as the set of those aspects of the prescriptive rules used in specifying a language that specify which marks, relative to some alphabet, constitute the elements of the language. Informally, the transfer syntax is the result of applying a particular encoding to a particular abstract syntax; the transfer syntax determines the particular bit patterns that will be used to represent information in transit between open systems. The ASN.1 Basic Encoding Rules Standard [15] specifies rules that may be used to encode the types and their values defined in ASN.1. Application of these encoding rules to ASN.1 constructs produces a concrete transfer syntax.

The Presentation Layer is responsible for representing the information exchanged between Application Entities in a manner that preserves it meaning. AEs are responsible for determining the abstract syntaxes to be used on a particular association and informing the Presentation Layer of these abstract syntaxes. Once given the set of abstract syntaxes to be used during an association, Presentation Entities serving the peer AEs agree on mutually acceptable concrete transfer syntaxes. The Presentation Layer mediates the choices of concrete transfer syntaxes to be used on a given presentation connection to support a given Application Association.

The Presentation Layer does this by negotiating transfer syntaxes and by transforming to and from the agreed transfer syntax.

The Presentation Layer protocol supports negotiation of transfer syntaxes, and the Presentation Layer contains the functions required for transformations between the transfer syntax in effect to the syntaxes used in the local systems [16]. The result of successful transfer syntax negotiation is a pairing of an abstract syntax with a compatible transfer syntax. This pair is called a *presentation context*. A presentation context, as viewed by the communicating AEs, represents an instance of use of an abstract syntax.

A. Information Transfer

Information exchanged between AEs is carried in user data parameters of presentation primitives. Information crossing the Application/Presentation service interface is structured as a list of typed data values. The data type identifies the presentation context applicable to the data value and the syntactic description of the data value within that context.

B. Definition of Presentation Contexts

The Presentation Layer provides the capabilities to define presentation contexts that support the information transfer requirements of AEs. One or more presentation contexts may be defined to support the requirements of communicating AEs.

The presentation services by which presentation contexts may be defined are P-Connect and P-Alter-Context. When a presentation context is defined it is placed in the defined context set. When a context is added to the defined context set it is available for immediate use. Contexts may also be deleted by the P-Alter-Context service.

If the defined context set is empty, the information is transferred in the default context. Information transferred using the P-Expedited-Data service is always transferred in the default context. The default context may be defined, but not subsequently redefined, using the presentation service.

C. Presentation Layer Facilities

This section presents a summary of the Presentation Layer facilities, as described in the Presentation Layer International Standard, which provide

1. Connection establishment.
2. Connection termination.
3. Presentation context management.
4. Information transfer.
5. Dialog control.

D. Presentation Layer Functional Units

The functional units visible at the Application/Presentation Layer boundary are the functional units provided by the Presentation Layer *and* the Session functional units defined in ISO 8326/8327 [17] and CCITT X.215/X.225[18] which are made directly available (by the Presentation Layer) to the Application Layer.

Session functional units that provide token management, synchronization, resynchronization, exception reporting, and activity management affect the state of the Presentation Layer. Therefore, in certain cases the Presentation Layer imposes additional restrictions on the use of the services provided by these functional units (e.g., in order to appropriately manage the state of the defined context set, as discussed below).

The functional units visible at the Application/Presentation Layer interface are thus the following:

a. *Presentation Layer Functional Units* as defined in ISO 8822/CCITT X.216

1. Presentation kernel functional unit.
2. Context management functional unit.
3. Context restoration functional unit.

b. *Session Layer Functional Units* as defined in ISO 8326/CCITT X.215:

1. Kernel functional unit.
2. Negotiated release functional unit.
3. Duplex functional unit.
4. Half duplex functional unit.
5. Expedited data functional unit.
6. Typed data functional unit.
7. Capability data exchange functional unit.
8. Minor synchronize functional unit.
9. Major synchronize functional unit.
10. Resynchronize functional unit.
11. Exceptions functional unit.
12. Activity management functional unit.

The presentation kernel is always available. The context management functional unit is optional and its use is negotiable. The context restoration functional unit is optional and its use is negotiable; the context restoration functional unit is only permitted to be selected if the context management functional unit is also selected for use during a presentation connection at connection establishment time.

E. Management of the Defined Context Set

The defined context set may be altered during a presentation connection only if the context management presentation functional unit is selected at presentation connection setup time.

The P-Alter-Context service request is used to alter the defined context set. The Presentation Layer is responsible for ensuring that the defined context set is identical at both ends of the presentation connection; thus P-Alter-Context is a confirmed service. It should be noted that certain destructive services may collide with or overtake the Presentation Protocol Data Units (PPDUs) for altering the defined context set. In particular, the

P-Resynchronize, P-Activity-Interrupt, and P-Activity-Discard are such destructive services.

F. Context Restoration

When the context restoration functional unit is selected at presentation connection establishment time, the presentation service remembers the defined context set at particular points in the information flow. If an Application Entity requests a return to one of these points, the defined context set is restored to what it was at that point.

The defined context set is remembered by the context restoration functional unit at the following points in the information flow:

1. Major Synchronization points.
2. Minor Synchronization points.
3. Activity Interrupt points.
4. The start of the presentation connection.

Table III lists the Presentation Layer service primitives and their parameters.

Table III. Presentation Service Primitives and Parameters

Service primitive name	Parameters
P-CONNECT request	Calling presentation address
	Called presentation address
	Multiple defined contexts
	Presentation context definition list
	Default context name
	Quality of service
	Presentation requirements
	Session requirements
	Initial synchronization point serial number
	Initial assignment of tokens
	Session connection identifier
	User data
P-CONNECT indication	Calling presentation address
	Called presentation address
	Multiple defined contexts
	Presentation context definition list
	Presentation context definition result list
	Default context name
	Default context result

(Continued)

Table III. (*Continued*)

Service primitive name	Parameters
	Quality of service
	Presentation requirements
	Session requirements
	Initial synchronization point serial number
	Initial assignment of tokens
	Session connection identifier
	User data
P-CONNECT response	Responding presentation address
P-CONNECT confirm	Multiple defined contexts
	Presentation context definition result list
	Default context result
	Quality of service
	Presentation requirements
	Session requirements
	Initial synchronization point serial number
	Initial assignment of tokens
	Session connection identifier
	Result
	User data
P-RELEASE request	User data
P-RELEASE indication	
P-RELEASE response	Result
P-RELEASE confirm	User data
P-U-ABORT request	Presentation context identifier list
P-U-ABORT indication	User data
P-P-ABORT indication	Provider reason
	Abort data
P-ALTER-CONTEXT request	Presentation context definition list
	Presentation context deletion list
	User data
P-ALTER-CONTEXT indication	Presentation context definition list
	Presentation context deletion list
	Definition result
	User data
P-ALTER-CONTEXT response	Presentation context definition result list
P-ALTER-CONTEXT confirm	Presentation context deletion result list
	User data
P-TYPED-DATA request	User data
P-TYPED-DATA indication	
P-DATA request	User data
P-DATA indication	

Table III. (*Continued*)

Service primitive name	Parameters
P-EXPEDITED-DATA request P-EXPEDITED-DATA indication	User data
P-CAPABILITY-DATA request P-CAPABILITY-DATA indication P-CAPABILITY-DATA response P-CAPABILITY-DATA confirm	User data
P-TOKEN-GIVE request P-TOKEN-GIVE indication	Token item
P-TOKEN-PLEASE request P-TOKEN-PLEASE indication	Token item User data
P-CONTROL-GIVE request P-CONTROL-GIVE indication	No parameters
P-SYNC-MINOR request P-SYNC-MINOR indication	Type Synchronization pint serial number User data
P-SYNC-MINOR response P-SYNC-MINOR confirm	Synchronization point serial User data
P-SYNC-MAJOR request P-SYNC-MAJOR indication	Synchronization point serial number User data
P-SYNC-MAJOR response P-SYNC-MAJOR confirm	User data
P-RESYNCHRONIZE request	Resynchronize type Synchronization point serial number Token item User data
P-RESYNCHRONIZE indication	Resynchronize type Synchronization point serial number Token item Presentation context identifier list User data
P-RESYNCHRONIZE response	Synchronization point serial number Token item User data
P-RESYNCHRONIZE confirm	Synchronization point serial number Token item Presentation context identifier list User data
P-U-EXCEPTION-REPORT request P-U-EXCEPTION-REPORT indication	Reason User data
P-P-EXCEPTION-REPORT indication	Reason
P-ACTIVITY-START request P-ACTIVITY-START indication	Activity identifier User data

(*Continued*)

Table III. (*Continued*)

Service primitive name	Parameters
P-ACTIVITY-RESUME request P-ACTIVITY-RESUME indication	Activity identifier Old activity identifier Synchronization point serial number Old session connection identifier User data
P-ACTIVITY-END request P-ACTIVITY-END indication	Synchronization point serial number User data
P-ACTIVITY-END response P-ACTIVITY-END confirm	User data
P-ACTIVITY-INTERRUPT request P-ACTIVITY-INTERRUPT indication	Reason
P-ACTIVITY-INTERRUPT response P-ACTIVITY-INTERRUPT confirm	No parameters
P-ACTIVITY-DISCARD request P-ACTIVITY-DISCARD indication	Reason
P-ACTIVITY-DISCARD response P-ACTIVITY-DISCARD confirm	No parameters

IV. An Example of Association Establishment

This section presents a simple example of how association establishment occurs between two AEs that each contain the association control ASE and the FTAM ASE. Connection establishment in the Presentation and Session Layers is also discussed. It is assumed for simplicity that the recipient AE is capable of supporting the Application, Presentation, and Session Layer capabilities requested by the originating AE.

When the application process initiates file transfer operations, the single association control function (SACF) in the AE of the initiating application process issues an A-Associate Request. The SACF ensures that the environment required by the FTAM ASE (e.g., application context, Presentation and Session Layer parameters) is reflected in the values of the parameters on the A-Associate Request primitive. In addition, the FTAM-specific parameters such as FTAM functional units requested are placed within the user data portion of the A-Associate Request primitive.

When the A-Associate Request primitive is issued, the ACSE issues a P-Connect Request primitive and sets the values of the Presentation parameters to reflect the Presentation requirements expressed in the Presentation parameters on the A-Associate primitive as well as its own Presentation requirements, and embeds the A-Associate Request APDU in the user data field of the P-Connect primitive.

When the P-connect request primitive is issued, the Presentation entity issues an S-Connect primitive, sets the parameters on the S-Connect Request primitive to reflect the AEs requirements, and embeds the P-Connect Request PPDU in the user data field of the S-Connect Request primitive.

The initiating Session entity then obtains a Transport connection from the Transport Layer and sends an S-Connect Request SPDU (over the Transport connection) to the destination Session entity.

Note that the Session connect request SPDU contains the P-Connect Request PPDU within its user data field and the P-Connect Request PPDU contains the A-Associate Request APDU within its user data field—which, in turn, contains the FTAM-specific parameters within its user data field. Thus, all of this information is conveyed in a single Transport Service Data Unit (TSDU).

Upon receipt of the S-Connect Request SPDU, the receiving Session entity examines the Session Layer protocol control information (PCI) in the header of the SPDU. If the Session entity can support the requested Session functional units and the values of other Session parameters, the Session entity passes the user data portion of the S-Connect Request SPDU (the P-Connect PPDU) to the receiving Presentation entity. If the receiving Presentation entity can support the requested Presentation functional units and the values of other Presentation parameters, the Presentation entity passes the user data portion of the P-Connect PPDU (the A-Associate Request APDU) to the receiving AE.

The receiving AE will accept the A-Associate Request if:

1. The association control PCI in the A-Associate Request is correct and acceptable to the ACSE; and
2. The requested environment (application context for FTAM) is acceptable to the receiving AE.

Item 2 above involves actions of the AE as a whole. Current Application Layer work suggests that the decisions required to be made in item 2 are modeled by the SACF.

If the recipient AE (SACF) accepts the A-Associate Request it issues an A-Associate Response accepting the proposed application context; the ACSE then issues a P-Connect Response, and in so doing sets the parameters on the P-Connect Response and embeds the A-Associate Response APDU in the user data field of the P-Connect response primitive.

When the P-Connect Response primitive is issued, the Presentation entity issues an S-Connect Response primitive and in so doing sets the Session parameters on the S-Connect Response primitive appropriately, and embeds the P-Connect Response PPDU in the user data field of the S-Connect Response primitive. The Session entity then sends the SPDU to the Session entity in the originating system, using the established Transport connection, again all within a single Transport Service Data Unit.

Upon receipt of the S-Connect Response SPDU the Session entity in the originating system passes the user data portion of the SPDU (the P-Connect Response PPDU) to the Presentation Layer, which then passes the user data portion of the PPDU (the A-Associate Response APDU) to the Application Layer.

In this way the Application Association, Presentation connection, and Session connection are established using one round trip exchange of embedded Session, Presentation, and Application PDUs. Thus, connection establishment in the upper three layers is performed in an efficient manner from the point of view of required PDU exchanges.

V. Future Directions

This chapter has dealt with the current state of OSI Application and Presentation Layer standards. In looking to the future of information technology standards, several key trends are emerging. Now that the basic OSI capabilities are in place, greater emphasis is being focused on the information processing aspects of interoperability. This includes the ability to communicate and share information at the semantic level and the ability to specify and obtain information processing functions required to support the activities of users. A major modeling effort called Open Distributed Processing (ODP) has been initiated in ISO. The purpose of this effort is to develop a model of information processing that describes the relationship among areas such as data base, communications, processing, and storage, and to determine what additional standards are required to move the work forward.

Another area of concern to information technology standards groups is the development of a model that describes voice, data, and integrated voice/data capabilities at all layers. One of the key questions here is, how do sophisticated voice capabilities such as those provided in an Integrated Services Digital Network (ISDN) environment manifest themselves at the Application Layer—i.e., how should voice and integrated voice/data capabilities be made available to application processes in the OSI environment?

It is the author's belief that architectural issues such as those mentioned above will represent much of the new work undertaken in the next five to ten years in the information technology arena and that there is an enormous potential for major advancements as we move into the era of information movement and management.

References

[1] ISO 7498, "Information processing systems—Open Systems Interconnection—Basic Reference Model," 1984. CCITT Recommendation X.200, "Reference model of open systems interconnection for CCITT applications," 1984 (updated expected in 1988).

[2] ISO DIS 9545, "Information processing systems—Open Systems Interconnection—Application Layer structure," September 1988.

[3] ISO TR 9007, "Concepts and terminology for the conceptual schema and the information base," 1985.

[4] ISO 8649, "Information processing systems—Open systems interconnection—Service definition for the association control service element," 1988. ISO 8650, "Information processing systems—Open systems interconnection—Protocol specification for the association control service element," 1988. CCITT Recommendation X.217, "Association control service definition for open systems interconnection for CCITT applications," 1988. CCITT Recommendation X.227, "Association control protocol specification for open systems interconnection for CCITT applications," final text December, 1987.

[5] ISO 8571, "Information processing systems—Open systems interconnection—File transfer, access, and management," Parts 1–4, 1988.

[6] ISO/DIS 9804, "Information processing systems—Open systems interconnection—Service definition for commitment, concurrency, and recovery," 1988 (text in SC 21 N 2573, March, 1988). ISO DIS 9805, "Information processing systems—Open systems interconnection—Protocol specification for commitment, concurrency, and recovery," 1988 (text in SC 21 N 2574, March, 1988). CCITT Recommendation X.237, "Commitment, concurrency, and recovery service definition," Draft Text, 1988. CCITT Recommendation X.247, "Commitment, concurrency, and recovery protocol specification, Draft Text, 1988.

[7] ISO DIS 9040, "Information processing systems—Open systems interconnection—Virtual terminal service—Basic class," 1988 (text in SC 21 N 2615, March, 1988). ISO DIS 9041, "Information processing systems—Open systems interconnection—Virtual terminal protocol—Basic class," 1988 (text in SC 21 N 2616, March, 1988).

[8] ISO DIS 9066-1, "Reliable transfer service, 1988 (text in SC 18 N 1408, March, 1988). ISO DIS 9066-2, "Reliable transfer protocol specification," 1988 (text in SC 18 N 1409). CCITT Recommendation X.218, "Reliable transfer: Model and service definition," 1988. CCITT Recommendation X.228, "Reliable transfer: Protocol specification," 1988.

[9] ISO DIS 9072-1, "Remote operations service," 1988 (text in SC 18 N 1410, March, 1988). ISO DIS 9072-2, "Remote operations protocol specification," 1988 (text in SC 18 N 1411, March, 1988). CCITT Recommendation X.219, "Remote operations: Model, notation, and service definition," 1988. CCITT Recommendation X.229, "Remote operations: Protocol specification," 1988.

[10] ISO DIS 9594, "Information processing—Open systems interconnection—The directory," parts 1–8, 1988 (text in SC 21 N 2751 through N 2758, April, 1988). CCITT X.500, "Series recommendations on directory," November, 1987.

[11] ISO DIS 10021, "Information processing—Text communication—Message oriented text interchange system," 1988 (text in SC 18 N 1487 through N 1493, May, 1988). CCITT X.400, "Series recommendations for message handling systems," 1988.

[12] ISO 8613/1–8, "Office document architecture and interchange format," 1988, awaiting publication. CCITT T.400, "Series recommendations for document architecture, transfer, and manipulation," 1988.

[13] ISO 8824, "Information processing systems—Open systems interconnection—Specification of abstract syntax notation one (ASN.1)," 1987; and ISO 8824/PDAD 1, "Information processing systems—Open systems interconnection—Specification for ASN.1:

Proposed draft Addendum 1 on ASN.1 extensions," 1988 (final text in SC 21 N 2341 Revised, April, 1988). CCITT Recommendation X.208, "Specification of abstract syntax notation one (ASN.1)," 1988.

[14] ISO 8822, "Information processing systems—Open systems interconnection—Connection oriented presentation service definition," 1988. CCITT Recommendation X.216, "Presentation service definition for open systems interconnection for CCITT applications," 1988.

[15] ISO 8825, "Information processing—Open systems interconnection—Specification of basic encoding rules for abstract syntax notation one (ASN.1)," 1987; and ISO 8825/PDAD 1, "Information processing systems—Open systems interconnection—Specification of basic encoding rules for ASN.1: Proposed draft addendum 1 on ASN.1 extensions," 1988 (text in SC 21 N 2342 Revised, April, 1988). CCITT Recommendation X.209, "Specification of basic encoding rules for abstract syntax notation one (ASN.1)," 1988.

[16] ISO 8823, "Information processing systems—Open systems interconnection—Connection oriented presentation protocol specification," 1988. CCITT Recommendation X.226, "Presentation protocol specification for open systems interconnection for CCITT applications," 1988.

[17] ISO 8326, "Information processing systems—Open systems interconnection—Basic connection oriented session service definition," 1987; and ISO 8326/AD 2, "Information processing systems—Open systems interconnection—Basic connection oriented session service definition—Addendum 2: Incorporation of unlimited user data," 1988. ISO 8327, "Information processing systems—Open systems interconnection—Basic connection oriented session protocol specification," 1987; and ISO 8327/AD 2, "Information processing systems—Open systems interconnection—Basic connection oriented session protocol specification—Addendum 2: Unlimited session user data protocol specification," 1988.

[18] CCITT Recommendation X.215, "Session service definition for open systems interconnection for CCITT applications," 1988. CCITT Recommendation X.225, "Session protocol specification for open systems interconnection for CCITT applications," 1988.

Message Handling System Standards and Office Applications

Ronald P. Uhlig

I. Introduction

The author of this chapter made the following prediction in 1977: "During the next 50 years...message systems will have as great an impact on the way business is done in our society as the impact that the telephone had on business practices during the last 100 years" [1]. That same article went on to describe some of the benefits of electronic messaging used for interpersonal interaction. Messaging was described in terms of an electronic post office box—a kind of data base where messages could be entered by a sender and retrieved at a later time by one or more recipients. This approach to messaging grew out of experiences on the U.S. Department of Defense ARPANET in the mid-1970s.

By 1976 a sizable community of ARPANET users were exchanging electronic messages with each other while conducting research. In addition, many senior managers in the Defense Department and its contractors were finding electronic messaging to be very beneficial in conducting business.

At that time, message systems were largely the product of computer scientists. Different managers wanted different special features, and by 1978 a significant number of incompatible electronic message systems had been created both within the USA and in other countries. They were incompatible in the sense that they could not receive messages sent from other systems.

RONALD P. UHLIG • Northern Telecom Inc., Richardson, Texas 75081.

Seeing a prospect for chaos, a number of people perceived the need for international standards. In September 1978 three persons, Ian Cunningham, John Pickens, and the author, wrote a proposal to the Data Communications Technical Committee (TC-6) of the International Federation for Information Processing (IFIP) to create a working group on messaging systems.

The proposal was accepted, and, following ratification by the General Assembly in 1979, IFIP Working Group 6.5 came into being. Through numerous meetings of the working group, both in Europe and the USA, an IFIP electronic mail/messaging model with wide international consensus was developed by mid-1981. This work was used as the basis for the new Rapporteur group created in CCITT in mid-1981. The Rapporteur group worked very hard over the next three years, turning the IFIP model into the X.400 set of recommendations for Message Handling Systems adopted at the October 1984 Plenary of the CCITT. A complete list of the recommendations is given in Table I.

Table I. CCITT Message Handling system (MHS) X.400 Series Recommendations

X.400	MHS: System Model-Service Elements.
	Introduces functional model. Describes naming and addressing. Defines service elements. Introduces layered model of MHS.
X.401	MHS: Basic Service Elements and Optional User Facilities.
	Overviews Interpersonal Messaging and Message Transfer Services and categorizes optional user facilities of each.
X.408	MHS: Encoded Information Type Conversion Rules.
	Specifies algorithms used by MHS for conversion.
X.409	MHS: Presentation Transfer Syntax and Notation.
	Defines the high-level language used to describe protocol data structures. Also defines how they are encoded.
X.410	MHS: Remote Operations and Reliable Transfer Server.
	Defines the mechanism used to structure Submission and Delivery protocol, P3, and describes the transfer mechanism used between peer entities. Also describes how P1 and P3 use presentation and session layer services.
X.411	MHS: Message Transfer Layer.
	Defines the conceptual "layer service" provided by the Message Transfer Layer and the peer protocols of that layer (P1 and P3).
X.420	MHS: Interpersonal Messaging User Agent Layer.
	Defines the conceptual operation of UA entities within the User Agent Layer (P2) and Simple Formattable Documents.
X.430	MHS: Access Protocol for Teletex Terminals.
	Specifies the protocol for Teletex Terminals to use in accessing MHS to provide Interpersonal Messaging Service to Teletex users.

II. Overview of the Message Handling System Model

CCITT Recommendation X.400, the first of the series, describes the Message Handling System Model and addressing principles. It defines all service elements and introduces the protocol architecture. Understanding the model is a key to understanding the remainder of the series of recommendations.

The model distinguishes functions associated with message transfer. The model has a User Agent (UA), which carries out all functions associated with message preparation and receipt, and a Message Transfer Agent (MTA), which handles "electronic envelopes" containing electronic messages prepared by a User Agent (UA).

Each user has a UA. There may be circumstances in which a user has more than one UA.

The User Agent provides all services needed for a user to prepare a message and submit it to an MTA. These include functions like prompting the user to fill in address fields for the message, a subject field, and the message contents. It is important to note that the standards provide for multimedia content. The message content may be text, voice, image, or graphics. The MTA will not normally examine the contents of a message. It simply treats the contents as a set of bits to be delivered to one or more receiving UAs.

The UA will provide assistance to the user in entering the contents of the message. For text content, a typical UA will provide some form of text editing support. For voice content, the UA will provide a mechanism for input of voice, and possibly, voice editing. The standards allow the contents to be mixed, with both voice and text (as well as graphics and image) in a single message.

The User Agent also provides everything a user needs to act on a message received from an MTA. This includes functions like "scanning," to get a summary of messages, displaying message content (text and graphics), playing voice content, and acting on messages (replying, filing, forwarding, deleting, ...).

A Message Transfer Agent receives messages submitted by User Agents. Normally a large number of UAs will interact with a Message Transfer Agent (MTA). The MTA examines only the addressing information supplied by the submitting UA. This may be thought of as addressing information on an "electronic envelope" containing the message. Depending on the address, the MTA may pass the envelope containing the message to another MTA, or direct to a receiving UA (if the receiving UA can connect directly to the MTA).

Three protocols have been defined in the X.400 Series of standards. These are: P1, which defines the protocol for interaction between MTAs; P3, which defines the interface protocols for a UA to submit a message to

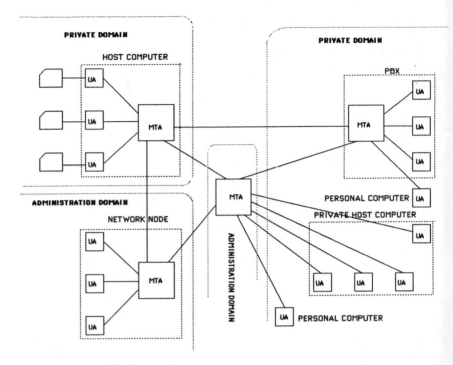

Fig. 1. Examples of possible implementations.

an MTA and for an MTA to deliver a message to a UA; and P2, which defines the message content protocol between cooperating User Agents that are providing what is known as an Interpersonal Messaging Service (IPMS).

The P1 and P3 protocols, which prescribe how the "electronic envelope" information must appear, are described in detail in Recommendation X.411. The P2 protocol for the Interpersonal Message Service is described in detail in Recommendation X.420.

To understand the enormous flexibility of the CCITT Message Handling System Recommendations and their implications for future services, it is important to understand some of the ways in which the MHS recommendations might be implemented. Figure 1 shows some of the many possible implementations.

III. MHS and Management Domains

The set of all MTAs and their interactions make up the Message Transfer Service. An MTA may be implemented as a public service or it may be implemented in a private system (host computer or PBX). A collection of equipment that is owned and administered by an organization is referred to as a Management Domain (MD). If the MD is operated by a public carrier or administration, it is called an Administration Domain (ADMD). Otherwise it is a Private Management Domain (PRMD).

The MDs are the administrative units that make up the Message Handling Systems. The MHS Recommendations define the interactions that must (required) or could (optional) take place between the MD and its subscribers.

Note that UAs may be colocated with an MTA or may be separate from it. A UA may be implemented in many different locations, including personal computers.

The recommendations cover three areas: (1) how an MTA in one MD interacts with an MTA in another MD, (2) how a UA interacts with an MTA, and (3) how UAs interact with other UAs to provide an Interpersonal Messaging Service through an MTA.

The recommendations do *not* prescribe how MTAs interact with other MTAs in the same MD. A service provider may choose to use the standard, or may add proprietary extensions, or may simply provide an interface to MTAs in other MDs. Similar considerations apply to interaction between UAs and between UAs and an MTA in the same MD.

IV. Addressing

An important aspect of Message Handling Systems is addressing. Within a Management Domain, the MD has complete freedom to define the form of address used to locate its users. However, the X.400 series deals with interdomain addressing. The mechanism is closer to the kind of addresses used in postal services than the mechanisms used in the telephone network.

Many early messaging networks used addresses related to a specific computer. For example, ARPANET used the name of an account coupled with the name of a particular host computer in the network. This form of address, although widely used in a number of messaging networks, has the disadvantage of being "user unfriendly."

Basic work on addressing began in IFIP Working Group 6.5. This work, carried out between 1981 and 1983, emphasized "User Friendly" addressing techniques. A part of the IFIP WG 6.5 work was adopted into the X.400 series. Later work, particularly the WG 6.5 recommendation, "A User Friendly Naming Convention for Communication Networks" [2], has been adopted by the CCITT Rapporteur Group on Directory Services as a basis for deliberations during the 1985–1988 study period. This will be discussed further below.

The current X.400 recommendations use a two-part addressing scheme. The first part identifies an Administration Management Domain (ADMD), which is assumed to be unique within a country. Some examples include MCI MAIL in the USA and ENVOY 100 in Canada. The combination of country name plus ADMD name is used to identify a particular messaging service.

The X.400 Recommendation provides for users of a messaging service to be identified by means of one or more attributes, which are selected from a standard list as follows:

- Personal Name (Surname, Given name(s), Initials).
- Organization Name.
- Organization Unit (e.g. Department, Division).
- Unique UA Identifier.
- Terminal (X.121) address.

The last item has been included to provide for interworking with Telematic Services.

There are a number of reasons why this approach to addressing was chosen. First, it provides for a degree of user friendliness by allowing the use of attributes that are reasonably common for addressing all over the world. At this stage, until current ongoing work on directory services is completed and implemented, there is no alternative to use of Domain names, which may not be particularly user friendly. It is hoped that the use of Domain names will become unnecessary in the future. A second important reason for this approach to message addresses is that it is largely independent of location and the electronic route used to reach a particular user. Some addressing schemes require specification of a route and/or a location, but these schemes have proved unwieldy in practice.

A third feature of the addressing mechanism currently provided in the X.400 series is that it provides considerable freedom to Management Domains. Different MDs can select different attributes and address forms. This is also true for Private Management Domains. It is important to note that the current recommendations require that a message be specifically addressed to the Administration MD to which a Private MD is attached. It is intended that the standard list of attributes will be sufficiently expanded in the future, so that it is no longer necessary to use domain defined attributes.

V. Relationship between MHS and Reference Model for Open System Interconnection

In understanding the significance of the X.400 series of standards to Office Systems and Office Applications, it is important to understand where the X.400 series of recommendations fits in with the Reference

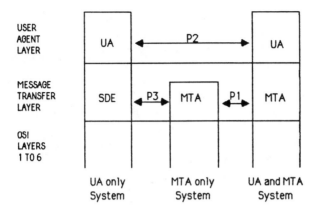

Fig. 2. Layered model.

Model for Open System Interconnection. This is shown in Fig. 2. The cooperating User Agent and Message Transfer protocols form sublayers of the Applications Layer. They use the services of the lower layers of the Reference Model. The MHS protocols are intended to work across an individual network or a set of networks which can be made to support the Connection Oriented Network Layer Service. This includes Local Area Networks, Packet-Switched Networks, and Circuit-Switched Networks. Standard protocols at the Transport Layer and Session Layer, plus a minimal form of presentation layer protocol, also support the UA and MTA layers. It is important to note that layers 4–6 can be quite general. They need not be implemented only for messaging. Indeed, it should be normal in the future for them to be used by other applications as well.

VI. The Interpersonal Messaging Service

As stated, P2 defines the protocol for the Interpersonal Messaging Service (IPMS). The IPMS is quite general in nature, and provides for a very wide range of message contents. This is important for office applications. Most proponents of message systems fall into two groups: voice messaging and text messaging. The IPMS bridges the gap between the two and provides more.

The IPMS provides for a basic message structure consisting of a header part and a body part. The header contains information that has become common in text messaging systems such as "TO," "CC," "SUBJECT," and "FROM." The recommendations provide for a number of different fields.

An indicator in the body part is used to inform a message recipient that part or all of the message content is encrypted. On detecting the indicator, the receiving UA could prompt the recipient to enter the key for decrypting the content. This assumes that the sending and receiving UAs are using compatible encryption and decryption schemes. This is an example of what the term "cooperating user agents" means.

While some UA service features are required, many features are optional, in that it is not mandatory that a UA make the feature available to someone originating a message. On the other hand it is expected that a receiving UA will respect the intentions of a sending UA. This is normally done by simply displaying the field for information, even if the receiving UA is not implemented to process the field.

The contents of the message body can be quite general. Figure 3 shows the message structure provided for in the IPMS. There may be multiple body parts to a message. The body parts may be text, digitized facsimile, digitized voice (soon), ... Table II shows the different body parts provided for by the standard. Most of these different kinds of body parts are defined by standards other than the X.400 series. Others remain to be defined in the future. Work is now underway to define the compression algorithm to be used for digitized voice.

One important area addressed by the IPMS was designed to overcome a common problem in text messaging systems. This is the problem of a user

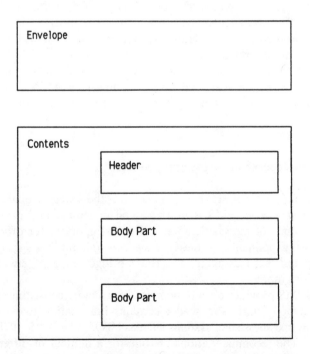

Fig. 3. Message structure.

Table II. Body Parts Supported by X.400 Series and Corresponding Standards

Body part	Standard
IA5 Text	CCITT T.50
Final Form Pages (Teletex)	CCITT T.60, T.61, F.200
Facsimile (Groups 3 and 4)	CCITT T.4, T.5, T.6, T.30, T.73
Telex	CCITT S.5, F.1
Mixed Mode (Teletex/Fax)	CCITT T.72, T.73
Videotex	CCITT T.100, T.101
Simple Formattable Document	CCITT X.410, T.61
Digitized Voice	Not yet defined
Encryption Indicator	X.420

generating a message that is sent to a terminal with different characteristics from the terminal on which it is generated (e.g., different line widths). To deal with this problem, Recommendation X.420 includes a new technique called Simple Formattable Document (SFD). This allows the layout of pages of text to be described independently of things like carriage control characters, line lengths, etc. The receiving UA is expected to know the characteristics of the receiving terminal. The receiving UA will use the SFD information to display the message content in a way that preserves the originator's format as much as possible.

A great deal of standards work in related areas is underway. SFD has been specified to be as compatible as possible with this ongoing work.

VII. The Message Transfer Service

While full implementation of many of the service elements of the IPMS is not mandatory, the same is not true of the Message Transfer Service. The Message Transfer Service is a very general store and forward distribution service. It is designed to support not only User Agents providing the IPMS, but many other (as yet undefined) kinds of User Agents as well. This is a main reason for separating the MHS standards into a Message Transfer Service and a User Agent Service. Virtually any office application that fits into the category of "store and forward" can use the Message Transfer Agent Sublayer of the MHS recommendations. Two examples include support for distribution of updates to copies of an electronic directory and support for routing of forms in an electronic forms service. The Message Transfer Service also provides a very powerful base for a general File Transfer Service, where the message is an entire file.

Much effort went into defining the Message Transfer Service, and it contains many service elements. The standards contain detailed descriptions of the elements, comprising many pages. No attempt will be made to repeat those descriptions here. The reader is referred to the CCITT Recommendations for these details. However, to amplify on the potential of the Message Transfer Service as a general base for store and forward applications in the office, a few major service elements will be discussed.

A Message Transfer Agent receives an "electronic envelope" from a User Agent. The UA asks the MTA to deliver the message to one or multiple addresses. The MTA must be capable of recognizing three priorities for delivery which the UA may request. The UA may also request that delivery be deferred until a later time.

The MTA must be able to provide acknowledgment to the sender concerning delivery of the message, if the sender requests this service. It must also be capable of providing notification of nondelivery, if it is not possible to deliver the message. Notification of delivery/nondelivery applies to any potential recipient.

An important related service allows the UA to issue a "probe" asking the MTA to determine in advance whether a message it intends to send is deliverable or nondeliverable.

Two important services are provided by the MTA which delivers messages to a receiving UA. These are (1) request MTA to hold messages to be picked up later by the UA, and (2) request MTA to deliver messages to an alternate UA. An example of the former could be a UA in a personal computer which connects periodically to an MTA to pick up messages waiting to be delivered. An example of the latter could be a UA designated to receive messages that are addressed to a particular organization, but there is no one in the organization with the name to which the message is addressed.

A third important set of service elements concerns content conversion. A user may request that the Message Transfer Agent convert the content of the message to some other form. For example, the message submitted may be text, but the addressee may have only a telephone set. The standard provides for the capability to request the Message Transfer Service to convert the text to voice. This would allow sending a text message to voice telephone sets. The sender might know that a particular recipient would be using a facsimile terminal to receive messages. Text to facsimile conversion could be requested for messages addressed to this user. Although this set of features may not be widely used initially, it is expected that they will greatly expand the number of different kinds of terminals that may be used in conjunction with messaging services.

VIII. Future Extension: Directory Services

As discussed above, the X.400 series has made important progress in the area of addressing; however, very significant work is now underway to extend this to general directory services. A directory for text messaging lists registered users of a particular messaging service. Similarly, a telephone directory lists the registered users of voice telephone services in an area. The listing in a voice telephone directory also provides a form of "electronic address" for each registered user. This allows a person who wishes to place a telephone call to look up the name and convert that name to the appropriate "electronic address"—namely, the telephone number. Although some individuals have proposed that messaging users be given numbers in much the same way that numbers are used for telephone calls, there is much stronger sentiment for providing the more user friendly form of directory service that specifies the kind of attributes described above (name, organization, etc.) and relies on a Directory Service to find the corresponding "electronic address."

The intent of the X.400 series of recommendations is to allow dissimilar message systems to exchange messages. Current efforts in CCITT and ISO and other bodies on Directory Services are likely to provide a broader set of allowable attributes that can be used in addressing, and to define mechanisms that will allow interaction between the Directories for different Management Domains.

Because message contents can be multimedia, existing directories need to be integrated so that a listing can provide information including telephone number, for delivery of voice messages, and display terminal address (e.g., on a PBX or local area network) for delivery of text and graphic content.

Work on Directory Standards is moving rapidly. For example, the CCITT Rapporteur Group on Directory Services intends to produce an interim recommendation by the end of 1986. Looking ahead a number of years, it is possible to imagine a worldwide directory service that will allow sending messages to users anywhere in the world, so long as the users are registered in some message system.

IX. Relationship Between IBM DCA / DIA / SNADS and X.400

IBM's Document Content Architecture (DCA), Document Interchange Architecture (DIA), and SNA Distribution System (SNADS) have many similarities to the MHS as defined in the CCITT X.400 series of recommendations, as well as some significant differences. Because there is no limit on the length of a message that may be distributed by an X.400 MHS, it can be used for document distribution as well as messaging. DCA/DIA/SNADS can be used for messaging as well as document distribution. Because of this, there has been considerable interest in the relationship between DCA/DIA/SNADS and X.400.

At the top level, DCA/DIA/SNADS and X.400 have similar structures. DIA/SNADS has a user layer and a transport layer. The DIA protocol for submitting a document from the user's environment to the transfer system is analogous to the P3 protocol of X.400. The SNADS protocol is analogous to the P1 protocol for transfer of messages between X.400 MTAs. The P2 protocol, which provides for compatibility between User Agents implementing the IPMS, is analogous to DCA.

These similarities have led to considerable discussion concerning building gateways between DIA/SNADS and X.400 MHS. Because both have sizable user communities, this is likely to happen. However, many details must be worked out. Despite the architectural similarities, there are some significant differences in the features of the two services.

One of the major differences concerns content types supported. Table II above lists the content types supported by X.400. DCA is intended to support an extensive set of document types, including text, digitized voice, facsimile, and graphics. But they are different from what is supported in X.400. For example, IBM supports EBCDIC, while the first item in Table II, IA5 text, is 7-bit ASCII. IBM's DCA Final Form Text and the X.400 final form text should be relatively simple to convert to each other in a gateway. X.400 final form uses the T.61 character set, which incorporates the 8-bit ASCII character set. Conversion of some of the other document types could be more complex.

The message structure of DCA and X.400 are similar. The structure supported by X.400 was shown in Fig. 3. DCA provides for a *document profile architecture*, analogous to the X.400 header, and a *document content architecture*, analogous to the body part. DCA also provides for a *message*. This can be up to 256 characters of EBCDIC text. Conceptually, this falls between the header and the body part of X.400.

X.400 has headers and IBM has base profile sets. Some of the elements correspond, while others do not. Because the intent of the two is different, this is not surprising. X.400 headers provide for an originator and recipient(s). Recipients may be designated as "primary" (for action), or "copy" (for information). IBM base profile sets provide for an author and a copy list, without any distinction between "primary" and "copy." Both support a "Subject" field.

A number of fields are unique to each, such as "In reply to," "Blind copy," and "Reply by" in X.400 and "Document Data," "Document Type," and "File Cabinet Reference," for IBM. These fields show the difference in emphasis between the two. X.400 is more oriented to the dynamics of mail distribution and handling, while IBM is more oriented to the document distribution and handling functions.

Focusing on the service element level of the Message Transfer Service of X.400, additional differences become apparent. We discussed a few of the many X.400 service elements above, including the capability for a UA to submit a single copy of a message to an MTA for Distribution to Multiple Destinations, Delivery/Nondelivery Notification, Probe, Hold

Messages for Delivery, and Deliver to Alternate UA. Some of these have IBM analogs while others do not. In addition, there are IBM features that have no correspondence in X.400.

Both support distribution to multiple locations, and delivery/nondelivery notification. The probe capability is unique to X.400. The "OBTAIN" command in DIA can be used to request delivery of queued documents, similar to the "hold for delivery" service element of X.400. "Deliver to alternate UA" has no analog in DIA/SNADS, but IBM provides a capability to deliver feedback to an alternate destination, something that is not supported by X.400.

DIA has the capability to carry out operations on a remote library such as file, delete, search, and retrieve. DIA also has the capability to do application processing on a remote host. SNADS has a transaction processing capability, which allows it to invoke transactions, using a distribution object as a parameter. None of these capabilities are supported by X.400.

There are significant differences in the area of naming and addressing. SNADS provides for a maximum of eight characters for the Element Name, e.g., R. UHLIG, and the Group Name, e.g., NETWORKS. This applies to both users and Distribution Service Units. X.400 provides for "Personal Name," "Organization Name,"...as discussed above. X.400 has placed considerable emphasis on "user friendly" addressing. In gateways, it should be fairly easy for X.400 to translate to IBM's Group Names and Element Names. The reverse may not be as simple, because of the additional flexibility available with X.400.

Given the X.400 emphasis on mail handling and distribution, and the IBM emphasis on document handling and distribution, the differences between the two are not surprising. X.400 is the base on which a global messaging system is being built by the many carriers who are implementing it. A high level of connectivity between X.400 and IBM DIA/SNADS can be expected to be available through gateways, in future years, because there is sufficient correspondence to make this feasible, and there is likely to be sufficient demand.

X. Message Handling Systems: A Base for Office Applications

The combination of Message Handling Services, as defined in the X.400 series, plus current work on Directory Services provides a powerful base on which to build other applications. To explore these in depth would be far beyond the scope of this chapter. However, we shall look at the basis for this and then provide one simple example.

Recommendation X.411 defines a "content type" parameter. The recommendation states, "A content type parameter is supplied by the originating UA and identifies the convention that governs the structure of the contents. *The only defined value identifies the P2 protocol for Interpersonal Messaging . . . ."* It is anticipated that there will be other protocols, corresponding to P2 for Interpersonal Messaging, which will provide other kinds of services that can be built on top of the Message Transfer sublayer.

As an example, we will discuss a few highlights of a possible forms service. Some key features of a hypothetical forms service might be as follows:

1. Form Filling—Prompts a user to enter information into each field on a form, with validation of contents by type.
2. Forms Routing—Allows a form to be sent, in sequence, to several persons, for review and approval.
3. Forms Authentication—Provides a mechanism for "electronic signature."
4. Forms Tracking—Allows the originator, the final recipient, or others to determine who has already acted on the form, and who has the form for action, at any time after the form has been sent by the originator.

A partial concept of a Forms Service built on the Message Transfer sublayer of X.411 is shown in Fig. 4.

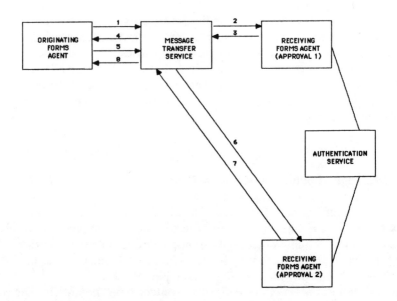

Fig. 4. Partial concept: Forms service built on message transfer sublayer.

The User Agent sublayer of Recommendation X.420 would be replaced by a Forms Agent. The Forms Agent would provide for Form Filling and would submit the form for approval using services of the Message Transfer sublayer. Forms Authentication is a new service not provided by the current MHS recommendations, but widely recognized as a needed service. It is closely related to encryption. As discussed above, some provision has already been made for encryption.

The capability to issue a probe to the Message Transfer Service, as discussed above, provides the basis for building Forms Tracking into a Forms Agent.

Forms Routing is a feature that would likely have to be built into a Forms Agent. After determining that an approval has been received, the Forms Agent would submit a copy of the completed form to the MTS for delivery to the next person in the approval chain. The Originating Forms Agent would likely use titles rather than names in consulting the Directory Service.

At best, this provides only a sketch of how one new service could be built on the Message Transfer Service defined in Recommendation X.411. It will be interesting to see how many new services are built using the base provided by the X.400 series of Message Handling Service Recommendations.

XI. Conclusions

A truly global message system is rapidly emerging, as electronic mail carriers, vendors of telecommunications equipment, and computer vendors all over the world implement X.400-based message systems. For some detail, see the paper by Joseph Pitteloud [3]. Some issues remain to be resolved, but implementation of X.400-based systems does not have to wait for the resolution of all the issues. Some of these issues include global interconnection of directories and routing, privacy and security across message systems, how users and receivers will pay for messaging services, particularly when international transport through public systems is required, and how to distribute the revenues collected across the various international systems through which a message may transit. Much hard work is going into the resolution of these issues, and there is no doubt in this author's mind that they will be resolved.

Acknowledgment

The author is grateful to Ian Cunningham for reviewing the manuscript and for many discussions, which have contributed to this chapter.

References

[1] R. Uhlig, "Human factors in computer message systems," *Datamation*, May 1977, p. 120.
[2] J. E. White, "Computer-based message services," in *Proceedings of the IFIP WG 6.5 Working Conference*, Nottingham, U.K., 1–4 May 1984, Amsterdam: North-Holland, 1984.
[3] J. Pitteloud, "Electronic message handling for the nineties," in *Proceedings of the 10th IFIP World Congress*, IFIP '86, Dublin, Ireland, 1–5 September 1986, Amsterdam: North-Holland, 1987.

VI

Network Architecture Examples

Xerox Network Systems Architecture

Abhay K. Bhushan and Dennis G. Frahmann

I. Introduction

This chapter describes the major characteristics of the Xerox Network Systems (XNS) architecture and relates it to the needs of end users that motivated its design. First, we will examine the basic design principles of XNS and compare its overall model of layers to that of the ISO Open Systems Interconnection (OSI) model. Within that model, we will move from the bottom layers up through the top layers—from the communications channel through support applications to the direct user applications.

II. XNS: A Distributed Architecture

Providing an architectural solution for networks is first a matter of design issues. The original researchers at the Xerox Palo Alto Research Center (PARC) developed an early distributed internetwork architecture named for PARC Universal Packets PUP(1). In working with these overall goals, several design principles emerged. They were as follows:

- Tie local area networks (LANs) together through internetwork routers to create a unified internetwork system.
- Use a distributed architecture based on replication of services.

ABHAY K. BHUSHAN AND DENNIS G. FRAHMANN • Xerox Corporation, El Segundo, California 90245.

- Provide paths, or gateways, from an XNS environment into non-XNS environments.
- Document and publish standards and make XNS an open system.
- Enable a long-term growth path from today's form of the XNS architecture to future forms.

Let us look at each of these design principles in more detail.

1. Distributed Architecture

XNS is a distributed network architecture, which means that it is nonhierarchical. Hierarchical systems are intended for applications where one or more mainframe computers dominate the resulting information system and its users. In such systems other elements are clearly subordinate to the large computers. In various ways, the network architecture underlying these systems are designed to create and reinforce this relationship.

Many networking applications prove to be ill-served by this model. In particular, many of the processes and activities in modern offices are essentially autonomous, including the creation, editing, storage, retrieval, and printing of douments. In most cases, document management requires dealing with a large series of autonomous processes, for which the centralized mainframe-oriented model of information flow and processing is a potential bottleneck. For this reason, XNS is designed to support autonomous processes, implemented by distributed, not centralized, processors.

XNS achieves a distributed architecture by placing an appropriate amount of computing power in each workstation to enable users to do their primary tasks by relying solely on one device's local resources. Those services that require infrequently used or expensive resources are provided through network servers, devices that provide their specialized services to a group of users. Print servers provide spooling and other printer-related services, for example.

XNS also depends on replication of information in databases and of services. One example is the authentication service that allows a user to access specific resources. A new employee is added at the local authentication server (Clearinghouse, to be discussed later); as this new information is shared with all the authentication servers in the network, the new employee's name and password have replicated themselves. Other services are replicated throughout the network, including local Time Servers.

2. Local Area Networks

To tie together workstations and servers in a local area network, XNS relies primarily on Ethernet, although many other products can perform at the physical and data link layers with XNS. XNS-compatible LANs use a variety of media with different functional limits, which will be discussed in detail later in this chapter.

3. Internetworks

Any transmission medium will have economic and physical limits on what it is able to tie together. Thus, if one wishes to design an architecture that allows a customer to grow from linking users at a single site to one that ties elements together through a company's worldwide offices, one needs to unite individual LANs into a large internetwork.

To meet this goal of an internetwork, one needs protocols that allow standard addressing across individual network boundaries. One also needs standards to support long-haul transmission between the LANs, if this is part of the internetwork. An internetwork might use the switched telephone system for long-haul transmission, for example.

4. Non-XNS Environments

As user demands and network environments grow more heterogeneous, the ability to provide a path for data to flow meaningfully from one environment into another increases in importance. In this regard, a full network architecture needs to worry not only about how to handle information generated within it, but also how to deal with information that flows from or to other environments, such as IBM's SNA or Digital Equipment Corporation's DECNET.

This includes not only considerations of physical transmission and logical addressing, but also the need for consistent formats and character codes for document interchange. A network architecture needs to develop standards and protocols to deal with these concerns.

III. Open Systems

The XNS architecture is an open system, able to provide services to all network citizens, whether produced by Xerox or not, and able to communicate, via gateways, with non-XNS citizens. Many gateway products now exist and many more are in development to make this interconnection with the non-XNS environments as complete as possible. It is also important, in the XNS view, that network standards be published and therefore easily available to developers and implementors of potential new network service products. A series of published standards is essential for the development of a truly open system.

Special development tools are also needed to ensure that a network is as architecturally open as possible. The Xerox Development Environment (XDE) allows programmers and other developers to use XNS's resources to support the program development and then integrate XNS services into the new programs so that end users can access them. The major components of XDE are tools for software development, including performance measurement facilities. XDE also includes an operating system for support of real-time interactive applications, including virtual memory management, process management and interfaces to network services, and application tools for general XNS end user support, including filing, mailing, printing, gateways, and other communications services.

IV. XNS and the ISO / OSI Model

Basic to most network architectures, including XNS, is the concept of layering. This means that the various functions supported by the architecture are divided into a series of layers. By convention, the tasks that are closest to the actual transmission medium are located in the lowest numbered layers while the higher layers provide services that are most visible to the users of the network. A useful model of this layered structure is the Open Systems Interconnection (OSI) model, first published as a standard by the International Organization for Standardization (ISO) in 1984 (ISO 7498) (see Chapter 2 for more information). In addition, Recommendation X.200, with the same content and in the same year, was published by the Comite Consultative International de Telegraphique et Telephonique (CCITT), one of the standardization arms of the ITU (International Telecommunications Union) of the United Nations.

In the OSI Model, the lower layers typically deal with data transmission matters, while the upper layers take care of end-to-end communication reliability, authentication and security, document coding and handling, and user applications, like filing, electronic mail, and data-base management.

The OSI Model in standard ISO 7498 is really a guide for the development of OSI standards. The CCITT, ISO, and ECMA (European Computer Manufacturers Association) are developing actual standards to achieve the functionality that the OSI Model lays out for each layer. Some of the existing standards provide nearly the same functionality as their related XNS protocols; these related protocols will be discussed in detail later in this chapter.

The OSI-compatible protocols are, by definition, created to be used in a heterogeneous environment. Because they are optimized around the greatest commonality, they are not necessarily best suited to every specific application. XNS protocols, because they are used in a homogeneous

ISO/CCITT Standards				Layers in OSI Model	XNS			
Information standards: Office Document Architecture (ODA), CCITT Group 4, and others to be defined				Layer 7: Application	Information standards: Office Document Architecture (ODA), Raster Encoding Standard (including CCITT Group 4), Interpress			
FTAM	Mail X.400	Virtual Terminal	Other		Filing	Mail	Virtual Terminal Gateway Access	Printing
Application Support: Directory, ISO 646, and others to be defined					Application Support: Clearinghouse, Authentication, Time, Character Code			
ISO/ASN.1		X.409/410		Layer 6: Presentation	Courier			
ISO Session				Layer 5: Session	Courier			
ISO Transport Class IV				Layer 4: Transport	Internet Transport Protocols			
ISO Internet Protocol		X.25		Layer 3: Network	Internet Datagram Protocol		X.25	
IEEE 802.2		HDLC/LAP B		Layer 2: Data Link	Ethernet Data Link Layer	Synchronous Point-to-Point	HDLC/ LAP B	
IEEE 802.3 CSMA/CD	RS232	Other		Layer 1: Physical	Ethernet/ IEEE 802.3 CSMA/CD	RS232	Future	
Cabling alternatives					Cabling alternatives			

Fig. 1. XNS and OSI layers.

environment, allow a higher level of functional integration than the collection of protocols that are emerging via compromise from the international standards bodies. In addition to the native protocols, XNS also includes protocols designed for internetworking XNS networks with outside devices and networks.

Figure 1 shows the basic structure of Xerox Network Systems, organized into a series of layers, approximately corresponding to the OSI Model layers. Since the XNS architecture predates the OSI Model, XNS does not have exactly the same number of layers; XNS uses fewer layers. Each XNS layer corresponds functionally to one or more of the OSI layers. Each of the protocols will be described in more detail as this chapter progresses.

V. Communications

At the lowest layer of the XNS architecture, we must consider data transmission between different devices of the network, corresponding to what the OSI Model calls the physical and data link layers. At the next higher layers, we consider the addressing, routing, and switching decisions that provide for reliable transport of information throughout the network (perhaps made up of connected subnetworks) so that we obtain end-to-end communication.

Within XNS, there are three major components: Ethernet protocols provide the OSI physical and data link layers' functionalities; XNS's Internet transport protocols provide the OSI network layer functions; the Sequenced Packet Protocol provides connection-oriented OSI transport layer functions.

1. LAN Architectures

The XC20 family of networks contains a network for almost every possible cabling preference. The XC24 uses Ethernet cabling and protocols, moving data at 10 megabits per second (Mbps). Ethernet was the model for the IEEE 802.3 LAN standard and interworks with 802.3 standard networks. The XC24 can use either standard Ethernet cable, 802.3 10 base 5, which is typically found in formal installations, in the wall or using drop cables (Fig. 2). It can also use Thin (RG58) Ethernet cable, 802.3 10 base 2, which is typically used for informal installations (Fig. 3). With standard twisted-pair telephone wire—either unused wire already in the walls or informal user installations—the XC22 network meets the specification of 802.3, 1 base 5, which is also used by the AT & T Starlan[tm] network. Up to ten workstations can be daisy-chained using the twisted-pair wire; every

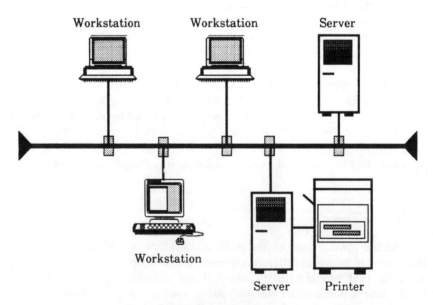

Fig. 2. Standard Ethernet (IEEE 10 bases).

PC   PC   Printer

Workstation   Workstation   Server

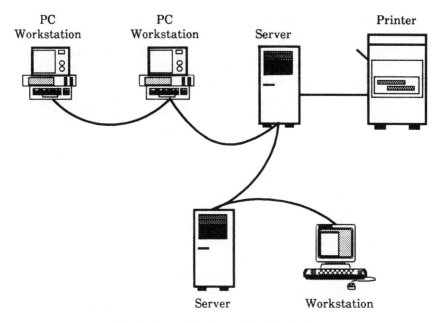

Server   Workstation

Fig. 3. Thin-cable Ethernet (IEEE 10 base 2).

eleventh workstation requires the addition of a Network Extension Unit. The XC22 network can move data at 1 Mbps.

Any member of the XC20 family can connect Xerox 6060 workstations, or IBM PC/XT/AT or compatible computers, running MS-DOS 3.1, into a functioning low-cost network. However, only the XC24 Ethernet-based networks can be upgraded into members of the XC80 family of networks, which support the full spectrum of XNS and OSI-compatible protocols.

Finally, the Astranet^{tm} member of the full-featured XC80 XNS family uses Ethernet protocols with either fiber optic cable or IBM Type 1/2 cable and combinations. Data moves at the Ethernet-standard 10 Mbps rate.

Chapters 3 and 6 contain more information about Ethernet.

2. Internet Architecture

The Xerox internet architecture offers a number of protocols and a rich addressing scheme, corresponding to the network and transport layers of the OSI reference model. The internet architecture enables Ethernets to be directly interconnected to each other in a variety of ways: directly, by telephone lines, through public data networks, or via other long-distance transmission media. It also allows the communication system to be reconfigured to satisfy the immediate and future requirements of the user.

The internet architecture makes use of several protocols that move information from source to destination in an organized and reliable manner. The internet protocols are as follows:

- Internet Datagram Protocol. This defines the fundamental unit of information flow within the internet—the internet datagram packet. It is in the OSI network layer.
- Sequenced Packet Protocol. This provides for reliable, sequenced, and duplicate-free transmission of a stream of packets. It is a connection-oriented protocol in the OSI transport layer.
- Packet Exchange Protocol. This supports simple transaction-oriented communication involving the exchange of a request and its response. It is a connectionless protocol in the OSI transport layer.
- Routing Information Protocol. This provides for the exchange and dissemination of internetwork topological information necessary for the proper routing of datagrams.
- Error Protocol. This standardizes the manner in which low-level communication or transport errors are reported.
- Echo Protocol. This diagnostic transport layer protocol is used to verify the existence and correct operation of a host, and the path to it.

The work done by the Internet layer involves interaction between one or more of the specialized protocols and the basic Internetwork Datagram Protocol. To simplify this discussion, functional distinctions between the operation of these various protocols have been ignored.

3. Datagrams

The XNS internet architecture uses as the fundamental unit of information flow in the OSI network layer an internet packet or datagram, shown in Fig. 4. A datagram is a packet (typically several hundred bytes of information) whose movement through the system is individually controlled. Other systems use the virtual circuit, in which a logical connection is constructed between source and destination prior to transmission, rather like the call setup involved in a telephone call. The virtual circuit is set up and packets flow from source to destination until the virtual circuit is taken down. Virtual circuits, however, have higher overhead costs associated with their use, including circuit setup and deletion, circuit maintenance costs, and loss of channel capacity for other uses. The datagram is a transaction-oriented transport mechanism. In the typical factory, office, or military LAN or company-wide network installation, the majority of the packets

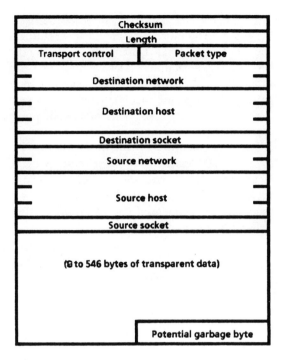

Fig. 4. Internet packet or "datagram."

involve simple transactions—a question and reply, for instance—and this type of traffic pattern finds datagrams to be the most cost-effective solution. A network that functions as a backbone, with a constant flow of large data files, would favor virtual circuits.

The internet packet fields fall into two categories: header and data. The first contains addressing fields, which specify the destination and source network addresses, and control fields, which consist of checksum, length, transport control, and packet type field. The data fields carry the data and consist of information that is interpreted only at the next higher layer, the OSI transport layer.

4. Internet Delivery and Routing

The internet architecture assumes the following topology: (1) within a LAN, any host can directly reach any other, (2) any number of LANs can be interconnected by Internetwork Routers, forming a graph whose nodes are LANs and whose arcs are transmission links between LANs.

A unique 32-bit network number is assigned to every LAN. An internet address includes both the network number and the unique 48-bit host ID used by the IEEE 802 standards. Internet routing is based on the network numbers only. The Internet Routers use routing tables and the network number in the Internet Datagram Protocol (IDP) destination address to determine the next immediately connected LAN to which the Internet Datagram Protocol packet should be sent. The Internet Datagram Protocol packet moves across intermediate LANs, from source to destination, up to a maximum of 16 hops. This maximum prevents indefinate looping of packets.

Use of datagrams means that the internet makes only a "best effort" attempt to deliver an IDP packet. Although most packets are delivered properly, a packet may not arrive, or it may arrive damaged and be discarded, or more than one copy of the packet may arrive. The next layer of protocols, especially the Sequenced Packet Protocol (SPP), correct these limitations of reliability and packet size.

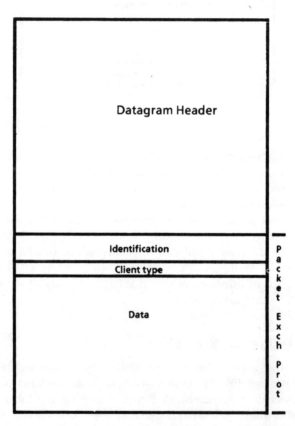

Fig. 5. A packet exchange protocol packet simply transmits a request and receives a response.

5. Message Integrity

It is complete messages that must be delivered to the user, not just isolated datagrams, so XNS needs a way of ensuring message integrity, the process by which the datagram building blocks successfully reassemble themselves into the original message.

XNS message integrity is provided through one of two protocols in the OSI transport layer. The Packet Exchange Protocol is designed for transaction-oriented communication, involving simple requests and responses. As shown in Fig. 5, there are only three fields in the packet. An identification field, which contains a transaction identifier, associates a request with its response. The client type field indicates how the data field will be interpreted at the higher protocol layers. The data field contains the data,

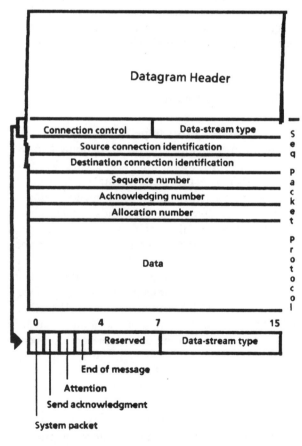

Fig. 6. Sequenced packet protocol packet allows successive transmission of internet packets.

specified by the needs of the higher-level protocols. For instance, this type of exchange might be used by a workstation trying to find an appropriate file server through the Clearinghouse resource-location service. For connection-oriented communications, the Sequenced Packet Protocol (Fig. 6) ensures message integrity.

6. Relation to the ISO Transport Layer Protocol

The OSI Transport Layer includes five classes of connection-oriented service, ranging from no error recognition or recovery to full recognition and recovery. These five classes are required to meet the full range of user requirements, whether the underlying network layer service is connection-oriented or connectionless.

The Sequenced Packet Protocol is a connection-oriented transport protocol that uses a connectionless network service and thus provides an extended conversation between two devices, called end systems in the OSI reference model. It is equivalent to ISO Transport Class 4.

VI. Remote Procedures / Courier

To XNS, an end system exists to perform useful work. The purpose of the communication infrastructure is to facilitate doing that work. Although other forms of information exchange are also used, a key part of the XNS architecture deals with how the exchanges take place between a service provider and a service consumer that cause work to be done at a location. This location may be at a distance of thousands of miles or within the same processor. To the architecture, there is no significant difference.

A special protocol called Courier (subtitled The Remote Procedure Call Protocol) defines the method by which directions for accomplishing work within XNS are sent and appropriate responses returned. Courier performs formatting like ISO Abstract Syntax Notation One (ASN.1) and remote operations like CCITT Recommendation X.410.

Courier facilitates the construction of distributed systems by defining a single request-reply or transaction discipline for an open-ended set of higher-level application protocols such as printing, filing, and mail. Courier specifies the manner in which a workstation or other active system element invokes operations provided by a server or other passive system element (see Figs. 7 and 8).

Courier does for distributed system builders some of what a high-level programming language does for implementors of more conventional, nondistributed systems. Pascal, for example, allows the system builder to think in terms of procedure calls rather than in terms of base registers, save

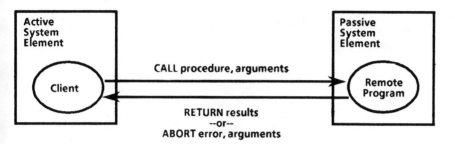

Fig. 7. Courier remote procedure call model.

```
SimpleFile Transfer: PROGRAM 13 VERSION 1 =
BEGIN
- types
Credentials: TYPE = RECORD (user, password: STRING);
Handle: TYPE = UNSPECIFIED;

- procedures
OpenDirectory: PROCEDURE (name: STRING, credentials:
    Credentials)
    RETURNS (directory: Handle) REPORTS (NoSuchUser,
    IncorrectPassword, NoSuchDirectory, AccessDenied) = 1;

    Store File: PROCEDURE (name: STRING, directory: Handle)
    REPORTS (NoSuchFile, InvalidHandle) = 2;

    RetrieveFile: PROCEDURE (name: STRING, directory:
        Handle)
    REPORTS (NoSuchFile, InvalidHandle) = 3;

    CloseDirectory: PROCEDURE (directory: Handle) REPORTS
    (InvalidHandle) = 4;

- errors
NoSuchUser:              ERROR = 1;
NoSuchDirectory:         ERROR = 2;
NoSuchFile:              ERROR = 3;
IncorrectPassword:       ERROR = 4;
AccessDenied:            ERROR = 5;
InvalidHandle:           ERROR = 6;
END.
```

Fig. 8. Example of Courier usage.

areas, and branch-and-link instructions. So Courier allows the distributed system builders to think in terms of remote procedure calls, rather than in terms of socket numbers, network connections, and message transmission. Courier also provides a rich set of predefined as well as constructed data types including Boolean, integer, cardinal, string, array, and record among others.

1. Relation to OSI Session and Presentation Layers

Courier uses an abstract syntax very similar to the ISO ASN.1, which is in the OSI Presentation Layer. The Remote Procedure Call mechanism used by Courier is very similar to Chapter 2 of the CCITT Recommendation X.410, which is currently being standardized in ISO as a Remote Operations Service (ROS). The emerging standard uses the Common Application Service Elements (CASE) kernel. Courier also includes the universal session functionality, similar to the Basic Common Subset of the OSI Session Layer standard.

VII. Applications Support

Although the OSI model ends in a single layer called Application, there are really two significantly different kinds of activities at this layer. XNS refers to these activities as application support protocols and application protocols. Applications support protocols define those application-level activities which support network activities as a whole. Within XNS, this includes a directory service protocol called the Clearinghouse, authentication protocols, and time protocols. In addition, XNS uses common data representation standards, including the character code standards, raster encoding standard, and the Interpress printing standard.

1. Clearinghouse and Directory Services

Every distributed system must have a way of identifying the location of system resources and users. For example, a workstation needs a way to gain access to a full-color printer in a distant location in the network, before it can print any documents on that printer. The challenge of locating specific resources is made worse by the constant movements, additions, and deletions of resources that occur in a typical network. For a large network, keeping track of addresses and key attributes of system elements is a major undertaking.

In XNS this problem is solved by Clearinghouse, a service whose purpose is to provide clients with the addresses of important objects. These addresses are used in the remote procedures through which clients get work done.

Clearinghouse is essentially a data base of objects. The entry for each object consists of a name and a set of one or more groups of data items that encode the object's properties. Clients use the Clearinghouse service by providing it with object names and properties, in return for which Clearinghouse provides the appropriate address information.

Objects in Clearinghouse are named unambiguously in a uniform manner with the same naming convention for every object regardless of whether it is a user, a workstation, a server, a distribution list, or something else. The naming is in a three-level hierarchy, in the form: LocalName: Domain: Organization.

This division into organizations and domains within organizations is a logical rather than a physical division. An organization will typically be a corporate entity, which can choose domain names to reflect administrative, geographical, functional, or any other type of divisions. Very large corporations may choose to use several organization names if their name space is extremely large. In any case, the fact that two addressable objects have names in the same domain or organization does not necessarily imply that they are physically close.

Clearinghouse allows any named object to use an alias. Names that are specified without a domain, Clearinghouse assumes to be of the client's domain, which makes assembling mail distribution lists easier. Since the Clearinghouse also associates a list of attributes with each named object, it is possible to request the names and network address of all objects with a specific attribute, for instance, all the laser printers.

Clearinghouse maps each name into a set of properties to be associated with each name. Each property is composed of a Property Name, Property Type, and Property Value. Values come in one of two categories: an Item, which is an uninterpreted block of data, or a Group, which is a set of names.

When a network object is referred to by name, the name must be bound to the address of the object. The later a system binds names, the more gracefully it can react to changes in the environment. If client software binds names statically, the software must be updated whenever the environment changes. On the other hand, binding takes time. Both static and early binding increase runtime efficiency since names are already bound at runtime. A useful compromise combines early and late binding, giving the performance and reliability of the former and the flexibility of the latter. Most XNS clients use early binding whenever possible and late binding only if any of these early bindings becomes invalid. Thus, software supporting printing stores the addresses of print servers at initialization and updates these addresses only if they become invalid.

Replication of data bases at multiple Clearinghouses means that information is always locally available to user workstations requesting service; it also means that important data bases will not become unavailable to the entire network when a single device fails.

2. Relation to the OSI Directory Standards

ISO and the CCITT are jointly developing a standard for a distributed Directory Service, using hierarchical multilevel naming. The XNS Clearinghouse is an instance of such a naming structure and its service operations are very similar to those now being standardized by ISO and the CCITT.

3. Authentication and Security

The advantage of an open, distributed system is that access to system resources, including files and facilities, is easy for any network user to obtain. Often this is a distinct asset, but at times only certain users should have access to a specific resource. The network's data bases and services must remain secure from unauthorized access.

XNS provides two security mechanisms: access control mechanisms designed into appropriate workstation and servers (e.g., filing, printing), and an Authentication Protocol, which helps clients and services determine each other's identity in a reliable and secure way.

For companies or agencies with a particular need for secure access control mechanisms, Xerox has Tempest-accredited versions of all the popular XNS workstations and servers. The Tempest security access service complies with NACSIM-5100A. The Tempest-accredited versions of the services and workstations work just like their unclassified counterparts, delivering mail, print, file, and document exchange services in a transparent manner.

The solution provided by the Authentication Protocol assumes a secure Authentication Service which all clients and services trust to know their specially encrypted passwords, and that the clocks in the system elements are reasonably well synchronized (see Time Protocol, below). The Authentication Protocol provides for both a strong and a simple level of security. The goal of strong authentication is to make it practically impossible for one user to impersonate another, whereas simple authenticatian merely makes it difficult.

Every user has two passwords, strong and simple (Fig. 9), while an XNS service has only a strong password. The password used depends on the workstation encryption capability and the security environment. Passwords are for human users to identify themselves to the system. When entered into a workstation, the passwords are immediately encrypted according to a specific algorithm—hashing for simple, and National Bureau of Standards (NBS) Data Encryption for the strong password—to form a strong key or a simple key, thus ensuring that a user's password is never transmitted unencrypted.

The authentication protection is more than mere encryption, however. A double authentication procedure between the initiator, the Authentica-

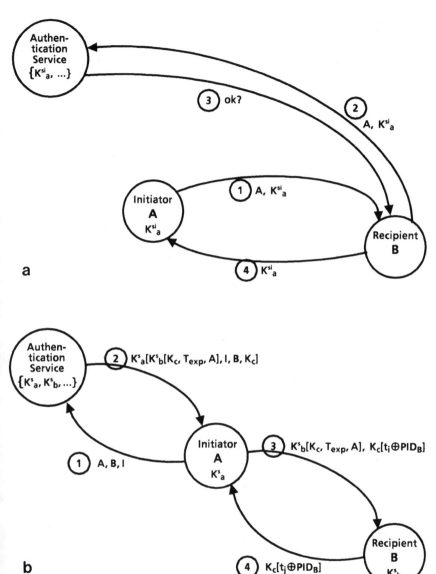

Fig. 9. The strong and weak model of authentication. (a) Weak: A, the fully qualified name of the initiator, A; K_a^{si}, A's simple key; ok?, a Boolean, which indicates whether or not K_a^{si} is A's hashed password. (b) Strong and simple: A, B, the fully qualified names of the initiator, A, and recipient, B; I, the nonce; K_a^s, K_b^s, the strong keys of A and B, respectively; K_c, the conversation key; T_{exp}, the expiration time for a set of credentials; $K_y[x]$, the value x encrypted with the key K_y; t_i, t_j, a time stamp obtained from the system clock at time i, j; PID_x, 48-bit processor ID or X, padded after the least significant bit with 16 bits of zero.

tion Service, and the recipient assures that no one is intercepting and reusing an encrypted password. In addition, the Authentication expires after a certain time period. This means that an unattended workstation, where the worker forgot to log off, cannot be used by an intruder because the valid user's authentication will have expired.

4. Time

In a distributed system, file services need to record the time when a file is read or written, electronic mail messages need time stamping, and time may be needed for authentication purposes. To facilitate the acquisition of time information in a reliable and unambiguous way while working with networks with a potential world-wide scope, a single time protocol and time standard is used throughout the XNS Network Systems.

The Time Protocol specifies the manner in which a Time Service makes the current time available to its clients, either workstations, terminals, or other servers. The Time Protocol is built upon the Internet Packet Exchange Protocol. The response packet from the Time Service returns the current time, along with the requestor's time zone, and information about when Daylight Savings Time is observed at its location. Since the Time of Day server is always a local system element, there is no measurable time delay between the time returned and the true local time. This global approach to the handling of time allows XNS systems to be implemented uniformly across geographical boundaries.

5. Character Code Standard

Information interchange on a worldwide internet requires a fundamental rethinking of the encoding of characters, the basic information element in written languages. In the United States and other English-speaking countries, 7-bit ASCII is widely used in all varieties of workstations, terminals and computers and provides only 128 different characters.

But 128 characters are hadly enough to deal with the printing needs of English, let alone other languages. To correct this situation, ASCII has been extended to 8 bits to define an additional 128 characters. OSI has adopted a similar 8-bit character code standard commonly referred to as ISO 646, but these 8-bit character codes are still inadequate. Special accents used in many European languages, Greek and Cyrillic alphabets, mathematical symbols and printing-oriented characters quickly exceed the 256-character capacity of any 8-bit code. In addition, Chinese and Japanese each have requirements for thousands of characters.

As a result, the ASCII and the ISO 646 standards permit only limited information interchange. A global information system must intelligently deal with all of these standards in a uniform and unambiguous way, while retaining compatibility with existing codes.

The XNS solution to this problem is a character encoding system that normally conforms to the ASCII and ISO 8-bit character codes, but expands to a 16-bit code when necessary. The 16-bit coding scheme permits 65,535 different character codes, which is sufficient for encoding all of the commonly used human languages. However, should future requirements warrant, mechanisms exist to expand the character code space beyond 16 bits.

The XNS multilingual Character Code Standard assigns a unique, unambiguous, and absolute numerical code to each semantically different character to permit efficient storage and processing while also ensuring proper interpretation of information. Each 16-bit character code may be viewed as consisting of two 8-bit bytes, the first of which is the character set code, and the second an 8-bit character code within that character set. The character codes are assigned so that characters within a single character set tend to be related to each other by traditional usage. Thus, all of the characters in Character Set 0 are for the Latin alphabet set. Some languages, such as Japanese Kanji, which involves well over six thousand characters, require many character sets.

To avoid the problem of having 16-bit codes take twice the storage and transmission time as 8-bit codes, the Character Code Standard also defines a string encoding technique which compresses the 16-bit codes into 8-bit bytes on a one-for-one basis (i.e., on 8-bit byte for each 16-bit entity), thus providing versatility with little or no loss in efficiency. Moreover, all sequences of 8-bit ISO 646 characters may be stored or transmitted as they are. For languages such as Japanese, which normally require the use of a 16-bit code, text may be stored and transmitted as a sequence of 16-bit codes.

6. Compatibility with Other Codes

Every effort has been made to keep the Character Code Standard compatible with the large number of national and international code standards. For example, the Character Set 0 assignment is fully compatible with ISO 646 and the ASCII standards, and the Japanese Kanji assignment is in accordance with the Japanese Industrial Standard Code JIS-C-6226.

VIII. Application Standards

We have now reached the top layer of the ISO model and that series of protocols and standards with which end users tend to have direct interaction. XNS has standard appications for mailing, document interchange, printing, filing, scanning, and access to other environments.

1. Document Management

The almost-universal end user wish is to be able to compose, print, mail, file, and exchange documents between different types of software programs and workstations. Since each workstation and software program has its own internal method of representing and storing documents, and the prospects of the entire world embracing a single standard word processing program are nil, the best way to achieve this document exchange goal is to have a common document architecture, with print, mail, and file standards that work in a homogeneous way.

2. Office Document Architecture (ODA)

It would be possible to have one document format standard, usable by documents in editable form as well as in final, printable form. That kind of strategy, however, leads to a standard that achieves neither of its goals particularly well. The Xerox developers, as well as the international standards bodies, have taken a different tack.

The standard for documents in revisable form should optimize editing efficiency, while the standard for documents in final, printable form must optimize printing efficiency, thereby minimizing the complexity of the printer interface. The revisable document standard concerns itself with logical structure and layout of the document, while the final form document standard concerns itself with physical placement of the images on the page and the placement of the pages within the total document. These two goals are separate but related.

3. Editable Document Interchange Standard

Interscript is the original Xerox standard for documents in revisable form. It says nothing about how a document editor works, but it does specify how a document will be exchanged between different document editors. The interchange standard externalizes a document into a script, which is then captured by the receiving editor and internalized into a form that the local editor can work with, as shown in Fig. 10. The script represents content, logical structure, and layout structure of the document and is capable of representing text, graphics, and virtually any kind of picture. It can include any kind of digital information, even digitized voice mail. The editor preserves without loss any information—graphics, for example—that it does not understand. After the editor edits the parts of

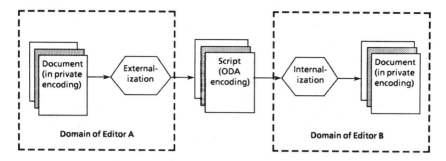

Fig. 10. Document interchange using ODA.

the document it does recognize, the unknown sections along with the edited parts can be reencoded into a new script and transferred elsewhere in the network.

When ANSI and the ECMA introduced their Office Document Architecture (ODA) standard, Xerox embraced ODA as its standard for revisable documents because it was clear that the two groups of developers had been working along the same lines; ODA and Interscript are very similar. Interscript ideas have been and will continue to be added to the XNS ODA, so that the XNS document architecture will remain compliant with the OSI standards while building upon them.

4. Printing (Final Form Document) Standard

A common desire for most users is the ability to share a range of printers—from simple workstation peripherals to high-resolution photo-typesetters—with a diverse base of computerized document creation devices, including workstations, mainframe computers, word processing systems, and input scanners. Computer-driven raster printers are capable of printing any imaginable combination of text, graphics, and pictures, because a raster printer can print anything at all simply by arranging the appropriate pattern of black and white (or colored) dots on the image.

Since raster printers are driven by computer software, and do not use device-specific control codes, it is possible to have a universal interface or interchange standard in which any document may be represented, and that can drive any raster printer, independent of its resolution (number of dots per inch) and other device characteristics. Interpress is the Xerox standard for final form document representation.

Interpress is more popularly called a page and document description language because it directs the printer to place specific marks on the page at specific locations and print pages in any specified order, signatures as well

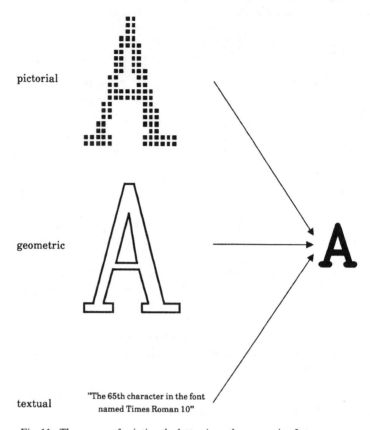

pictorial

geometric

textual

"The 65th character in the font
named Times Roman 10"

Fig. 11. Three ways of printing the letter A on the page using Interpress.

as standard pages. Interpress represents a series of instructions to the printer and the printer follows these instructions just as a computer follows the instructions found in a program. This technique makes possible the printing of substantially more complex documents than would be possible with static format specifications that are used with character printers.

To understand the Interpress approach, consider Fig. 11. The pictorial representation of the letter "A" takes several thousand bits at typical printer resolutions, the geometric representation takes several hundred bits, and the textual representation requires only 8 bits once the font has been specified. While Interpress permits all three approaches, the textual approach is preferred in most applications. The geometric approach of outline characters is useful when unique character sizes are desired or when the characters need to be rotated at unusual angles.

The appearance of material printed by a raster printer is generally limited by the skill of the artist, by the quality of the printer, and by the

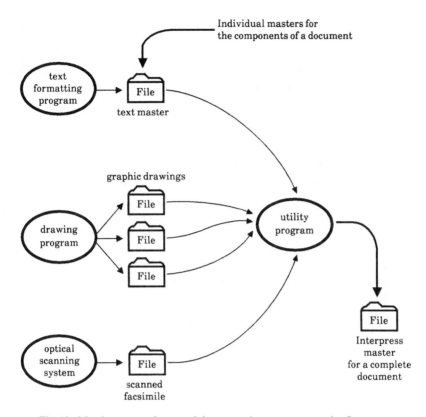

Fig. 12. Merging separately created documents into one master using Interpress.

ability to communicate just what it is that the printer should print. The design goal for Interpress was that it should not have any built-in limitations on its expressive power. Interpress is therefore, represented as a program (called an Interpress Master) written in the Interpress programming language, that is executed by the printing machine to produce the finished document. Although most masters consist only of simple statements, such as text and vectors, the full power of the programming language is available for complex applications. Programming is useful if the master is to adapt to the various properties of printing device (such as page size, order of page printing, or color or black and white) and for changing Interpress masters in complex ways (Fig. 12).

Describing pages is not enough. What is communicated to the printer usually is a document, which is a collection of one or more pages. A printing standard, therefore, must be able to describe documents (i.e., how the pages are put together) as well as the pages within it. Interpress design, therefore, includes an extensive set of printing instructions which enable the user to control the printing of documents (e.g., to invoke two-sided printing or special finishing such as stapling). Printing instructions also provide information necessary for multiuser environments (e.g., who printed the document, its name) and enable the declaration of resources required for the printing of the document (e.g., additional files, fonts, and font sizes).

5. Printing Protocol

XNS clients use the Printing Protocol to cause documents to be printed on a Print Service. This protocol assumes that the document is encoded in Interpress. Documents not encoded in Interpress can be printed in an XNS system only if special provisions are made by private arrangement between the client and service.

The Printing Protocol model assumes an abstract printer service that has three distinct processing phases: spooling, formatting, and marking or printing. In the spooling phase, the Interpress master is queued in a special holding areas (which may or may not be in the print server) for subsequent printing. In the formatting phase, the Interpress master is converted to a form suitable for rendering by the specific printer marking engine. In the marking phase, the document is actually printed on the physical medium.

6. Filing Standards

In XNS, the interaction between clients and file services is defined by the Filing Protocol. The Filing Protocol is both a guide for using a file service and a specification for the implementation of such a service. The protocol provides a general filing facility to support a wide variety of applications.

The Filing Protocol follows a session-oriented model in which a client interacts on behalf of a user, either a human or another service. A session begins when the user logs on to the file service and ends when the user logs off the service. The session can last just long enough to perform one operation or it can include several operations, with inactive periods between them. The user can have several sessions operating simultaneously; they may or may not have all been established by the same client.

The Filing Protocol may be viewed as being composed of four sublayers, one above the other, which progressively provides additional functionality. These four sublayers, from lowest to highest, are as follows:

- The Session sublayer provides a context within which action requests from the client occur.
- The Files sublayer is responsible for the file data structure and, within the context of a session, for implementing those operations that deal with the individual file.

- The Directory sublayer organizes the files into the typical hierarchical tree.
- The Search sublayer can identify a whole set of files with related file names.

7. The Filing Subset and OSI's FTAM

File Transfer, Access, and Management (ISO DP8571) (FTAM), the OSI Model's filing standard, is a filing protocol that moves data between dissimilar systems. It ensures that the data stream is transferred such that if its retransmitted to its original sender it will return in exactly the same format that it was sent in. The XNS Filing Subset performs a similar function, but in a more document-oriented way. The full Filing Protocol also provides information about the file without actually accessing it, as well as security integration, authentication, interlocks to prevent simultaneous writes to the same file while allowing simultaneous access and directory service, among other services.

8. Mailing Standards

The Xerox mailing standards consist of two protocols and a format standard. The Mail Transport Protocol and the Inbasket Protocols address the functions of sending and receiving mail, and the Mail Format Standard defines the format of messages transported using the two protocols. Together these three standards specify a layered mailing architecture.

The relationship of the two Mailing protocols can best be explained in terms of the CCITT X.400 architectural model, which defines two layers: Message Transport and User Agent, as shown in Fig. 13. The model envisions an originator and a recipient user agent, ordinarily a human/ workstation client although entirely automatic processes might also originate or receive mail.

9. Mail Transport Protocol and Inbasket Protocol

The XNS Mail Transport Protocol corresponds to the boundary between the Message Transport layer and the User Agent layer, and provides operations for sending and receiving mail using posting and delivery slots. Accessing the XNS Mail Service according to this protocol thus equates the physical boundary between the workstation and the server with the architectural boundary between the two layers. However, there are pragmatic reasons for departing slightly from this model. A typical user agent (client) will transfer incoming mail from the delivery slot to an "inbasket" container (mail file) for perusal by the recipient.

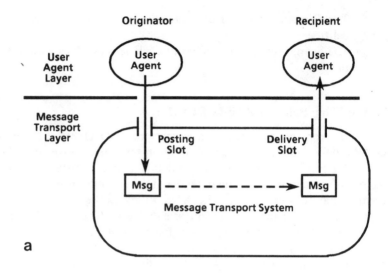

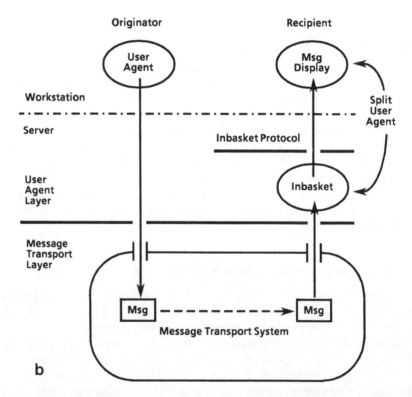

Fig. 13. Message transport. (a) The CCITT X.400 architecture model. (b) XNS Inbasket Protocol provides an internal interface for user convenience.

Using the Mail Transport Protocol for delivery implies that the mail is transferred to an inbasket container residing on the client machine, almost always a workstation. Since users may wish to gain access to their mail from any of several workstations, it is preferable that the inbasket resides on the server, making it globally accessible. The Inbasket Protocol, which corresponds architecturally to an internal interface of the user agent layer, accomplishes this goal, approximating the function of the P3 protocol of the CCITT's X.400 standard.

The interface between the workstation and the service thus consists of both protocols: The Mail Transport Protocol is used for posting messages, while the Inbasket Protocol is used for receiving them.

To send a properly formatted message, the originator invokes the Mail Transport Protocol, handing the message through the "posting slot." The Mail Transport Protocol is responsible for calling the appropriate Courier procedures that will deliver the mail to the intended recipient(s) through "delivery slots." The Mail Transport Protocol then delivers the mail to a holding facility managed by the Inbasket Protocol. This protocol acts at the request of the receiving user by fetching messages that have arrived at the holding area since the last access.

Together, the two mail protocols provide a transport medium for all variety of electronic information that is totally transparent to the content and format of the messages.

10. Mail Format Standard

While the transparency of the Mail Transport system provides layering and flexibility, most usage of Mail Transport will adhere to a single format for the content. This format is defined in the Xerox Mail Format standard, which is functionally aligned with the CCITT X.420 P2 specification for interpersonal messages. The presence of such a message is signaled by the appearance of the corresponding Content Type on the envelope. Messages of this type are encapsulated within the outer level defined by Mail Transport, and provide, in turn, a second level of encapsulation by defining a two-part structure consisting of a message heading and message body.

The Mail Format standard is transparent with respect to the format of the encapsulated message body. Examples of message body types would include document formats for various document preparation systems (e.g., the Xerox 8010); these correspond to subtypes of the Nationally Defined body type of the CCITT X.420 model. In all cases, the heading information, including originator and subject, is represented in a uniform way, independent of the format of the body. At the innermost level, the mail message body must be interpreted by appropriate format-specific software, selected according to the type indicated in the body Type field of the heading.

11. Scanning

Another common customer application is transforming existing images into an electronic form through scanning processes.

Within XNS, the Raster Encoding Standard (RES) is a general-purpose encoding and Image format design that permits an image to be used as a document by itself, or as parts of documents subsequently handled by Interpress.

RES describes a digital representation of all raster images, such as those created by scanning devices or by software that outputs rasters. The representation accommodates images of different sizes, resolutions, intensity gradations, encoding, and compression attributes.

RES uses Interpress encoding rules and Interpress terminology to simplify printer software that handles raster image files specified by Interpress masters. Since RES uses the Interpress encoding rules, it is also possible to include a raster image file in an Interpress master, and execute it. To limit the possible effects of executing a raster image file with its complexity, RES allows only a subset of primitive operators specified by Interpress. RES image descriptions are like Interpress masters: they are not intended to be read directly by people, so no attempt has been made to encode the data in human readable form.

In RES, each raster image, whether binary, continuous-tone, or color, is described as a single file. Raster image data, which can become very large, is described in compressed form with provision for several standard compression schemes. The compression options are Packed; CCITT-4 (Group 4 facsimile); Xerox-standard Adaptive; and a simpler version called Compressed.

12. Scanning Services

The XNS architecture has two common models for how scanning devices can be integrated into the network. In the graphic input model, the scanner is an XNS systems element with its own user interface. In the peripheral device model, the scanner is a peripheral to a workstation that is either directly part of an XNS network, or is connected to an XNS network through a gateway service. (Gateway services are the subject of the next section.)

The graphic input model is best suited for production-oriented applications requiring line graphics or photographs. The scanner digitizes a hard-copy image, then the image (in RES format) can be stored in a file server or to a printer, using the Printer Subset of the Filing Protocol. The scanner image can be combined with text or graphics to form a new composite image, either at the workstation or at the printer, using Interpress commands.

The peripheral device model is best suited for applications where the user will need significant interaction with the image, such as preparing high-quality original artwork from photographs. The appropriate workstation for this model has bit-map display and image editing capability.

13. Gateway Access

The basic functions performed by the Virtual Terminal Circuit Protocol (VTCP) are to move information, support other communication models, and support terminals. The OSI Model has spawned a Virtual Terminal service specification that is functionally similar to the VTCP, shown in Fig. 14.

VTCP is an application-level protocol that makes use of the Courier and Internet Protocols (specifically, Sequenced Packet Protocol) to interconnect an XNS system with a non-XNS system or device. The Gateway Access Protocol (GAP) uses these protocols to issue or receive customized command and data sequences that exactly replicate the command and data sequences used by the target system or device. Courier establishes a session with the target and the Sequenced Packet Protocol handles transmission and receipt of the appropriate bit and character sequences.

VTCP creates a logical appearance of communication compatibility with the other system or device. For interconnecting with a remote mainframe system, this appearance emulates particular devices, such as TTY, VT-100, or IBM 3270 terminals, which the mainframe supports. VTCP also makes it possible for non-XNS terminal devices to interconnect with an XNS system and access XNS services.

VTCP supports many communication models, including the document transfer model for electronic mail, transaction-oriented model for remote data base access, and an interactive model for interface to a mainframe data processor. VTCP does not deal with the content of data; the protocol provides information transfer but not content translation. If translation is needed, either party can provide it or the files can move via a separate conversion service, such as a document interchange service in gateway or

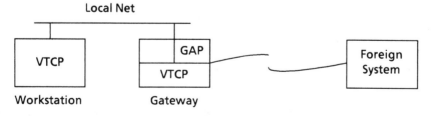

Fig. 14. VTCP and gateway services.

workstation products. In supporting other communication models, VTCP adapts to the other protocols if they are not contained in XNS. Thus many of the commonly used communication protocols are supported by the XNS gateway services.

IX. Future Trends in the XNS Architecture

1. Engineering Information Systems

Engineers involved in circuit design and development already have several graphic design engineering workstation available to them, fully integrated into the XNS architecture. Combined with the latest color plotters, these workstations include not only document creation and editing, but also schematic design, actual logic simulation, and component placement. The future forms of these workstations will add more graphics design functionality, while maintaining full integration with XNS.

2. Data Base Management Systems (DBMS)

Data base management systems are important in any size network, and new data base management products are high on the list for future expansion of service offerings through XNS. New data base servers will be particularly important in department-level networks with heavy data access demands. When demand is heavy, speed of access becomes an important measure of overall functionality and special high-speed data base servers improve the user perception of network performance.

3. Color

The heyday of monochrome in business is over. Color makes business information come alive—makes it more understandable, more dramatic, and more satisfying—so Xerox has made a strong commitment to color technology. New products for XNS networks will include advanced color applications in Interpress, advanced color capabilities in workstations, advances in color printing technology, including new ink-jet and laser printers, as well as color plotters, and advances in color document representations.

4. New Gateway Services

The newest gateway products allow XNS network users (XNS calls them citizens) to communicate with the IBM SNA world, including IBM

377X Remote Batch Service, 3270 interactive terminals, and the IBM Distributed Office Support System (DISOSS). Users will then be able to access documents stored in IBM's Document Interchange Architecture (DIA) and Comment Content Architecture (DCA). New gateways also allow XNS networks to interconnect with other network architectures: the IEEE 802.4 Token Bus and General Motors' Manufacturing Automation protocol, and the IEEE 802.5 Token Ring. Higher-speed links to PBXs will make the network-to-PBX traffic easier to manage for XNS users.

5. Toward an Integrated Future

As international standards are approved and emerge into the market, Xerox will support them in products and will work to integrate with other vendors who implement these standards.

The future of all network architectures is the same—ultimate interconnection through international standards such as the OSI model and the Integrated Services Digital Networks (ISDN). The XNS architecture, with all the advantages that internal homogeneity confers, will be part of this integrated future.

Reference

[1] D. R. Boggs, J. F. Shoch, E. A. Taft, and R. Metcalfe, "PUP: An internetwork architecture," Xerox Palo Alto Research Center, July 1979.

16

IBM's Systems Network Architecture

Diane P. Pozefsky, Daniel A. Pitt, and James P. Gray

I. Introduction

A network is more than simply a collection of machines and communication lines. A properly designed network serves a particular purpose for a particular user or class of users. In order to design networks for different users and purposes while minimizing design effort, one can employ a network architecture. In this chapter we look at the services provided by a network, examine what an architecture is, and then look at Systems Network Architecture (SNA),* its design principles and how it enables a network to provide the required services. Our discussion of the architecture appears in two major sections, transporting data and distributed programming. We have attempted to minimize the amount of jargon in this description of SNA. Necessary new terms are introduced in italics. After looking at the architecture, we conclude with a discussion of how SNA has applied the underlying principles.

A. Networking Services

1. History

In discussing the services provided by a network, it is helpful to remember how computers, computer usage, and computer complexes have

---

*The architecture description is by nature very general. No specific implementations are described except as explicitly noted.

DIANE P. POZEFSKY, DANIEL A. PITT, AND JAMES P. GRAY • IBM Corporation, Research Triangle Park, North Carolina 27709.

grown. In the early days of computers, stand-alone use of the computer was common. Such usage was inefficient of computer resources and did not encourage the sharing of data and programs among users. As usage of the computer grew, batch and spooling operating systems were developed. These operating systems were concerned with resource-sharing at three levels: input devices, processor cycles, and output devices.

With the advent of direct-access storage devices, another type of resource sharing became important: sharing of files—whether the files were programs, such as compilers, or data files, such as data bases. Along with this type of resource sharing came the requirement to be able to locate resources, and a directory was added to operating systems.

As new types of input and output devices were introduced to the computer system, the operating system took on the function of giving the user a higher-level interface for input and output. As multiprogramming was introduced, the virtual machine concept shielded the user from many of the hardware nuances such as interrupts, program relocation, and page swapping.

When computer usage grew beyond the capabilities of a single machine and multiprocessor configurations were introduced, the user expected to be shielded from the appearance of multiple machines. If there was a requirement for a specific job to run on a specific processor (e.g., because of special hardware such as a vector processor that was attached to only one machine), the operating system was expected to assign the job to the correct processor.

When networks of independent computers were created, the user expected the same sort of services. To the user, the computing system is still a single entity and the services provided should hide the location of any resource that it needs just as it hides all concerns about access to a printer, disk drive, or tape drive. The ultimate goal of a network is to make distributed programming available to the user.

2. Services

The above description has focused on the resource-sharing supported by an operating system and clearly this is the prime responsibility of an operating system. In the case of a network, there are further services that are provided that assume responsibility for hiding from the user the distributed nature of the computer facility being used and for guaranteeing fairness in the resource sharing.

The primary additional resource that a network must allow users to share is the communication line. Sharing of a communication line requires that users be given orderly access to the line and that no users be allowed to consume more than a fair share of a single line or route.

The directory function is much expanded from that provided in an operating system. Not only does the directory need to find the appropriate user, file, or device on its own system, but it needs to be able to find them on other systems as well. This requires either replicated directories or a distributed directory and the ability to maintain and query the directory.

Once a resource is located on a remote system, the network needs to give the requesting user access to it. In a nondistributed environment this is simply a matter of assuring that the proper access (e.g., exclusive access for updating) and addressability are given. In a network, a communication path must be established as well.

Just as an operating system enhances the reliability of input and output devices with retries and allows the user flexibility in assuring that a program can tell exactly what data have been updated in case of error (e.g., by identifying the data write that failed), a network must perform the same sort of function for remote updates. This function is performed by means of protocols that return positive or negative responses and allow partner programs to establish synchronization points—common backout points in case of failure in the next program segment.

Another type of function that operating systems provide is management facilities: the ability to identify errors and help in problem determination, the maintenance of accounting data to allow fair charging to users, and the gathering of data to allow the tuning of a system and planning for future enhancements. All of these functions take on much more complex aspects in a networking environment. In order to provide problem determination, the path of any remote accesses must be known. Statistics must be maintained on all intermediate components in order to charge users appropriately, to recognize components that have intermittent problems before hard failures occur, and to identify those parts of the network that are reaching capacity limitations. When enhancements are to be made in a network, control of release levels is required to maintain knowledge about the current status of all network components and to assure that changes that need to be coordinated are installed at appropriate times.

Finally, the function of an operating system in raising the level of the user interface to the machine becomes crucial in a networking environment. The user needs to be insulated not only from the different operating systems and hardware that are available throughout the network, but even from the difference between local and remote accesses. In order to provide this function, a network provides generic interfaces for communicating between programs. These interfaces have a common set of functions supported by all implementations and can then have additional functions that are not required for basic communication but can be used when both partners support them. Such functions can include asynchronous distribution of data, data access and update, and document interchange—both the actual interchange and the contents of the document.

B. How Do Different Systems Work Together?

While a few networks may be truly homogeneous, most are heterogeneous since differences in configuration, release level, and service level generate important divergence in even apparently homogenous networks. In order to allow diverse systems to cooperate and become a network, there must be agreements on both the data that flows between the systems and the action taken based on the data received or sent. The data flows are referred to as formats and the actions taken are the protocols. The formats and protocols together constitute an architecture.

For us to understand what an architecture is, it is easiest to contrast it to an implementation. The architecture defines the external appearance, whereas an implementation defines how that external appearance is constructed internally to a product. In IBM hardware, the S/370 architecture defines the instruction set [1] and models 3090 and 9370 are radically different implementations. In networking, the architecture defines the formats and protocols, but each product is a different implementation. Thus the SNA architecture for type 2.1 nodes* defines the external interfaces that any T2.1 node must meet, and S/38 [2] and AS/400 [3] are two implementations of T2.1 nodes: the architecture will function the same way with any T2.1 implementation. Because the external interface of each implementation is the same, a product attached by a communication line to a T2.1 node does not need to distinguish among the possible implementations. The concept of an architecture inherently includes the important information hiding principle of separation of internal formats from external formats. This is key to all forms of heterogeneity.

C. Network Architecture Scope

A network architecture is a general service that is intended to cover a large range of network configurations and applications. An architecture differs from a single-instance network such as TYMNET[†] [4] or TRANSPAC [5] in that the protocols and techniques that the architecture applies must be applicable to an "arbitrary" network—i.e., any one that a customer wants to configure. Single-instance networks have many of the same requirements as an architecture: range of applications, provision of an agreed-to service level, deadlock prevention, fair access, security, and management facilities. The added requirements for an architecture are the range of configurations that must be supported, the arbitrary mix of products and release levels, and the inability to rely on the addition of hardware or pricing to solve congestion or failure problems.

---

*A type-n node is often referred to as a Tn node, thus "type 5" and "T5" are equivalent as are "type 2.1" and "T2.1."

[†]TYMNET is a registered trademark of Tymshare, Inc.

We briefly review the range of applications and configurations that must be supported and discuss their implications for an architecture. A major concern of an architecture is the co-residence of different types of applications, some of which might require high security (such as banking) and others that might be "high-risk" unauthorized users (e.g., personal computer owners using their own programs). In a network with some applications that require highly secure transmission facilities (e.g., business financial data) and some that do not (e.g., the stock market prices), the network must be sure not to use all of its secure resources on nonsecure data. Different applications require different priorities: interactive traffic should not be delayed by large batch jobs. Some messages require minimal guarantees of delivery (the junk mail of networking) while others require specific knowledge of what data have or have not been received and processed (such as data base updates). Some messages must be delivered while the origin and destination are simultaneously connected to the network (synchronous delivery) while many forms of electronic mail and document delivery can employ delayed or asynchronous delivery.

Thus an architecture must support different levels of security, protection from intruders, appropriate load distribution to prevent overuse of resources in short supply, synchronous and asynchronous communication, priority levels, and levels of coordination from "send-and-hope" to complete synchronization of distributed updates. The range of options that must be provided adds considerably to the complexity of an architecture, and the requirement for efficiency means that the low-function applications must not pay the overhead for functions they are not using.

An architecture must support a wide range of configurations and transmission media to meet the needs of diverse networks. Configurations will range from tree structures, rings, and stars to complete mesh networks —and everything in-between. The media supported will include 1200-bit per-second telephone lines, T1 lines, I/O channels, satellite links, packet-switched networks, and local area networks. Decisions as to network topology and media will be affected by the requirements for availability, traffic, response time, cost, reliability, security, and management organization. An architecture must also support changes in a network; constantly changing, a network experiences change in nearly every aspect over the years of its existence.

D. SNA's Design Principles

Before looking at the specific concepts of SNA, we briefly look at some of the design principles that have driven the architecture. After looking at the details, we will revisit the design principles and see how they have been applied. Here we introduce four of the key design principles—those concepts that have guided how the architecture was built. These key principles are layering, subsettability, evolution, and consistency.

Layering is a well-understood principle originally borrowed from operating system design [6] and introduced to network architectures by ARPANET. It is widely used in network architectures today (e.g., the OSI model [7] and DNA [8]). It is network architecture's form of information hiding or shielding. Layering improves all aspects of architecture work—design, change, validation—and helps products in all phase of their development cycles, from design through test and maintenance.

The requirement for subsettability of an architecture is the ability to accommodate a wide range of product costs and functions while maintaining connectivity. Connectivity can easily be achieved by requiring all products to implement exactly the same thing, but this either requires the reduction of supported functions to a least common denominator or makes the cost of an architecture implementation prohibitive for small systems. The better solution is to define a base of function that all products implement. Use of additional functions can be negotiated.

Just as networks continue to change as if alive, a networking architecture can be likened to a living, growing organism. With rapid changes in technology and requirements, an architecture must continue to grow to fit the changing environment. It must evolve through time rather than undergo revolutions. An evolutionary architecture is one that permits products to grow in reasonable steps and allows networks to add components with new capabilities in small increments—as small as one at a time. Both product developers and users require evolutionary growth. Products are constrained by cost, schedule, and other limits on the size and frequency of changes in their design. Users need to maintain their networks' operation while testing new products and, of greater importance, to preserve their investments in current applications and equipment. An evolutionary architecture allows products implementing one version of the architecture to work through many architecture releases.

Everyone wants consistency—for design principles, technical principles, and encodings. Why is consistency hard to achieve? There are three basic problems: (1) an architecture is built by many different people; (2) as networking continues to mature, we learn better ways to do things; and (3) external constraints often conflict—consider, for example, the immediate needs of customers for functions and the measured pace of standards development.

II. Major Concepts

Before discussing specific functions, base concepts are introduced: the layered architecture, the physical and logical elements of an SNA network,

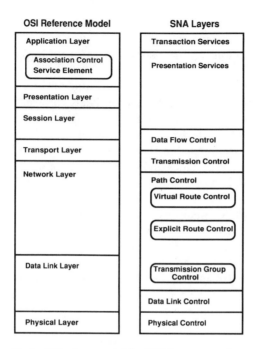

Fig. 1. OSI reference model—SNA layer comparison.

the naming and addressing of elements, and how independent SNA networks can be connected.

A. Architecture Layers

SNA is a layered architecture containing seven layers, whose approximate mapping to the OSI reference model is illustrated in Fig. 1.

Briefly, the seven SNA layers, their services and functions are as follows:

- *Physical control* provides connectivity to the transmission facility. It is concerned with the electrical and mechanical characteristics of the transmission facility and is the same as OSI's Physical Layer. It is responsible for encoding and decoding between bits and transmission symbols.
- *Data link control* provides reliable data transmission between adjacent nodes. (SNA has no concept of an unreliable data link control.) It is responsible for scheduling data over the link and for error detection recovery between the adjacent nodes: checking of data integrity, lost messages, and retry logic when errors occur. The data link control manages the sharing of multiple access facilities (e.g.,

multipoint lines or token rings). This is the lowest layer of an SNA node that participates in network management.* The layer supports error conditions and statistics that can help in problem determination.

- *Path control* provides a reliable route between a source and a destination. It routes data appropriately, provides first-in/first-out end-to-end delivery of messages, and manages the shared data link resources of the network. Path control may change the size of the units that are shipped in the network: it may segment messages into smaller pieces or block multiple messages together as appropriate. Path control recognizes if a route becomes unavailable and reports that fact to its users.

    In subarea nodes (see "Nodes and Links" for a description of the different types of nodes defined in SNA), this layer is divided into three sublayers: *Transmission group control* creates a logical link from one or more physical links between a pair of nodes. *Explicit route control* manages the physical route between nonadjacent subarea nodes. *Virtual route control* selects a transmission priority for and controls congestion on a route between nonadjacent subarea nodes.

- *Transmission control* provides a metered, secure end-to-end connection. It provides pacing, cryptographic services, and facilities to aid in error recovery. Pacing allows a node to control the rate at which it receives request data from its partner in order to match its processing capacity. Cryptographic services allow the end-to-end enciphering of all data or selected data. Transmission control is responsible for the protocols to reset pacing and sequence number processing in those error cases where they can be reinitialized without destroying the connection.

- *Data flow control* provides the logical grouping of end users' data. There are two types of grouping in SNA: *chains*, which identify the smallest recoverable unit for error recovery, and *brackets*, which keep together a related conversation (or dialogue) for application purposes. Data flow control also corrclates requests and responses and manages the protocols that regulate the sending of responses.

- *Presentation services* provides data transformations to map data into formats and representations more appropriate to the applications. It

---

*SNA's network management interacts with transmission services (e.g., modems and PBXs) through direct means (e.g., modems with sophisticated problem determination capabilities) and indirect means (e.g., equipment coupled to the SNA network management focal point, such as through NetView/PC [9]).

is also responsible for the coordination of resource sharing and for synchronization processing. The interface to presentation services is available to the user; for LU 6.2, it is a well-defined interface and is available as the communications interface of Systems Application Architecture (SAA) [10].

- *Transaction services* provides application services such as distribution services [11] or distributed data management [12]. Transaction services are privileged transaction programs; they run like other transaction programs except that they may have access to privileged commands and they provide architected functions. Several transaction services, such as SNADS and DDM, have been identified as SAA protocols.

B. Nodes and Links

In architectural terms, a network consists of *nodes* and *links*. Nodes are the architectural home of the protocol implementations, the *protocol machines*, and the end points of one or more layers of protocol.

Different node types exist in SNA to reflect differing capabilities to provide network services. *Subarea nodes* use unique network-wide ("subarea") addresses and provide the full complement of network services including intermediate routing and address mapping between local addresses and network-wide addresses. Addresses are described in more detail in "Names and Addresses." Node types 4 (T4) and 5 (T5) are subarea nodes. Virtual Telecommunications Access Method (VTAM) and Network Control Program (NCP) are T5 and T4 nodes, respectively [13]. Subarea nodes can communicate with other subarea nodes or peripheral nodes.

Peripheral nodes (e.g., 3174, Application System/400, Transaction Processing Facility (TPF) [14], Series 1, or the IBM Personal Computer) use local addressing; they connect to subarea nodes through *boundary functions*. Nodes that use local addresses are further classified as type 1 (T1), 2.0 (T2.0), and 2.1 (T2.1). New SNA products in this class support node type 2.1. S/36's and AS/400's Advanced Peer-to-Peer Networking feature (APPN) [15, 16] adds network services such as routing and directories to type 2.1 nodes. In APPN, there are two roles that a node can take: that of a *network node* or that of an *end node*. A network node performs intermediate routing of data, provides network services to end nodes, and supports its own end users. An end node only provides services for its own end users. Figure 2 shows an SNA network consisting of subarea and peripheral nodes and Fig. 3 shows an APPN network.

There is no architectural association between node type and the kind of hardware device that implements it. One commonly thinks of hosts as

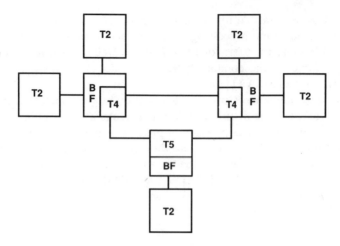

Fig. 2. Example of a subarea network. T2, T4, and T5 are type 2, 4, and 5 nodes, respectively; BF is the boundary function.

type 5 nodes, communication controllers as type 4 nodes, and everything else as type 2 (meaning either 2.0 or 2.1) nodes, but that is only because historically the developers of these products have chosen to implement these node types (often with good reason, of course, such as processing and storage capabilities).

Links in SNA can be any transmission facility that connects two SNA nodes but that does not contain intermediate SNA nodes (or at least not

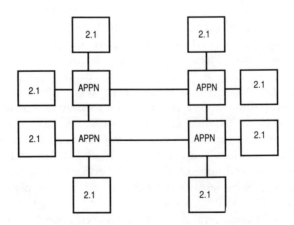

Fig. 3. Example of an APPN network. 2.1, T2.1 node; APPN, APPN node.

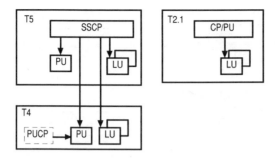

Fig. 4. Control points and the NAUs they control.

any in their same SNA network*). Thus two nodes connected by a link are considered to be adjacent. A link is also defined by the ability of two attached nodes to operate, with no intermediate protocol partners, a data link control procedure between them. Carried on this link can be traffic originating in many different nodes and destined for many different nodes.

C. Network Addressable Units and Sessions

Nodes and links are the building blocks of the physical SNA network; *network addressable units* (NAUs) are the logical users of the transport network. There are three types of NAUs: the *logical unit* (LU), which is the access port of the end user into the network, the *physical unit* (PU), which controls the attached links and other resources of the node, and the *control point* (CP), which is responsible for managing the node and its resources and possibly for supporting other lower-function nodes. Thus, for example, the PU is the component that understands how to activate a link and the control point is the component that determines that a link should be activated.

The external appearance of PUs and control points differs by node type. In T5 nodes both the PU and the control point are separately addressable and the control point is referred to as a *system services control point* (SSCP). In T4, T2.0, and T1 nodes, the PU is addressable and the control point, the *physical unit control point* (PUCP), is not. In T2.1 nodes, the PU has been merged into the control point. Figure 4 illustrates the various control points and the NAUs that they control.

*As an example of an adjacent SNA node not in the same network, an SNA network can be built to appear as an X.25 network using the X.25 SNA Interconnection product on NCP [17]. In such a case, the SNA link between the two SNA nodes acting as X.25 DTEs (terminals attaching to an X.25 network) is actually over an entire SNA network.

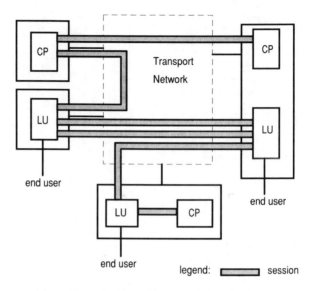

Fig. 5. Network addressable units (NAUs) and sessions.

NAUs communicate with other NAUs in other nodes over logical connections called *sessions*. Sessions are established with a specific mode. The mode identifies session parameters such as whether encryption is to be used and maximum packet size and identifies the route characteristics to be used by the class of service. (See "Route Selection" for more information on class of service.) Functionally, a session roughly equates to an OSI transport layer connection.

Figure 5 illustrates the concept of NAUs and sessions. It illustrates that the logical units (LUs) may have multiple sessions with the same or different LUs and that a control point may communicate with LUs in its own node or, as is the case for the SSCP, may have sessions with LUs in other nodes. Control points may also have sessions with other control points. The transport network shown in the figure is discussed in "Transporting Data."

D. Names and Addresses

All NAUs have names, where a name is a character string and is unique within a network. These names are used at the presentation services layer and above. All networks also have names, and network names are unique.* Thus a fully qualified network name, NETID.NAUNAME, is

---

*Actually, the names are unique within the world of the user to whom the networks belong; there is currently no universal name-administering authority.

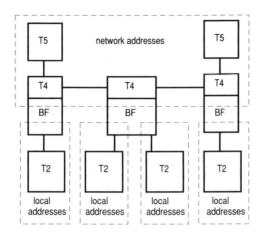

Fig. 6. Scope of address spaces in subarea network. T2, type 2 node; T4, type 4 node; T5, type 5 node; BF, boundary function.

universally unique. A user who wants to establish a session with a remote partner provides the name of the logical unit wanted. SNA uses the name to determine a location for the logical unit. Within the subarea, the location is identified by an address. As the following discussion will point out, there is no global addressing scheme in T2 nodes and therefore the name of the control point is used to denote the location of these nodes.

The addressing schemes in a subarea node and in a peripheral node differ. In the subarea, an address must be unique within a network, not just within a single subarea node, and is assigned to a NAU when the NAU is activated. The address is a fixed length, consisting of two parts: 15 bit element, 31 bit subarea (although not all 31 bits are currently supported). The subarea address identifies the subarea within the network, and the element address identifies the element within the subarea.

There are two peripheral node schemes that currently exist, those for a T2.0 node and those for a T2.1 node.* In both schemes, the scope of a NAU address is local and is a single field. In a T2.0 node, an address refers to a NAU and is statically assigned to the NAU. In a T2.1 node an address is dynamically assigned for the duration of a session, identifies the session (not the NAU), and is unique for a specific transmission group. Figure 6 shows the scope of different addresses spaces in a subarea network and Fig. 7 shows the scope of different addresses among T2.1 nodes in a APPN network.

A subarea is defined as a subarea node and the peripheral devices attached to it. All NAUs in the subarea share a common subarea address

---

*T1 nodes are obsolete.

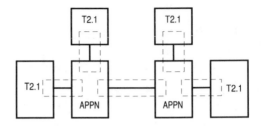

Fig. 7. Scope of address spaces in APPN network: All address spaces are local.

and each has a unique element address. The element addresses are assigned when the network is brought up or at dynamic reconfiguration except for resources accessed over switched links and LUs engaging in parallel sessions with the same partner. Since addresses used for subarea resources are unique in the network, subarea path control uses them to define the location of the resource.

Logical units in peripheral nodes attached to a subarea network have both a network address that is known in the subarea nodes and a local address that the peripheral node uses. Since T2.0 nodes are accessible only from the subarea, the subarea address is used for global identification.

The address used in a T2.1 node is a seventeen-bit value that is called the *local form session identifier*. Because two T2.1 nodes initiate sessions that flow over the same transmission group, it is convenient to divide the address space in half. One bit of the address is an address space divider; the value that it takes is negotiated when the link is brought up (see "Link Activation" for further details) and each node will always set that bit to its negotiated value when it assigns the address for a session.* In a T2.0 node, there is an origin address and a destination address in the routing header, each address being a local eight-bit address. In order to retain the header format and allow migration of T2.1 nodes from roles played by T2.0 nodes, the remaining 16 bits of the local form session identifier are encoded in the origin and destination address fields.

A single path control instance runs in a specific address space. There are several places in SNA where a session traverses multiple address spaces; when this occurs, a session connector component is used to bridge the different address spaces that meet at a single node. There are three types of session connectors in SNA: boundary function, SNA network interconnect gateway, and APPN intermediate routing function.

Boundary function is the connection that is used to connect a NAU in a peripheral node with a NAU in the subarea and therefore connects a T2 address space with the subarea address space. Between the subarea NAU

---

*For the reader familiar with SNA headers, the address divider bit is the ODAI (OAF'-DAF' Assignor Indicator) field in a FID2 header.

and the boundary function, the subarea address is used; the boundary function maps the headers and addresses onto the T2 address space on the other side.

SNA network interconnect gateway is the component that accepts data packets from one network and transmits them to the appropriate destination in the other network. The end-point NAUs are unaware that they are communicating with a partner in another network as cross-network resources are given alias names and addresses in the partner's network. (For further discussion of the alias function, see "Network Interconnection.") It is the responsibility of the SNA network interconnect gateway to map the names and addresses appropriately as the data packets cross the network boundary.

APPN session connectors are used for intermediate routing within an APPN network. Because each transmission group has its own address space, an APPN session connector bridges two T2.1 address spaces as data packets move from one transmission group to another. APPN session connectors are discussed further in "Routing in T2 Nodes."

E. Network Interconnection

As communication functions have grown, network users have found a need to interconnect networks without giving up the autonomy of the networks. SNA network interconnection [18] permits this function. Figure 8 illustrates three interconnected SNA networks and shows the different components: the *gateway SSCP*, which is responsible for locating and setting up sessions to partners in different networks, and the *gateway node*, the component that actually connects the two networks.

The general principle of network interconnection is that a user need not be aware that his or her partner is in another network: the differences between an intranetwork session and an internetwork session are all handled by the gateway components. Gateways also provide LU name aliasing and hiding (the ability to give an LU in one network a name and

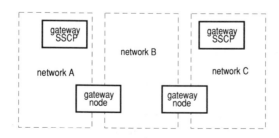

Fig. 8. Interconnected SNA networks.

appearance in another network) and customer exits usable for security. These gateway functions allow programs written prior to network interconnection (and therefore with no knowledge of network-qualified names) to communicate across a gateway. They also provide strong isolation of the two networks: one network need not be aware of changes in the other network's structure or names and a network that wants to keep its internal structure or names unknown can easily do it.

III. Transporting Data

In this section we examine how a network of SNA nodes moves data. We describe the overall concepts of data transportation in SNA, the variety of transmission facilities from which a customer may choose in setting up an SNA network, and the key networking functions of intermediate nodes.

The transporting of session data from one session end point, or half-session, to the other is the responsibility of what is sometimes called SNA's *transport network*. The transport network provides reliable service to the sessions, indicating only failures from which it cannot recover. SNA's transport network covers the functions defined in the OSI model's physical layer, data link layer, and network layer, as well the reliability sometimes provided in the transport layer [7]. Governing data transport are the formats and protocols in the layers designated in SNA as the physical layer, data link control, path control, and part of transmission control. Together these layers hide from the sessions the details of the underlying transmission network, including its topological arrangement, the characteristics of the transmission lines, the public or private administration of the lines, the security of the lines, and the cost of transmission. Any interest the sessions have in these details is manifest in the primitives exchanged between layers. For example, if a session requires a secure transmission path, it can request a certain level of security and the transport network either establishes a path that meets that level or turns the request down because such a path is unavailable.

The functions performed by the transport network and the control points in support of the transport network include the following:

- Management of data links.
- Activation of links.
- Locating of requested partners.
- Selection of a route between partners.
- Activation of a route in the subarea.
- Congestion control within the network.
- Segmenting and blocking of messages to match link characteristics.

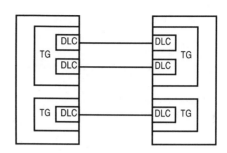

Fig. 9. Examples of transmission groups: Two parallel transmission groups—one multilink with two links and one single-link.

Data link controls in SNA provide reliable service to path control. Reliable service is defined as the delivery of frames in the proper order, without error, and without duplication. Each type of link provides its own form of checking; a common technique is a cyclic redundancy check, which is purely an error-detecting code whose probability of undetected error decreases with the number of check bits.

Two adjacent nodes (except T2.0 nodes) are allowed to maintain more than one link between them, to provide backup, increase capacity, or improve performance. Transmission on each link is governed by a distinct data link control. The links can each be used independently (any session choosing exactly one to use) or in subarea nodes they can be bundled together into a multilink *transmission group*. (A single link also constitutes a transmission group.) Coordination among the data link controls of the multiple links of a transmission group is performed by the path control layer in SNA. Figure 9 illustrates parallel and multiple transmission groups.

When a link fails, a link failure is reported immediately to network management. If all links of a transmission group fail, the transmission group failure is reported to the session end points, so that users are not left hanging. Frames that are lost or that arrive with errors are retransmitted over the transmission group at the first available opportunity. Retransmitted frames consume bandwidth on only one link, not on all links in a path as would end-to-end detection and retransmission. Link control parameters can be carefully tuned to the characteristics of each specific link, providing lower delay and better response time than end-to-end tuning [19].

During data transfer, each frame carries an unambiguous sequence number and is stored at the sender until it receives an acknowledgment for that sequence number. If a frame is lost, the sender retransmits the frame. After a given number of retransmissions have failed, the link is considered to have failed and the connection is terminated. An acknowledgment timer is used to detect missing frames; this timer is long enough to allow for normal transmission delays but short enough to detect true losses as soon as possible. The variance in link delay is substantially smaller for single links than for concatenated links, allowing single-link protocols to run very tight timers. Timers that are unnecessarily long delay recovery when frames are lost and require the transmitter to buffer more frames.

The speed of the link, the value of the acknowledgment timer, and the number of bits in a frame determine the number of frames that a transmitter can have outstanding and unacknowledged at a time. If this number exceeds the size of the transmit window (the allowable number of sequence numbers that can be sent unambiguously), the transmitter will be unable to transmit continuously, since at some point it will be waiting for acknowledgments. This problem first appeared with satellite links because of their combination of high speed and long delay. Compensation for this delay takes the form of an increased sequence number space. Early SDLC allowed only three bits for sequence numbers, but that has been increased to seven bits. The 3737 channel-to-channel extender uses a 16 bit sequence number.

Every SNA node terminates the data link control operating on each link to which it attaches. When data link control releases (to the path control component in the same node) a packet received on a link, the node decides whether to keep the packet itself or retransmit it on another link. Subarea nodes and APPN nodes route packets between links and within the node; T2.1 nodes route packets only within the node. Flow control is performed on a link-by-link as well as an end-to-end basis to prevent congestion in the network. Link stations (the protocol machines that run the link protocols) can inform one another when they are unable to accept frames because they are congested. Link stations on a local-area network can unilaterally detect congestion on their links (in either their partner stations or intervening bridges) and modify their transmit windows until the congestion subsides [20]. Routing and both network- and session-level flow control are functions of the path control and transmission control layers.

A. Data Links

Links between SNA nodes may be point-to-point, multipoint, or multiple access. These types of links are discussed in detail in Chapter 3. In SNA, the three categories of multiple-access links are satellite links, local-area networks, and public packet-switched networks. Despite their broadcast nature, satellite links are used point-to-point. There are no connection-oriented broadcast protocols in data link control (or any higher layers) and the polling needed for satellite multipoint links would degrade throughput compared to the performance of a point-to-point connection. On a local-area network, any two SNA nodes are considered adjacent. Point-to-point link controls are used, although broadcasting may be employed within the medium access control sublayer as part of the generic local-area network access protocol. Nonetheless, while stations on a local area network transmit one at a time in some scheduled or random order, the mediation is performed transparently to the link control procedure so that full duplex link controls are used. Likewise on a public packet-switched network, point-to-point data link controls are used.

While SNA runs on a variety of LANs,* the local-area network architecture described in SNA is that of the IBM Token-Ring Network. Conforming to ANSI/IEEE Standard 802.5-1985 [22] and ISO International Standard 8802-5 [23], the token ring allows multiple stations to share the same ring network. Under SNA, multiple token rings at a single site can be interconnected to form a *bridged local area network* [24]–[26]. Each ring retains its own token, but all the addresses in the bridged network must be unique. A bridge that connects two rings observes only the token ring protocols, not the data link control protocols between the communicating stations. Any interconnection topology is permitted so long as no two rings are separated by more than five intervening rings [27].

SNA also allows nodes to be connected via public packet-switched networks. SNA nodes act as DTEs and communicate to the network across an X.25 interface. The X.25, LAPB, and X.21 or X.21bis protocols operate between the SNA node and the X.25 network. SNA protocols at the path control layer and above operate between the SNA nodes themselves, which are viewed within SNA as adjacent. Since X.25's LAPB operates only between the DTE and the network, SNA nodes operate an additional data link control (Physical Services Header, QLLC, or ELLC [28]) between themselves, below path control but above X.25. Otherwise, there is no end-to-end (across the X.25 network) protocol for control functions, error detection, and recovery. Public packet-switched networks that exhibit a high rate of signaled failures† use ELLC, with its checksum and retransmission capability.

Finally, SNA includes a very different type of link, the mainframe channel. These channels provide high bandwidth, up to 3 Mbytes/s over distances up to several thousand feet using multiwire cables or optical fibers. The channel protocols apply to all channel-attached devices (hosts, communications controllers, terminal controllers, and printers).

We briefly describe the prominent characteristics of the data link controls supported by SNA; additional details can be found in the references and in Chapter 3.

1. SDLC Normal Response Mode

Synchronous data link control (SDLC) [29] is a bit-oriented link protocol that allows arbitrary bit patterns of user data in the information field of packets called frames. When running in normal response mode

*For example, RT/PC supports SNA on IEEE 802.3 LANs [21].
†Signaled failure rates of tens per day per virtual circuit have been experienced.

(NRM) or in normal response mode-extended (NRME), type 2.1 nodes connected in a point-to-point fashion may negotiate which is to be the primary.

2. X.25 and LAPB

An X.25 network between two SNA nodes is treated as a link between them, with X.25 as the access protocol [28]. The X.25 packet layer protocols apply between the X.25 DTE and the X.25 network, to which the X.25 DTE attaches via an X.25 DCE. These protocols provide a method for multiplexing virtual circuits on a single access line. They carry no end-to-end (DTE-to-DTE) significance, with the possible exception of the D-bit when it is set to one, which some users have employed for higher layer acknowledgments.

While two SNA nodes connected by an X.25 network are considered adjacent, X.25 alone does not constitute a complete link control between them. An SNA node requires certain protocol exchanges, communicated by data link control, with the adjacent node. These are the common SDLC functions of identification exchange (XID frames),* link testing (TEST frames), operational mode selection (SNRM frames), and link disconnection (DISC frames). None of these are provided by X.25. Three varieties of logical link control may thus be used over X.25. All provide the required services and each is distinguished by a unique value in the first byte of the call user data field.

Physical Services Header. The physical services header, the first logical link control provided for SNA nodes attached to X.25 networks, has its own encodings for the physical services commands. It contains provisions for data segmentation that are independent of X.25's segmenting capabilities. It also provides data packet sequence checking but no recovery, only notification, when frames do arrive out of order. The physical services header type of logical link control depends on acceptable rates of signaled failures and residual errors from the X.25 network.

QLLC. Qualified logical link control (QLLC), more recent than the physical services header, also depends on acceptable rates of signaled failures and residual errors. It employs the Q-bit,† when set to one, to encode the physical services commands and allows either asynchronous balanced mode or normal response mode of operation. Data packets are sent as normal, unqualified X.25 data. If segmenting is required, X.25's

---

*The use of XID by higher-layer functions is discussed in "Link Activation."
†The Q-bit is a bit in an X.25 data packet that allows higher-layer protocols to distinguish between its control and data packets.

M-bit facilities are available. (The M-bit is used to indicate that multiple packets have been sent and need to be reassembled.) No sequence number checking is provided and the D-bit is not used.

ELLC. Enhanced logical link control (ELLC), used over networks whose rates of either signaled failures or residual errors are unacceptable, provides complete link recovery facilities. It detects lost, duplicated, and corrupted packets (the last with a cyclic redundancy check) and retransmits them, thus providing the same quality of service to path control as SDLC does, all without using the Q-bit. It allows recoverable call clearing, virtual circuit restarting, and interface restarting. A connection type indicator distinguishes between initial connection requests and those resulting from a recovery attempt. An explicit connection identifier is included in the call setup packet. There is a two-byte address field (of which one bit distinguishes commands from responses), a two-byte frame check sequence, and a two-byte control field that allows 7-bit sequence numbers. ELLC employs the elements of procedure of asynchronous balanced mode.

3. IEEE 802 LAN LLC

Logical link control for local area networks also employs the elements of procedure of asynchronous balanced mode (with 7-bit sequence numbers) but contains many novel features present in none of the other data link controls used by SNA [30]. Conforming to the international standard ISO IS 8802-2 [31], logical link control operates in conjunction with any of the standard medium access controls for local area networks, which offload the tasks of frame delimiting and error checking from logical link control. The two (destination and source) six-byte addresses of medium access control combine with the two (destination and source link service access point) one-byte addresses of logical link control to allow multiple simultaneous logical links to be maintained through one physical port. Half of the link service access point address space is reserved for definition by standards organizations (currently the IEEE) according to the notion that a link service access point uniquely identifies a higher-layer entity or architecture (such as the ISO network layer). SNA uses link service access points from the user-assignable half of the address space, and while the architecture specifies a default link service access point, SNA implementations may choose additional values.

SNA requires Class II logical link control stations, which incorporate both connectionless (type 1) and connection-oriented (type 2) services. Connectionless service does not exist in the other data link controls, and the simplicity of its protocol is inadequate justification for the low quality of service it would provide to SNA's path control. In addition to sequence number checking and retransmission, the connection-oriented service includes two types of flow control. Basic flow control, present in normal response mode and asynchronous balanced mode as well, allows a receiver to notify a transmitter when to stop, and restart, sending data. Dynamic window flow control [20] allows a transmitter to detect congestion in the transmission path, including the receiver or any intervening LAN bridges, and modify its transmit window as the congestion subsides. It operates on the principle that when congestion is greatest, throughput is maximized with the smallest transmit window, and when congestion is absent, throughput is maximized with the largest transmit window.

4. S/370 Channel

The System/370 channel predates SNA but its data link capabilities can be examined comparably to those above. The channel in essence uses direct memory access for parallel transmission of data on a bus. The bus's parallelism allows certain wires to be dedicated to certain signals, so the mapping of data link control functions onto the channel is not one-to-one. Channel command words, written to and read from the channel, satisfy the requirements for XID, Test, connection initiation (Contact and Restart Reset), and connection termination (Discontact). The transmission of a block of data is delimited by a Write Start command and an Ending Status byte. An explicit length field accompanies each data word in the block. The equivalent of a one-bit sequence number, frame-by-frame acknowledgment, and parity checking provide assurance of reliable delivery.

There are several configurations in which the channel can be used. When it supports devices such as communications controllers, terminal controllers, and printers, it has a multipoint appearance. In this configuration, access to the channel is governed by a loop arrangement in which a single primary station selects the others, which are secondaries. A one-byte address designates the secondary in all commands and responses. Unlike an SDLC link, the channel allows a secondary station to send an attention signal to the primary to ask for servicing; this feature relies on loop-style wiring of several channel signal wires. The channel also provides VTAM's channel-to-channel support. In this function, it can be configured as a point-to-point connection to one other mainframe or it can take on a multiple-access appearance by connection through a 3088 switch. Each 3088 switch allows up to 16 processors to be connected.

B. Routing Data

There are three portions of the routing problem: locating the session partner, selecting a route between the session end points, and routing the data. We first describe how directory services are used to locate a partner, then we look at how routing is done in subarea and APPN networks, and finally we examine how the route is selected.

1. Directory Services

The directory support needed in a network is to map from resource name to location. As described above, the location is identified by the address in the subarea network and by control point names in T2.1 nodes. In both cases, the location is represented by the control point that is sought; in the subarea this is represented by the subarea address and in T2.1 nodes by the control point name. Since the location is used to select a route, identification by control point or subarea correctly defines the physical end point that is sought while reducing the number of end points in the route selection task. For a logical unit located in a peripheral node attached to the subarea, the subarea address yields the correct route because a peripheral node is connected to the subarea network by only a single link.

Not all control points are involved in the directory function. System services control points (SSCPs), located in T5 nodes, perform directory services (and all other session services as well) for resources in T4 subareas. An APPN node's control point performs directory services for resources in attached T2.1 nodes. Thus a directory control point is a control point in a T5 node or in a T2.1 network node.

In early SNA, directory services required the customer to define the location of all resources that a directory control point might want to use. Each control point defined all the resources in its domain and all resources in other domains with which it required sessions. When SNA network interconnection was introduced, a gateway SSCP became capable of search-

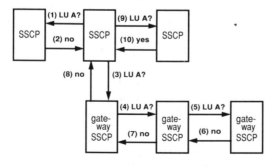

Fig. 10. Directory search among subarea nodes.

ing its adjacent gateway SSCPs for a resource and in fact this search could cascade through any number of gateway SSCPs to find a particular resource. Figure 10 illustrates the directory function performed among subarea nodes. As a search is cascaded across a gateway, name translation may be performed if aliasing is in use.

In APPN, there is no requirement for control points of communicating logical units to be in session with one another, so the search for a resource cascades throughout the network until the resource is found. The search creates a spanning tree across the network and each node retains information about the search until all of its children reply. APPN does not support network interconnection, so the search is always within a single network.

The directory protocol (as with all APPN control point protocols) is built as LU 6.2 transaction programs. While these functions are using the presentation services layer, they are part of the control point, and not a user's LU. The use of 6.2 is a matter of reusing an existing, generalized function, rather than inventing a new, specialized one.

Figure 11 illustrates the directory search in an APPN network. When a node responds positively to a query, it still propagates the request to its neighbor and the crossing of queries is treated as a negative reply.

Node A broadcasts a query for LU X to both its neighbors, B and C.

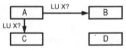

Node C continues the broadcast to node D.

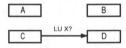

Node D has the resource and reports positively to node C. It also continues the broadcast to its neighbor, node B. Node B likewise continues the broadcast to node D. The crossing of queries is interpreted as a negative reply by both nodes B and D.

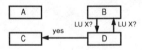

Node C forwards the positive response to node A. Nodes B and D return search completion indicators.

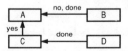

Node C will forward the search completion to node A, which now knows that the search has completed.

Fig. 11. Directory search among APPN nodes.

Reference [32] describes how resources in end nodes can be learned: For very low function or untrustworthy nodes (e.g., some personal computers), the resources are defined by an operator to the network node. Trustworthy nodes may choose to register their resources with the network node or may be willing to process directory queries for their own resources.

Once the location of the two end points of a session is determined, a route needs to be established between the two. We now look at the routing mechanisms used in subarea networks and in APPN: how routing is done, how routes are selected, and how they are managed. To support the sharing of any resource, the network must assure fair access to the resource for all users. For the sharing of transmission channels, this management is the congestion control function.

2. Routing and Congestion Control

One of the important functions of a network is to provide to its users fair access to all resources. Providing NAUs and the sessions they use fair access to the communication lines requires congestion control.

Both at the session level and in the transport network, SNA uses *pacing* to prevent congestion. In all SNA pacing, a request is sent for permission to send the next pacing window and a response gives the permission. Figure 12 illustrates the general concept of pacing; the specific pacing protocols in the transport network and in sessions are described in later sections. In SNA, the next pacing window is always requested when the current window is started. This prevents the sender from waiting long periods between windows. It also means that the receiver has at times agreed to receive almost two complete windows. (Since the pacing request flows on the first data of a window, the maximum is one less than two windows' worth.)

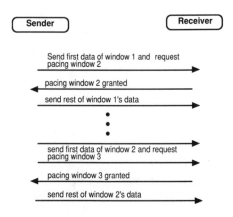

Fig. 12. General SNA pacing protocol.

The basic requirement of routing is to move data from the origin to the destination. In SNA, traffic for a session travels a fixed path. As described in "Architecture Layers," the path control layer has three sublayers: transmission group control, explicit route control, and virtual route control. Transmission group control is common to both subarea and APPN routing, but APPN does not define the multiplexing of sessions onto (explicit and virtual) routes as does the subarea. After discussing transmission groups, we discuss the routing among subarea nodes and the routing among T2 nodes.

a. Transmission Groups. Transmission groups are the logical connection between adjacent nodes that add multiple link procedures and transmission priorities to the services provided by data link control. A transmission group may contain one or more links and may support different transmission priorities. The Network Control Program (NCP) [13] supports both of these options. T2.1 nodes do not support multilink transmission groups (since the transmission header does not have space for the required fields); AS/400 [16] supports transmission priority by carrying the transmission priority on the BIND and saving the priority with other routing information. (See "APPN Route Selection" for further discussion of setting up the route.)

Multilink transmission groups are used to bundle links together into a single link appearance. Traffic that is directed over a transmission group can traverse any of the links in the transmission group. The user perceives two main benefits from transmission groups: improved reliability, because the failure of a single link will not affect a route or the sessions on it, and increased bandwidth. Because messages can flow on different links and arrive at the receiving node out of order, a transmission group sequence number enables the receiver to resequence messages and eliminate duplicates after each hop. The multilink transmission group function is similar to that defined by the ISO standard IS 7478 for multilink procedures for X.25 [33], but it is applicable to all data link control types, and, in fact, a multilink transmission group could have links supporting different data link control types.

There are three transmission priorities available for user data: high, medium, and low. At each transmission group, messages are queued according to their respective transmission priorities, with the higher-priority messages being transmitted first. An aging algorithm prevents lower-priority messages from being permanently blocked.

b. Subarea Routing. Just as each hop between adjacent nodes has two levels of protocols for it (data link control and transmission groups), the path between two subareas has two levels of protocols: explicit routes and virtual routes. An explicit route is a physical connection between two subarea nodes; it can be defined as a series of transmission groups and serves to carry user data in one direction. Separate explicit routes (ERs) are

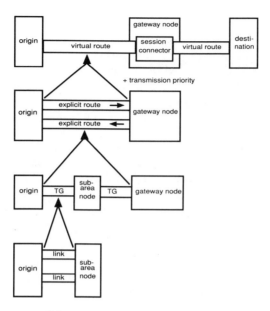

Fig. 13. Virtual routes, explicit routes, transmission groups, and links: The four figures show the composition of a virtual route from its underlying constructs. TG, transmission group.

used in each direction to simplify the ER numbering task [34]. A virtual route is a two-way logical connection between two subarea nodes; it flows over an explicit route and the reverse explicit route (that follows the same physical route). A virtual route is defined by a virtual route number, a transmission priority, and its origin and destination.* Virtual routes are constrained to stay within the boundaries of a single network. If a route is required across SNA networks, there is an independent virtual route in each network and the SNA network interconnect session connector bridges the two virtual routes. Figure 13 shows the relationship of these various components.

The primary additional services provided by virtual routes beyond those provided by explicit routes are transmission priority and global flow control.† The transmission priority defined for the virtual route determines

---

*Two virtual routes with the same number and different priorities are completely independent entities and can flow over different explicit routes.

†Virtual route sequence numbers are also carried on messages. Because resequencing and duplicate suppression are done at every transmission group, the function of the sequence number checking is improved protection against residual link errors (e.g., on single link transmission groups, TG sequence number processing need not be done).

which queue is used at each transmission group. The use of transmission priority permits interactive data and batch data to use the same transmission facilities with interactive traffic traversing the network more quickly. Global flow control is provided by virtual route pacing.

When a virtual route is activated, minimum and maximum pacing window sizes are defined based on the number and types of links in the underlying explicit route. Initially the pacing window is set to the minimum value. The pacing window is increased by one with each additional window up to the maximum value, unless a node on the underlying explicit route recognizes that it is congested. Depending on the degree of congestion, the node can cause the window to be decremented by one (down to the minimum) or slammed all the way to the minimum. The result of these protocols is that the virtual route will keep as large a window as the network can afford to provide within the range that has been defined as acceptable.

c. Routing in T2 Nodes. T2.1 nodes do not use virtual routes, and instead establish paths on a session basis. In *low entry networking* [34], two adjacent T2.1 nodes are able to communicate directly. The only decision such nodes need to make is which of the parallel transmission groups that might connect the two nodes should be used. Peripheral nodes attached to a subarea node maintain only a single connection to the subarea node and hence need to make no routing decisions.

Routing becomes a question of interest in the intermediate nodes of APPN. Recall that the address space used in T2.1 nodes is applicable to only a single transmission group and that a single path control element runs on only a single address space. Therefore each intermediate node in APPN contains a session connector to bridge the two path controls. The session connector uses a different local form session identifier on each transmission group that it couples and therefore swaps the appropriate values into the transmission header. Figure 14 illustrates the session connector and address swapping. "APPN Route Selection" discusses how the session connectors and mappings are created.

Just as flow control procedures in subarea routing are limited to a single virtual route and hence a single address space, so is flow control in APPN administered on a transmission group by transmission group basis. Because there is no multiplexing of sessions onto a single object such as a virtual route, flow control is handled on an individual session basis. Thus there is very good control and adaptiveness in APPN flow control: each intermediate node can adjust window sizes on each individual session.

Global flow control is produced by *back pressure*. As destinations start reducing pacing window sizes, intermediate nodes reduce their window sizes to prevent a large collection of messages from being buffered in their nodes. This backward pressure will eventually reach the source of the data, unless

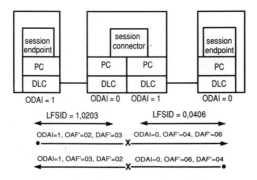

Fig. 14. Example of address swapping in an APPN session connector. PC, path control; DLC, data link control; LFSID, local form session identifier; ODAI, address space divider as carried in transmission header; OAF', half of LFSID as carried in transmission header; DAF, other half of LFSID as carried in transmission header.

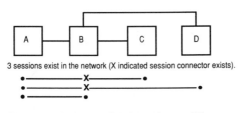

3 sessions exist in the network (X indicated session connector exists).

The current pacing counts with Node A as receiver are as follows:

Node A becomes congested and reduces its windows.

Node B cannot afford to buffer extra messages so it reduces its windows.

The result: all flows in the network are reduced.

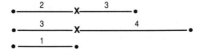

Fig. 15. Example of pacing back pressure.

there is adequate capacity in the network to handle the queued messages. Figure 15 shows the pacing stages and how flow control is achieved.

The pacing used in APPN is exactly session-level pacing. Whether a session connector is adjacent to the session end point or to another session connector, it uses the same protocols. Session end points need not know whether the adjacent node is the session partner or a session connector. Session-level pacing in APPN is an adaptive pacing scheme; unlike virtual route pacing, though, the value of the pacing window can be changed arbitrarily—from any value to any value. In case of congestion, a node can request its partner to return the unused portion of the current, already given, pacing window and to change the size of the next window that has been authorized (to zero if congestion is bad enough).

This increased flexibility is important because APPN intermediate nodes are required to reserve buffers before they authorize a window. Adaptive pacing and buffer reservation are important parts to APPN's self-running, self-adjusting design philosophy—a philosophy based on the expectation that many APPN networks will be installed in environments without systems programmers or network maintenance organizations.

3. Route Selection

From the user's perspective all routes in SNA are chosen based on *class of service*. The class of service defines the type of transmission facility that is to be used for a specific session, according to characteristics such as its availability, security, and response time.

a. Subarea Route Selection. In the subarea, the determination of what routes meet a specific class of service is the responsibility of the network designer. The customer defines the classes of service, the virtual routes that they map to, and the transmission groups that the underlying explicit route traverses. Figure 16 illustrates a sample definition where the customer has defined alternate routes for backup on failures or load balancing of session traffic.

b. APPN Route Selection. In APPN, the least weight path is dynamically calculated using the characteristics of transmission groups and nodes that are exchanged among all APPN nodes. (See below for more information on the route computation.)

Each APPN node is responsible for reporting changes in its local topology—the characteristics of the node itself and the characteristics and operative status of its link. As Fig. 17 illustrates, when a link fails, two nodes are responsible for reporting the event.

Each link is represented as two unidirectional links so that only one node is responsible for reporting about any resource. This prevents conflicts of different reporting by the two end points; this is important when a node fails since a node failure often appears as link failures to all its neighbors

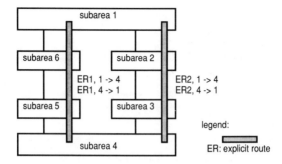

Class of Service Definitions between subareas 1 and 4

Class of Service	Transmission Priority	Virtual Route Number
BATCH	low	1
	low	2
INTERACTIVE	high	1
	high	2

Virtual Route Definitions between subareas 1 and 4

Virtual Route Number	Transmission Priority	Explicit Route	Reverse Explicit Route
1	low	ER1	ER1
1	high	ER2	ER2
2	low	ER2	ER2
2	high	ER1	ER1

Fig. 16. Class of service and virtual route definition.

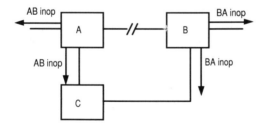

Fig. 17. Outage notification in APPN.

and the failed node does not report any change to its links, leaving inconsistent information in the two directions.*

*This inconsistency does not affect the correctness of routes that are selected: when a node fails, all links into it will be declared inoperative and therefore the node is not reachable and no links out if it will ever be used.

All nodes have sequence number 10 when new information arrives at Node A.

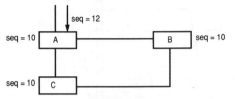

Node A recognizes it as new information and passes it to Nodes B and C.

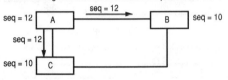

Nodes B and C accept the data and pass it on.

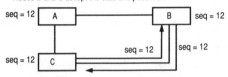

Nodes B and C recognize this data as duplicate and discard
the message.

Fig. 18. Topology distribution in APPN.

Topology information is passed on until all APPN nodes have received
it. When a node receives data that it already has, it stops forwarding the
information. As shown in Fig. 18, the nodes recognize duplicate information because every resource update carries a monotonically increasing
sequence number that is set by the originating node: equal sequence
numbers indicate duplicate information.[†]

Through these protocols, a correct, consistent topology is known
throughout the network.

AS/400 uses the topology information with a class of service definition
that directly describes the characteristics of the route requested. The class
of service definition defines for nodes and links the acceptable range for
each characteristic and a function to calculate the weight. When a link or
node is being considered for inclusion on a route, all characteristics are
compared to the acceptable ranges and if they are all in the acceptable

---

[†]If sequence numbers are the same but the data are not, an error has occurred and protocols
exist to return the information to the originating node in order to rectify the situation.

range, the weight is calculated. Instead of using a simple shortest path algorithm, the least weight route is calculated.

An SNA session always flows over a fixed route. Once the route is selected, a definition of that route is built and appended to the session setup message, the BIND. As the BIND progresses through the network, each node saves the TG it was received on, the LFSID used on that TG, the next TG to send it on (from the route definition), the LFSID that it chooses for that (outgoing) TG, and the transmission priority for the session. With this information, it builds a session connector and now the data for the session will be properly routed.

C. Message Repackaging

As another service provided by the network, path control can repackage packets into different sizes in order to improve link performance or insulate the user from specific link or intermediate node characteristics. *Blocking* is the combining of packets; it yields significant performance gains on the S/370 channel, its only use in products. *Segmenting* and *reassembly* treat the case of limited buffer space in smaller nodes and high block error rates on some links. If path control receives from a session end point a message that is too large for the next link or partner to handle, it segments the message into pieces that are sufficiently small to permit transmission.

The actual segmenting and reassembly are straightforward; the difficult part is the negotiation of packet sizes that can be sent, particularly on an APPN route that passes through intermediate nodes, with possibly different buffer requirements and different segmenting and reassembly capabilities. When a link is activated, both partners learn the maximum packet unit that can be received by one another over that link. Segmenting and segment reassembly are options not supported by all nodes. When a session setup message is processed, each node negotiates the maximum data packet that can be sent in each direction in order to assure that no data will be segmented that cannot be reassembled.*

D. Link Activation

The last service provided by the transport network is the activation of links when needed in order to establish routes. During the link activation process, a series of message exchanges allows the two end nodes to learn

---

*A similar problem of different frame sizes being supported on different segments in a bridged local area network is treated by the inclusion, in a frame, of an explicit indication of the maximum frame size supported by a path through a series of local area networks [25]. No segmentation or reassembly is provided in data link control.

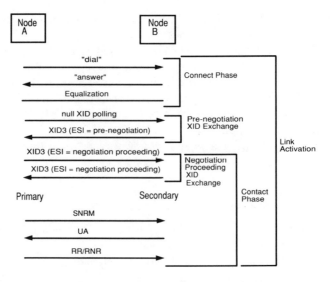

Fig. 19. Link activation for a switched connection between T2.1 nodes.

about each other's basic capabilities and to negotiate others. For example, when an asymmetric data link protocol is being used (e.g., SDLC Normal Response Mode) the two stations can negotiate the role of primary; T2.1 nodes negotiate the setting of the address space dividing bit described earlier. T2.1 node negotiation is based on comparison of control point names; subarea nodes use subarea addresses. Values that are learned from each other include segmenting capabilities for session setup messages and maximum message sizes that can be received. This procedure of negotiating and dynamically learning each other's capabilities removes a great deal of coordinated system definition and makes network maintenance that much easier.

Figure 19 illustrates the bring up of a switched line connection between two T2.1 nodes. Link activation for T2.1 nodes is composed of up to three phases:

- Connect phase.
- Pre-negotiation XID exchange.
- Contact phase.

The connect phase, which is only needed for certain link types, allows initial establishment of communication between nodes. "Dial" and "answer" establish physical layer connection on switched facilities. It is applicable to connections over the IBM Token Ring or X.25 networks, as well as between modems. "Equalization" is the transmission of training sequences that occurs between two modems. Once the connect phase has completed, the two nodes are able to exchange and establish node characteristics via XID commands.

The prenegotiation phase is begun with a null XID poll. A null XID is an XID with an I-field of zero length, and is used to poll an adjacent node. Polling is performed in order to determine if the adjacent link station is active; a null XID is used when the polling node does not know if the polled node is a T2.1 node, i.e., that it can accept an XID format 3 (XID3) poll.

During the initial link-activation XID exchange, before link station roles have been determined, the link station roles may be primary, secondary, or negotiable. Primary or negotiable stations may send a null XID. XID collisions that may occur are resolved with a randomized retry algorithm. In this example, node B, receiving a null XID, responds with an XID3 whose Exchange State indicators field (ESI) is set to "Prenegotiation."

The contact phase begins with XID negotiation. XID negotiation is performed by T2.1 nodes to establish the primary and secondary roles of the link stations, as well as other characteristics of the link. The primary–secondary role determines which link station will have control of the link, and is also used in setting the value of the ODAI bit in the LFSID.

The negotiation-proceeding XID exchange finishes when link station roles have been established as complementary, i.e., one link station is primary and the other is secondary. The primary link station, node A, sends the mode-setting command, SNRM.

Once the mode-setting command and UA have been sent, and RR or RNR has flowed on the link, the contact phase and link activation are over. See [35] for more detail on link activation between T2.1 nodes.

IV. Distributed Programming

The transporting of data, described in the previous sections, is only a means to make distributed programming possible. We now examine how, given the services of a transport network, SNA allows distributed resources to interact in a meaningful way for the end users that access the network through logical units, or LUs.

Functionally, SNA's LUs are execution environments for programs. Each program has access to one or more local resources, such as files, data bases, memory, or I/O devices. Each one also cooperates with one or more remote programs through a shared resource, illustrated by Rs in Fig. 20.

The use of a shared resource to coordinate the processing of cooperating programs (or processes) is seen in many areas of computing. The shared resource can take many forms: semaphores, queues, files, pipes, memory

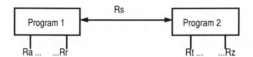

Fig. 20. Cooperating programs.

areas, monitors, and so on, which have all been used in nondistributed programming. SNA's distributed processing is distinguished from other forms of cooperative processing by the shared resources that are made available to programs, and by restrictions that exist on the forms of cooperation that are supported.

Figure 21 shows the basic form of cooperation that is supported. Each program can have one or more connections, called *conversations*, to other programs. A conversation is a serial reuse of a session. The session is the connection between two LUs, the execution environments, while the conversation is the connection between two programs. The programs can be in the same or different LUs. Conversations, which are created as a side effect of creating an execution instance of a program partner, are destroyed at the request of either partner. Conversations are never shared by more than two programs, but programs can have multiple conversations with many other programs either simultaneously or sequentially. Programs, once created, terminate themselves, and can outlive the program that created them as well as any that they may create.

Because the lifetimes of conversations are under the control of the programs that use them, it is possible for trees of cooperating programs to divide into several trees and continue computing. In Fig. 21 this would happen if the conversation between P1 and P3 were broken. It is not possible for trees to recombine directly because a new conversation will create a new instance of its target program; similar computational effects can be achieved by local sharing of resources other than conversations.

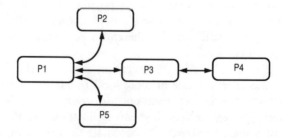

Fig. 21. A distributed processing tree.

A. Conversation Properties

Conversations appear to the programs that use them very much like a telephone conversation appears to two people. There is a beginning, an end, and a period in between that consists of alternating periods of speech, with an occasional interruption. Interruptions are of several types. Sometimes the speaker pauses and then corrects a recent utterance. Sometimes the listener rejects what has been said without letting the speaker finish. Sometimes the listener requests a chance to speak. There are other parallels as well. The conversation can be long or short, and can consist of anything from a one-way message from the caller to the called party (almost analogous to conversations with answering machines) to an extended two-way exchange. In principle, if not in practice, a conversation could last indefinitely, continuing until forced to end by an equipment failure.

Some of the properties of conversations are not so obvious in the telephone analogy. For instance, while both telephones and conversations deliver messages in the order of their transmission, conversations nominally deliver them without error; the actual residual error rate, while nonzero due to undetected link-level errors and other residual sources of errors, is so small that it can usually be ignored by comparison to other sources of errors, such as operator-induced data errors. Speed matching is provided by SNA's conversations, while telephone conversations require the sender and receiver to be generating and consuming information at the same rate.

B. Alternatives to Conversations

Why are SNA conversations the way they are? To be sure, there is some similarity between SNA conversations and OSI sessions, but since OSI sessions were defined many years after SNA conversations, that does not explain SNA's conversations. We can explore this question by examining various alternatives to conversations.

• Sessions—Programs could be built to share sessions rather than conversations in order to cooperate in their processing. SNA sessions are, in some respects, more similar to OSI sessions than are SNA conversations since both types of sessions are suitable for long periods of connection. Since creating an SNA session is a relatively expensive operation (see, for example, Fig. 24), it is not desirable to create and destroy sessions each time a short transaction is to be executed. This means that their direct use is limited to long periods of cooperation.

Conversations actually are an abstraction—time-multiplexed portions of sessions, with the multiplexing algorithms controlled at the two LUs that share the session and within which reside the programs that wish to cooperate via a shared conversation. The SNA bracket protocol performs this time multiplexing.*

• Queues—Queues are often used for cooperation between and among programs in a single node, and they can be used for distributed cooperation as well. For instance, the START command in IBM's Customer Information and Control System (CICS) [37] can be used to queue data to a named CICS program for later execution. This command is even implemented so that it works for local or remotely located programs and queue names. In contrast to the internal-to-the-session queues used by conversations, one can imagine literal shared queues as the shared resources used by all programs that wish to do distributed processing. For examples of this approach, see [38] and [39].

SNA prefers conversations over shared queues for several reasons. Firstly, queues are not primitive constructs. That is, since the network layer needs to provide long-term connections for outage notification, for streamlined performance, and for buffer (as well as bandwidth) management, the insertion of messages into a remote queue needs to be performed after transporting messages over the approximate equivalent of a session. Secondly, they are not easy to use in a distributed environment. That is, when used in a local environment, they usually are only a portion of a larger interface between the cooperating programs, so that the cooperating programs are also sharing data areas, or files. Thirdly, they are not easy to use for any application that needs to correlate multiple messages. Conversations, on the other hand, provide the natural correlation of the half-duplex conversation protocol whereby the speakers take turns. This shows up as simple linear code in the programs themselves.

A more complete discussion of the queue alternative would distinguish between queues-in-memory and queues-on-disk, showing how the difficulties described above apply to both types of queues but with different emphases, and would discuss those applications where queues-on-disk can be used to advantage, such as a simple message-switching application, where the disk queues might simplify the application logic.

• Storage—Why not apply the shared storage metaphor to cooperating programs? Shared storage is not primitive when shared memory is not provided by the hardware nor is it efficient to simulate shared memory across the range of network speeds and delays that are supported by SNA, even though high-speed, low-delay connections do allow shared memory to be synthesized with reasonable results, at least for virtual memory pages or for files.

---

*See [36] for detailed definition of the bracket protocol. Briefly, each session is polarized so that one end is the contention winner. Each new conversation begins with a *begin bracket* indicator; the contention winner's conversation is accepted in the event both sides try to start a new conversation at the same time.

• Full-duplex data flow—Making the conversation full-duplex appears, at first glance, to be an attractive generalization. However, very few applications truly use a full-duplex data flow, which tends to be transport oriented, not computation oriented. There is a high cost implied in storage management within the LU for buffers to support two flows as active, even though only one is truly active. Seen this way, the half-duplex service provided by conversations is for the aid of the LU. Half-duplex service also allows efficient conversation termination, and allows the cooperating programs to have a simpler internal structure (linear single processes), as mentioned earlier.

• Simultaneous multiplexing—Since conversations are multiplexed over sessions, why not do this multiplexing at a buffer or packet interleave level rather than dedicating a full session to a single conversation at a time? The answer lies in two facts: First, this would force the creation of another multiplexing layer on top of the session layer, certainly not needed in SNA since virtual routes provide multiplexing efficiency underneath the session layer. Second, parallel sessions are available for simultaneous communication and the interleave multiplexing of a single session does not remove the need for parallel sessions since a single session can have only a single class of service. The serial reuse of sessions by conversations is the simplest solution.

• Remote procedure call—Some efforts have been made to use location transparent procedure calls as the basis of distributed computation [40, 41]. Since SNA's conversations can be represented as a set of local procedure calls [36], the discussion is really one of contrasting a generalized location-transparent procedure call mechanism with conversations. Other than generality of semantics, the largest distinction is the degree of serialization of the calling and called programs. Conversation partners can execute in parallel or be interlocked, while the usual method of transparent procedure calls requires serial execution, the characteristic of a local procedure invocation.

Another area in which the SNA approach to distributed processing has been tuned carefully is that of *transaction programs* as contrasted to processes. The programs that use conversations in SNA are transaction oriented because they come into existence with the arrival of the first message in the conversation that they share with their parent program. By contrast, some distributed processing tries to provide cooperation between long-running processes, so that a connection can be built to a program or process that already exists simply by providing its identifier.

SNA uses transactions for several reasons. One is that the name of a transaction program is the name of a generic function. If a state is needed as well, the state data or name can be provided to the program instance in the data stream. Anything that can be done with a (long-running) process can therefore be done with a transaction. Another reason is that it is easier to implement a transaction program model of computation in each node, especially if the node already has such a model. If the node has only long-running programs, a transaction model is easily created merely by delivering to the program the unique name of the incoming conversation. For example, the IBM Series/1 implementation of LU 6.2 provides such a capability as one of several alternatives. Finally, the use of transaction programs reduces network protocols since process identifiers do not have to be resolved by a network service; state names are shared only between the programs that directly need to know and manipulate them.

C. Convergence on LU 6.2

Conversations are provided in several forms by SNA. One form, defined by LU 6.2, is universal in that it encompasses the facilities of conversation on other LU types and extends them to be a service capable of universal application [42]. Thus, over time, the applications that use earlier SNA conversation and session protocols will either stabilize or move to LU 6.2 in order to take advantage of the broader connectivity and greater function of LU 6.2.

One important benefit of LU 6.2 is the connectivity provided by its approach to function subsets. As discussed further in "Subsettability," a common core of function, known as the *base*, is defined for implementation by all products, thereby ensuring connectivity whenever the programs that use the LU are connectible. Thus an LU 6.2 printer cannot meaningfully connect to another LU 6.2 printer, but an LU 6.2 printer can attach to any LU 6.2 product that contains a program that can create its datastream and protocol. Additional functions, called *option sets*, can be used by mutual agreement of connected LU 6.2 session partners. Like the base, an option set must be implemented in its entirety.

D. Sessions and Session Management

The LU contains three major layers, as shown in Fig. 22. The session layer, consisting of one or more half-sessions, is controlled and managed by the *session manager* component. The session manager sends and receives BINDs, which are the commands used to create sessions, and UNBINDs, the commands used to destroy them. When both sides send a BIND and only one can be accepted because of agreed-to limits between the LUs, the session manager uses a race-resolution algorithm to accept one and reject others. The session manager also receives session outage signals from the transport network when sessions fail before they are unbound.

Each LU keeps a table of the LU names of partners and the *mode names* that can be used with each partner. The mode name describes the quality of service that a given set of sessions will provide, directly determin-

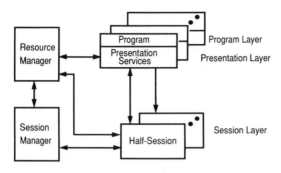

Fig. 22. Components of the LU.

ing the security of the session (e.g., by using encryption or not) and the capacity (e.g., packet size and packet window width*). Mode name also selects the class of service name for the session, and so indirectly determines such transport network characteristics as path delay, path bandwidth, path security, and path availability.

The session manager component verifies passwords on the BINDs exchanged with partner LUs that support the option set for security. The password algorithm uses the U.S. Data Encryption Standard to protect passwords against observation by wiretapping. The session manager also attempts to rebind sessions that have failed, up to a limit on the number of sessions upon which the LUs have agreed. The command to change the session limit can be issued at any time by either LU's control operator. The simplest LUs (e.g., printers) do not implement this function because they have a fixed session limit of 1.

One of the sessions layer's main functions is end-to-end flow control. The goal of flow control is to keep the program that is consuming data running at its maximum rate without using more buffers than required. SNA's algorithm for doing this is pacing, the sender being paced to the speed of the receiver, and operates at two levels. One level, discussed in "Subarea Routing," limits the buffer occupancy of the transport network. The other, session pacing, limits the buffer occupancy of the receiving half of the session. In order to keep the receiving session from being sensitive to the physical configuration of the network, especially the delay-bandwidth product of the path used by the session, intermediate session pacing spots are placed at specific points in the network. This principle of multiple pacing stages has been enhanced in APPN. The stages' adaptive session pacing also improves flow control efficiency and fairness by automatically

---

*The student of SNA will recognize these fields as request unit (RU) size and pacing count.

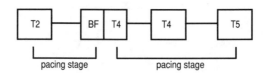

Fig. 23. Example of pacing stages.

controlling buffer allocations and adjusting to the level of relative demand placed on links by each session. Figure 15 illustrates multiple pacing stages in an APPN network and Fig. 23 illustrates them in the subarea.

E. Session Services and the Control Points

The *session services* component in the node is not able to build sessions entirely by itself; it needs help from control points. There may be one or more control points involved, the first in the node containing the LU, or in the node that controls and owns the LU's node. Cooperating among themselves, the control points search the distributed directory of LU names, as described in "Directory Services," when asked to do so by an LU. When the desired LU name is found, the address of the LU is determined (or, in some cases, assigned dynamically) and the path between the two LUs is selected. APPN computes this path dynamically [15, 32]. Other SNA configurations select the first available of several precomputed paths. If requested by the LU, the control points will queue the directory request until the target LU is able to support another session (when the LU has a session limit of one), until a path to its control point is available, or until the target LU becomes active (such as when its node powers on). A session encryption key is computed if session encryption has been requested, and a suggested set of values for various BIND fields is extracted from a table. The requesting LU (for sufficiently capable requestors) or the target LU (for simple terminal requestors) is then provided with all the information needed to enable it to send a BIND and create a session.

Figure 24 shows the kind of cooperation required by the control points to establish a session across a gateway. The figure illustrates a *secondary LU* (SLU), the BIND receiver, requesting the *primary LU* (PLU), the BIND sender, to initiate a session. While LU 6.2 protocols are symmetric, allowing either partner to send BIND, older LU types were defined such that each partner had a specific role and only one side could be the BIND sender. For more information on this flow, see [18].

Session initiation for an LU–LU session is mediated by one or more SSCPs. The two LUs may be supported by the same SSCP, in which case only one SSCP is involved. If the two LUs are supported by different SSCPs, both SSCPs are involved, exchanging *cross-domain* session services

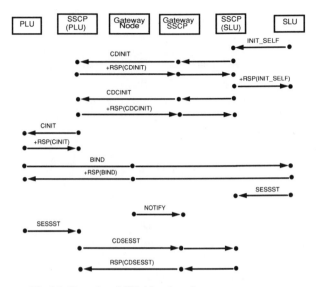

Fig. 24. Secondary LU initiated session across a gateway.

RUs. For cross-network LU–LU sessions, two or more SSCPs may be involved, including at least one gateway SSCP.

The initiating LU in this sequence, the SLU, sends its SSCP an INIT_SELF requesting that a session be established to it.

The SSCP of the SLU (SSCP(SLU)) sends a CDINIT request to the SSCP of the PLU. This is the initial directory request. In this case, a positive response is returned, indicating that the resource has been located. In general, a negative response could be received and the SSCP (SLU) would query other SSCPs to try to find the desired LU. When the CDINIT passes through a gateway SSCP, it will translate the name if aliasing is being used.

The SLU receives a response from its SSCP after its SSCP receives the positive CDINIT response.

The SSCP of the SLU sends a CDCINIT containing session parameters for the BIND image. The SSCP of the PLU responds to the CDCINIT and sends a CINIT request with the BIND image to the PLU indicating that it should BIND a session.

After a successful exchange of BIND request and response, the SLU and the PLU each send SESSST RUs to their SSCPs. The SSCP of the PLU sends CDSESSST to the SSCP of the SLU and the SSCPs update their session awareness records to show that the session is active. After it has seen the BIND and BIND response, the gateway node notifies the gateway SSCP that the session has started and identifies the virtual route used for the session in both networks.

LUs can request several enhanced services of session services in the control points. For example, if LU A has a session with LU B, it can pass its session end point to C, creating a session between B and C. Session services will also automatically create sessions between LU pairs when the LUs are active and the path between them is available. In configurations where two SNA networks are interconnected, the session services at the gateway control points authorize access from one network into the next one and provide LU name translation when desired.

F. Conversations and Conversation Management

Just above the session layer is the presentation layer, in which each conversation is represented by a presentation services component. Presentation services presents the current state of the conversation to the using program and allows the program to send data, receive data, and control the state of the conversation. Like other local resources (e.g., files or queues), conversations are managed by a local resource manager; the resource manager for conversations is a component of the LU.

The allocation of conversations, however, is specified by SNA. When a program asks that a conversation be built for its use, the resource manager tries to use an existing active session, first checking for *contention winners* (the sessions over which it will win a race to be the data sender) and then bidding on a *contention loser* session (one of the sessions over which it will lose a race). If no sessions are available, a new session will be bound (with the help of LU network services) if possible. If no session can be added, and none is available, the conversation request will be denied if immediate allocation is requested or queued if delayed allocation is permitted.

Figure 25 illustrates the verbs issued by transaction programs in order to generate a conversation and the line flows that those commands generate. The example shows a conversation between transaction programs on two T2.1 nodes that is requested when no session exists. Partner-LU verification is needed for the session and the conversation itself is a simple exchange of data. For more detail on this function see [36, 43, 44].

The first transaction program issues an ALLOCATE indicating that it wants to begin a conversation with a partner transaction program (TP). The presentation layer requests use of a conversation from the resource manager, which requests the session manager to begin a session.

In general, the LU provides three functions to assist in providing security: partner-LU verification, partner end-user verification, and session cryptography. Partner-LU verification is a session-level security protocol; it involves protocols at the time the session is activated. Partner end-user verification is a conversation-level security protocol, taking place at the time a conversation is started. Session cryptography is another session-level protocol that allows data to be encrypted.

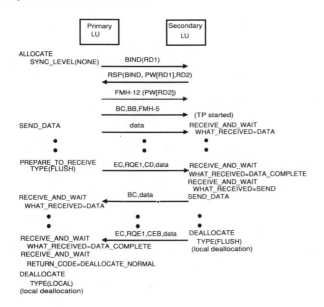

Fig. 25. Example of an LU 6.2 conversation. RDi, Random data ($i = 1|2$); PW, LU–LU password; PW[RDi], RDi enciphered using PW as cryptography key; BC, begin chain; EC, end chain; BB, begin header; CD, change direction; CEB, conditional end bracket; FMH-12, header for encryption password; FMH-5, header with name of TP to attach; RQE1 indicates that no response is required.

In this example, partner-LU verification is used. Partner-LU verification is done by a three-flow exchange between the two LUs. During session activation, random data (RD1) are sent in BIND from the primary LU to the secondary LU. The secondary LU enciphers these random data using the LU–LU password and the random data as input to the DES algorithm. The secondary LU returns the now enciphered random data (PW[RD1]) to the primary LU along with its own randomly generated data (RD2) in the BIND response. The primary LU compares the received enciphered random data with its own copy of the random data that it enciphered using its LU–LU password and the DES algorithm. Since the two versions of the enciphered random data compare equal, the primary LU has verified the identity of the secondary LU and LU–LU verification continues.

Using the LU–LU password and the DES algorithm, the primary LU enciphers the random data received from the secondary LU and returns it in a Security header (FMH-12) to the secondary LU. The secondary LU compares these enciphered random data with its own enciphered version. Since the two versions compare equal, verification is complete and the session is established.

An Attach header (FMH-5) is now sent indicating the name of the transaction program that the conversation is to be established with. The flows indicate the beginning of the conversation with Begin Bracket indicator and that this is the beginning of the allocator's data with a Begin Chain indicator. On receipt of the FMH-5, the destination TP is started.

The allocating TP now sends as much data as it has, the flows indicating that it is in the middle of the chain. When it has sent all the data it wants, it issues a PREPARE_TO_RECEIVE. This indicates that the data are finished (End Chain), that the other TP may now send data (Change Direction) and, in this case, that there is no need for a response.

The other TP has been issuing RECEIVE_AND_WAITs to get the data being sent. After the PREPARE_TO_RECEIVE has been sent, it will receive indications that all the data have been sent and that it may now send data.

The two transaction programs now reverse roles. When the TP now sending data finishes, it issues a DEALLOCATE, indicating that it wishes to end the conversation. This is indicated in the flow by the Conditional End Bracket. Once the TP has issued the DEALLOCATE indicating that no response is needed, the conversation is terminated.

G. Deadlock Detection

Since the conversations are allocatable resources, allocation deadlocks can develop. If only local resources are involved, deadlocks can be detected by the LUs locally. Distributed deadlock is detected by a *dead transaction timer*. Each LU implements a local timer that is used to identify programs that might be in an allocation deadlock or hung due to a logic error. The deadlocked programs are then terminated locally, and the resulting conversation outage signals are propagated to all conversation partners. More extensive algorithms are possible, but do not seem to be warranted for most environments [45], since they require the overhead of a generalized distributed lock manager.

H. Synchronization Points

While LU 6.2 does not currently define a fully distributed lock manager, it does include a fully distributed synchronization point manager. The synchronization point, or *sync point*, service supports commitment and backout of changes to local and remote protected resources. If LUs, programs, or conversations should fail, the distributed unit of work is resynchronized and consistency is established if possible, using global commit or global backout. Achieving full consistency may be impossible after massive failures, so the protocol includes the ability to detect damage to the unit of work. LU control operators are informed about damage when it is detected, and are responsible for application-level repair actions.

I. Efficiency and Datagrams

LU 6.2 includes a number of features that are designed to improve the efficiency of the network. For instance, while the transaction program is continuing to send, multiple SEND DATA requests are usually packed together* to create packet sizes that are efficient for the network path of the session and the LUs sharing the session. Parallel sessions make conversation allocation very inexpensive of flows; on contention winner sessions, for example, allocation involves no additional messages.

For certain kinds of transaction environments, datagrams are suitable. LU 6.2 includes a form of datagram known as a one-way conversation. The sending program allocates a conversation and names the target LU and partner program that is to run in the target LU. It follows immediately with the data to be sent and then deallocates the conversation with a flush option. The flush option allows an immediate ending without waiting to receive any feedback from the target LU or partner program.

This sequence produces a one-way flow of packets. If the amount of data is small enough, the result is a single packet or datagram. This datagram is fully flow-controlled since it travels on a session. It can be lost without notice if the session fails, if the receiving program fails to operate properly, or if the operator of the receiving LU has failed to enable the named receiving program for execution. However, a reliable datagram protocol can be constructed for certain kinds of applications. APPN, for example, contains two such protocols, one used for broadcast of topology information about the network, the other for directory searches [32]. At a different extreme, the recovery from loss of datagrams can be pushed all the way back to the terminal or the terminal operator: if no reply comes, the operator resubmits the previous input.

V. Transaction Services

Sitting above the logical unit (LU) are user applications and transaction services. Transaction services are architected IBM applications that customers or other transaction services can use. They represent important services that the writer of distributed programs can use. They provide extensions to many operating systems functions to enable the functions to be provided over communication lines.

We begin with a very simple picture of an application (Fig. 26), where there is a single application supporting a terminal and accessing data.

---

*This action by LUs is different from blocking by path control, which involves collecting packets from multiple sessions into one DLC transmission.

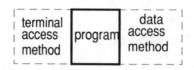

Fig. 26. Simple nondistributed program.

There are two ways in which this program could be distributed: the data could be moved to a remote location (Fig. 27) or the terminal could be made remote (Fig. 28).

In Fig. 27, Distributed Data Model (DDM) [12] is used to access remote files. DDM supports record-mode data access across LU 6.2 connections. Once it has located the requested file, DDM simply forwards the request to the system on which the file resides. The system that receives the DDM request accesses the data and sends them back to the requesting node. The application program receives the data as if they were retrieved locally.

In Fig. 28, the data and application are co-resident, but the terminal is remote. The data and screen-handling commands need to be transported between the host system and the terminal. SNA uses the EBCDIC character set and the SNA 3270 data stream [46], running on LU type 2, as the device control data stream. The 3270 data stream includes controls to describe the screen format, field attributes (e.g., which are highlighted, which are alterable, and which colors are used), and partitions of the screen into active and reference areas (the active area is the only part that can be changed). SCS (SNA Character String) [47] and IPDS (Intelligent Printer Data Stream) [48] provide similar device control support for printers.

But not all programs follow the model of supporting a terminal that accesses data. There are many program-to-program functions performed in a network as well. An important class of these programs do not require

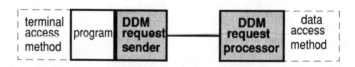

Fig. 27. Remote data access.

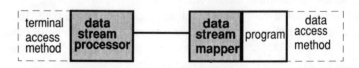

Fig. 28. Remote terminal.

synchronous communication between the partners. Just as spooling in an operating system permits job input or printing to be decoupled from the processing, so SNA Distribution Services (SNADS) [11] allows transmission of data to be decoupled from the sending and receiving procedures. There are many uses of such a function: currently, usage focuses on office applications such as electronic mail, but the service can be used for many types of file transfer, such as distribution of changes to different systems in the network or transporting of job streams from one system to another.

VI. Application of Design Principles

Having looked at some aspects of SNA, we can now discuss how the principles of its design have been applied. We focus here on the four design principles, those that guide how the architecture as a whole is built, described in "SNA's Design Principles"—layering, subsettability, evolution, and consistency. We also give an example of how technical principles, those that guide the technical decisions that are made, are used in SNA.

A. Layering

The layering of SNA has been a major contributor to the smooth enhancement of SNA's capabilities. For example, when SNA introduced alternate routing capabilities, the only effect on layers above path control was the exposure of a new service, namely, the ability to classify routes by their performance or security characteristics and to identify which class of service, the representation of these services, is desired for a session. The addition of the token ring to data link control was transparent to the higher architectural layers as well.

It is sometimes argued that layering can introduce performance penalties. That has not been our experience. Performance is an implementation issue; there are many examples of the same SNA protocols implemented very efficiently in one product and less efficiently in another. The architecture of the implementing product and the ability of the product developers to focus on performance bottlenecks are what affect the performance of the product. Indeed, we assert that the ability to do both those tasks well is enhanced by a well-structured, well-layered architecture.

B. Subsettability

SNA's early approach to subsettability was the creation of logical unit (LU) types. There are seven types of LUs, some designed to support

communication between programs and workstations and some designed for
program-to-program communication. The use of LU types introduces sub-
setting into SNA by reducing, for example, the number of different options
at the transmission control and data flow control levels that any one
product needs to support. There are several dozen options that are identified by the LU type, including use of pacing for normal requests, use of
cryptography, use of full duplex or half duplex communication, and whether
responses to requests are required, available upon request, or not even
allowed.

The use of a set of LU types worked well with the technology of the
1970s. The world of computer communication was a world of fixed function—fixed function workstations communicating with known applications.
With notable exceptions (including the first SNA remote products, the 3600
banking system), most workstations were not programmable and program-
to-program communication was limited. Connectivity was designed into
products and the connectivity requirement was to avoid a proliferation of
different LU types that needed to be supported by host systems.

With the advent of intelligent workstations and more sophisticated
program-to-program communication, the LU types began to be stressed in
two directions: (1) the range of valid functions within LU types was so wide
that two implementations of the same type were not necessarily compatible,
and (2) the desired couplings became much more dynamic and numerous.

Because of the growth in desired communication partners, it became
necessary to design a single universal LU that all products could implement. With the growth of programmable workstations and the implementation of many workstations moving to software, program-to-program, rather
than program-to-device, protocols became widespread. SNA satisfied program-to-program communication requirements with LU Type 6.2: Advanced Program-to-Program Communication (APPC) [36].

APPC was designed to continue the ability for simple products to build
reasonable implementations at an acceptable cost while giving a wide range
of functions to those products that wanted them. What was new in the
APPC design was a way to support arbitrary connectivity. The design is
based on two technical principles: (1) base and option sets and (2) negotiation. Base and option sets reflect the simple idea of a base set of functions
that all products must implement. The base must be complete enough that
useful functions can be performed by two base products and all base
products can communicate. Thus part of the base is the ability to negotiate
such values as pacing windows and message sizes. Without this ability, such
simple parameters divide the product set into classes of products that
cannot communicate. Once the communication channel is established,
well-written applications can learn what facilities are supported by the two
partners and negotiate the functions they will use.

C. Evolution

An evolutionary architecture allows products implementing one version of the architecture to work through many architecture releases. Generally, the products that must be the most stable are workstations; the intermediate routing nodes still require smooth migration, but only between limited releases. Thus the 3600 banking system that was introduced in 1974 is still supported by today's SNA networks.

Naturally, upgrades permit functional additions. For products where regular upgrades are expected—such as the host and communication controller software products, VTAM [49] and NCP [50]—the SNA objective is that products can always work in a mixed environment with architectural levels that are one behind them or one ahead of them. The objective of moving forward without losing touch with the past presents a challenge for architects. Too much concern with migration prevents new and innovative steps; too little concern delays or prevents product implementation.

In some cases, evolution is simple. Addition of a new data link control is a simple evolutionary step: because of the layered nature of SNA, the new data link control is strictly additive. Once two components have been upgraded to support the data link control (which need not be done at the

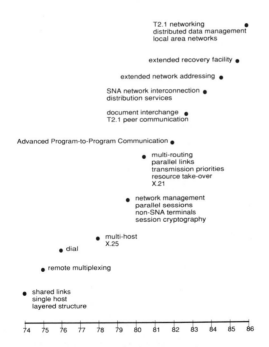

Fig. 29. Evolution of SNA, related architecture, and SNA product function.

same time), the user is free to add the new transmission facility between them.

Throughout its history, SNA has continued to increase the variety of transmission facilities it supports. When introduced in 1974, SNA supported only leased SDLC links and S/370 channels; SDLC dial was introduced in 1976, X.25 support in 1977, and X.21 in 1980. In 1985, SNA added local area network support for ISO standards 8802-2 (IEEE 802.2) and 8802-5 (IEEE 802.5) [22, 23, 30, 31, 51].

Figure 29 illustrates the major evolution of SNA, related architectures, and its products' functions. There are several trends besides the data link control growth that should be noted:

- Increasing range of supported configurations and attachments.
- Growth of peer-oriented function.
- Growth of architected transaction services.
- Increasingly comprehensive and sophisticated network and system management.

1. Configuration and Attachments

Figures 30–33 illustrate the growth in configuration complexity. No step was revolutionary; each step helped evolve SNA to a more general networking facility.

When introduced in 1974, SNA supported a simple tree configuration: a single host connected to terminals through a communications controller. Within a year, remote communication controllers were added. Multiple hosts were added in 1977 with single links between communication controllers. In 1980 full mesh connectivity between hosts and controllers became available with multiple routes, parallel links between nodes, and channel-to-channel connection of hosts.

Attachments to an SNA network have grown from T2.0 nodes to T2.1 nodes, ASCII and Bisync terminals through protocol converters, and X.25 PAD terminals.

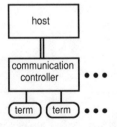

Fig. 30. 1974 SNA configuration capabilities.

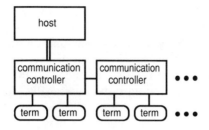

Fig. 31. 1975 SNA configuration capabilities.

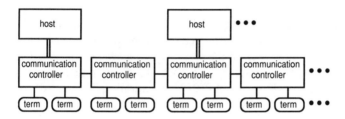

Fig. 32. 1977 SNA configuration capabilities.

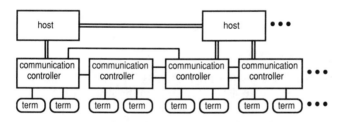

Fig. 33. 1980 SNA configuration capabilities.

2. Peer Functions

Since 1979* when parallel sessions were introduced in program-to-program sessions between application subsystems such as CICS [37] and IMS [52], SNA and its products have made considerable improvements in peer-oriented functions. The 1982 introduction of APPC enhanced program-to-program function and low entry networking [35], first introduced

---

*Single-session peer communications appeared even earlier: JES–JES communication was introduced in 1977 and CICS–CICS in 1978.

in the S/38 [2] in 1983, gave smaller systems the ability to communicate directly. AS/400's Advanced Peer-to-Peer Networking [16] further extends the capabilities of such systems to communicate through intermediate nodes and to learn about the topology and partner locations dynamically. APPC/PC (1986) and OS2 EE (1988) extended APPC to desktop personal computers and the AIX operating system uses APPC for high-performance distributed services [53, 54].

APPC represented the introduction of a new LU type, LU 6.2. Adding new LU types is a straightforward process; a product can add the new LU as an independent component, and nodes with the new capability can be added to the network independently. When two nodes have added the function, transaction programs can be written on the two sides to use the new facility.

Peer-to-peer networking of T2.1 nodes has developed in several steps. T2.1 nodes introduced the first architected method of direct communication between two nodes without the assistance of a system services control point. The ability of two adjacent T2.1 nodes to communicate directly is called low entry networking. In APPN, two additional levels of peer-to-peer networking are introduced: a low entry networking T2.1 node can attach to an APPN node and a session can be established to an LU in a remote T2.1 node. The low entry networking node believes that the partner is in the adjacent node. Finally, an APPN node acquires additional function such as being able to choose which route is preferred when there are multiple links that it can choose from.

3. Transaction Services

Architected transaction services were first defined when program-to-program protocols were developed for communication between application subsystems. General models were defined for DL/I [52] access, scheduling of processes, and handling of queues. CICS also shipped product-defined protocols for transparent file access. In 1982, SNA introduced additional transaction services. With these architected services, connectivity and user function are considerably enhanced. In 1982 Document Interchange Architecture [55] and Document Content Architecture [56] extended support for office applications. SNA Distribution Services [11] introduced an asynchronous file delivery service to SNA in 1983. Distributed Data Management [12], announced in 1986, facilitates data sharing.

Because transaction services run like any other transaction program on an LU 6.2 session, addition of these services is a simple evolutionary step: they can be added to each node independently and a network of communicating services is built up as more nodes acquire the transaction services.

4. Network Management

From a beginning with virtually no network management, SNA has evolved to include a sophisticated set of network management tools covering many aspects.

Network operators use the services of the control points to view and control the session in the network. In large, System/370-based networks, the functions of the control point are implemented by two cooperating products, Virtual Telecommunications Access Method (VTAM) and Netview [13, 57]. The control points keep lists of active sessions, and files of previously active sessions can be maintained if desired. The mapping of sessions to routes can be displayed, and both sessions and routes can be deactivated if desired. Performance and accounting data can be kept with the historical session file entries: accounting can be done via user exits and performance via the Network Performance Monitor product. In addition, response time for terminals can be measured with the Response Time Monitor.

Network operators also need the ability to recognize errors in components in the network. Nodes send alerts when they recognize problems [57]. SNA modems are able to diagnose errors in the modems, lines, and adjacent nodes and, at operator request, to return this information to a central operator [58]. Netview/PC [9] provides customers the ability to manage non-SNA entities, such as PBXs.

Network management is a very large topic and the interested reader is referred to [57, 59–62].

D. Consistency

SNA has been defined by many people (currently about 100 architects) with products from sites around the world over almost two decades. When large decisions are being made, communication is good; the smaller the issue becomes, the less it is exposed and the higher the risk of missing similarities to other functions. As large as SNA has grown, it is not surprising that inconsistencies have appeared. Examples occur in some of the SNA formats: there are two different techniques used for defining the length and type of a control vector (a common piece of data that is optionally attached to several different message units) and there are several different types of data used to identify sessions. SNA has been able to keep such problems to a minimum by procedures that require widespread reviews and regular design meetings.

The second reason for inconsistencies is harder to handle. Like the dichotomy between migration and new functions, a decision often has to be made between continuing to use a current technique and switching to a newer, better one. This type of problem occurs when a function is being introduced that is related to an existing function but is simpler than or otherwise different from the existing function. Is it better to add the function to an existing message unit, with fields that are no longer necessary, or to design a new, sleeker message? Such a decision will be based on the ability of products to reuse code, the complexity of the parsing and processing, and migration and coexistence considerations. If new components will be processing the messages, the improved version might be chosen. If some components will need to process both versions and they are both complex structures, the tendency would be to use the old format. There are no "correct" answers—only a range of trade-offs to be explored.

The technical principles discussed in the next section are further examples of consistency in SNA.

E. Technical Principles

Technical principles are those principles that influence technical design decisions and are applicable throughout the architecture in a variety of different situations. There are many such principles in SNA—the use of sessions being an obvious one; base and option sets and negotiation as described in "Subsettability" earlier are others. While a lengthy discussion of these principles is beyond the scope of this paper, we briefly discuss one important principle—the avoidance of timers—in order to show how a technical principle affects the architecture.

There are a few timers in SNA, but the principle is to use them only when absolutely necessary and at as low a layer as is possible. For example, data link controls use an inactivity timer to recognize outages and an acknowledgment timer to protect against lost messages or acknowledgments. The use of DLC timers is critical to SNA: because SNA requires a reliable delivery by data link control, it can avoid higher-layer timers. This is important in that it is easier to properly set a single-link timer than an end-to-end timer [19]. The ability to tune timers more easily at the link layer is one of the benefits of using connection-oriented link protocols (such as IEEE 802.2 type 2) instead of connectionless link and connection-oriented transport protocols. Session outages, on the other hand are not recognized by timers, but by specific protocols. Hung resources are sometimes recognized locally by timers, but the partner is then notified by a specific protocol.

Why avoid timers? Probably the most significant reason is that any timer that is architected needs to be set to some value and appropriate values are difficult to choose. What is the proper timeout for a session partner? What would you assume is wrong if the timer expires? Timer-based solutions require that problem determination begin with little information.

Because an architecture cannot easily specify timer values, the burden frequently falls on the user—a burden most users prefer to avoid. Because of the problem with setting them, timers are a major source of user-reported problems. Historically, timers were also avoided in order to save cycles. In larger nodes, timer-driven protocols implied many timers that needed to be running and this would become a drain on what in 1974 was a critical resource—cycles. Today, cycles may not be as critical as they once were, but customers purchase a computer to run applications and they want to maximize the percentage of machine resources used for their applications; therefore saving cycles remains an important principle. Further, timer-driven protocols are very difficult to validate.

We now look at how the principle of timer avoidance has affected SNA routing and session outage. Without timers, a method must be available to recognize when an outage has occurred, either on the path or at one of the partners. Consider first the handling of link outages.

In SNA, all traffic between logical units flows over a session with a path that is defined when the connection is established. Because the path that the session uses is known, an outage on the path can be translated into a session failure. While a full path is not known at each intermediate node, the intermediate nodes do have a local view of the path. Because all paths in SNA are bidirectional (i.e., the traffic in both directions follows the same path), the intermediate nodes can route in both directions. When a link fails, the nodes adjacent to the link report the outage for each path that was using the link in the direction opposite the failure. When the notification of the path failure reaches the path ends, all LUs using that path are notified of those sessions that have failed.

For failures of the partner node, SNA uses different techniques. If the failure is of the entire node, the node adjacent to it will detect the failure as a link outage and proceed as described above. If only the logical unit (or its subsystem) fails, another component of the node will generate a message requesting the remote partner to break the session. The message is sent on the path that session traffic was using; this ensures that either this message will arrive at the partner or a message indicating that the path has broken will arrive there.

It is interesting to recognize that a number of other principles are closely related to the timer avoidance principle:

- Use of sessions (a connection-oriented binding between partners) instead of datagrams.
- Use of the same path by session partners.
- Use of the connection path for outage notification.
- Hierarchical generation of outages.
- Protocols that minimize cycle usage.

VII. Summary

Systems Network Architecture provides the user with the services needed in a distributed operating system: sharing of communications lines, directory services, connection establishment, improved reliability, management services, and an appropriate interface and set of facilities for distributed programming. It provides these services within the scope of a layered architecture that has evolved significantly from its simple beginnings to the many functions and products that it supports today.

There are many topics and details that were necessarily omitted from this discussion. For example, we have not discussed the many aspects of network management in SNA or the capabilities of SNA modems, which are capable of diagnosing errors in the modems, lines, and adjacent nodes and at operator request returning this information to a central operator. We also have omitted discussion of Systems Application Architecture (SAA), which can be viewed as an extension of SNA's connectivity-oriented network services to include support for application portability at the source code level, the provision of additional distributed services (e.g., the presentation manager), and specifications of standards for user interactions. For more information on SAA, see [10, 63, 64].

For more detailed information on SNA, the interested reader is directed to references [36, 35, 61, 65–79].

References

[1]   "System/370 principles of operation," Order No. GA22-7000, available through IBM branch offices.

[2]   "System/38 data communication programmers guide," Order No. SC21-7825, available through IBM branch offices.

[3]   "IBM AS/400 communications: User's guide," Order No. SC21-9601, available through IBM branch offices.

[4]   G. P. Ravikumar, "The role of private networks in communications today," IEEE Midcom Conference, Dallas, Texas, Dec. 1, 1982. Reprinted in TYMNET Technical Papers 1971–1983, Tymshare, Inc.

[5]   A. Danet, R. Despres, A. LeRest, G. Pichon, and S. Ritzenthaler, "The French public packet switching service: The TRANSPAC network," Proc. Third ICCC, pp. 251–260, 1976.

[6]   E. W. Dijkstra, "The structure of the 'THE' multiprogramming system," Commun. ACM, vol. 11, No. 5, pp. 341–346, May 1968.

[7]   International Organization for Standardization, International standard 7498, Information processing systems—Open system interconnection—Basic reference model.

[8]   "DECnet DIGITAL network architecture (Phase IV): General description," Order No. AA-N149A-TC, available through DEC sales offices.

[9]   "Netview/PC: Application program interface/communications services reference," Order No. SC30-3313, available through IBM branch offices.

[10] "Systems application architecture: An overview," Order No. GC26-4341, available through IBM branch offices.

[11] "Systems network architecture—Format and protocol reference manual: Distribution services," Order No. SC30-3098, available through IBM branch offices.

[12] "IBM distributed data management architecture: Reference," Order No. SC21-9526, available through IBM branch offices.

[13] "Network program products general information," Order No. GC23-0108, available through IBM branch offices.

[14] "Transaction processing facility: General information manual," Order No. GH20-7450, available through IBM branch offices.

[15] "S/36 Advanced peer-to-peer networking (APPN) guide," Order No. SC21-9471, available through IBM branch offices.

[16] "IBM AS/400 communications: Advanced program-to-program communications and advanced peer-to-peer networking user's guide," Order No. SC21-9598, available through IBM branch offices.

[17] "XI General information manual," Order No. GH19-6575, available through IBM branch offices.

[18] "Systems network architecture—Format and protocol reference manual: SNA network interconnection," Order No. SC30-3339, available through IBM branch offices.

[19] A. Syed and J. A. Field, "Performance analysis of error control alternatives in local area networks," Proc. IEEE Infocom '86, pp. 503–509, Miami, Apr. 1986.

[20] W. Bux and D. Grillo, "Flow control in local area networks of interconnected token rings," IEEE Trans. Commun., vol. COM-33, No. 10, pp. 1058–1066, Oct. 1985.

[21] Institute of Electrical and Electronics Engineers, ANSI/IEEE Standard 802.3-1985: "CSMA/CD media and media access control."

[22] Institute of Electrical and Electronics Engineers, ANSI/IEEE Standard 802.5-1985: "Token ring access method and physical layer specifications."

[23] International Organization for Standardization, International Standards 8802-5, "Information processing systems—Local area networks—Part 5: Token ring access method and physical layer specification."

[24] D. A. Pitt and J. L. Winkler, "Table-Free Bridging," IEEE J. Selected Areas Commun., vol. SAC-5, No. 9, pp. 1454–1462, Dec. 1987.

[25] D. A. Pitt, K. K. Sy, and R. A. Donnan, "Source routing for bridged local area networks," in Advances in Local Area Networks, J. O. Limb and K. Kummerle (eds.), pp. 517–532. New York: IEEE Press, 1987.

[26] J. A. Berntsen, J. R. Davin, D. A. Pitt, and N. G. Sullivan, "MAC layer interconnection of IEEE 802 local area networks," Comput. Networks ISDN Syst., vol. 10, No. 5, pp. 259–273, Dec. 1985.

[27] D. A. Pitt and F. Farzaneh, "Topologies and routing for bridged local area networks," Proc. IEEE Infocom '86, Miami, Apr. 1986.

[28] "The X.25 interface for attaching SNA nodes to packet-switched data networks general information manual," Order No. GA27-3345, available through IBM branch offices.

[29] "IBM synchronous data link control general information," Order No. GA27-3093, available through IBM branch offices.

[30] "IBM token ring network architecture reference," Order No. SC30-3374, available through IBM branch offices.

[31] International Organization for Standardization, International Standards 8802-2, "Information processing systems—Local area networks—Part 2: Logical link control."

[32] A. E. Baratz, J. P. Gray, P. E. Green, J. M. Jaffe, and D. P. Pozefsky, "SNA networks of small systems," IEEE J. Selected Areas Commun., vol. SAC-3, No. 3, pp. 416–426, May 1985.

[33]   International Organization for Standardization, International Standards 7478, "Data communication—Multilink procedures."

[34]   "Network design and analysis: User's guide for the VM environment," Order No. SC34-4061, available through IBM branch offices.

[35]   "Systems network architecture—Format and protocol reference manual: Architecture logic for type 2.1 nodes," Order No. SC30-3420, available through IBM branch offices.

[36]   "SNA format and protocol reference manual: Architecture logic for LU 6.2," Order No. SC30-3269, available through IBM branch offices.

[37]   "Customer information control system/virtual storage (CICS/VS): General information," Order No. GC33-0155, available through IBM branch offices.

[38]   H. C. Forsdick, R. E. Schantz, and R. H. Thomas, "Operating systems for computer networks," *Computer*, vol. 11, No. 1, Jan. 1978, pp. 48–57.

[39]   L. D. Whittie, "A distributed operating system for a reconfigurable network computer," Proc. First Int. Conf. on Distributed Computing Systems, October 1979, pp. 669–678.

[40]   A. D. Birrell, and B. J. Nelson, "Implementing remote procedure calls," *ACM Trans. Comput. Syst.*, vol. 2, No. 1, Feb. 1984, pp. 39–59.

[41]   E. C. Cooper, "Replicated procedure call," *Proc. Third Ann. ACM Symp. on Principles of Distributed Computing*, Aug. 27–29, 1984, pp. 220–232.

[42]   J. P. Gray, P. J. Hansen, P. Homan, M. A. Lerner, and M. Pozefsky, "Advanced program-to-program communication in SNA," *IBM Syst. J.*, vol. 22, No. 4, 1983, pp. 298–318.

[43]   "SNA transaction programmer's reference manual for LU Type 6.2," Order No. GC30-3084, available through IBM branch offices.

[44]   "Systems network architecture formats," Order No. GA27-3136, available through IBM branch offices.

[45]   R. Obermarck, "Distributed deadlock detection algorithm," *ACM Trans. Database Syst.*, vol. 7, No. 2, pp. 187–208, June 1982.

[46]   "3270 data stream programmer's reference," Order No. GA23-0059, available through IBM branch offices.

[47]   "Systems network architecture—Sessions between logical units," Order No. GC20-1868, available through IBM branch offices.

[48]   "IPDS architecture reference manual," Order No. S544-3417, available through IBM branch offices.

[49]   "ACF/VTAM: General information," Order No. GC27-0642, available through IBM branch offices.

[50]   "ACF/NCP/VS (Network control program, systems support programs): General information," Order No. GC30-3058, available through IBM branch offices.

[51]   Institute of Electrical and Electronics Engineers, ANSI/IEEE Standard 802.2-1985: "Logical link control."

[52]   "IMS/VS general information manual," Order No. GII20-1260, available through IBM branch offices.

[53]   C. H. Sauer, D. W. Johnson, L. K. Loucks, A. A. Shaheen-Gouda, and T. A. Smith, "RT PC distributed services overview," *Operating Syst. Rev.*, vol. 21, No. 3, pp. 18–29, July 1987.

[54]   "IBM RT personal computer AIX operating system technical reference manual," Order No. SA23-0806, available through IBM branch offices.

[55]   "Document interchange architecture: Concepts and structures," Order No. SC23-0759, available through IBM branch offices.

[56]   "Document content architecture: Revisable–Form–Text Reference," Order No. SC23-0758, available through IBM branch offices.

[57]   "Systems network architecture format and protocol reference manual: Management services," Order No. SC30-3346, available through IBM branch offices.

[58] D. H. Thoenen, "Communications Systems Bulletin: Network problem determination application version 3 release 2," January, 1986, Order No. GG22-9108-00, available through IBM branch offices.

[59] L. J. Cole, "Network management as described in systems network architecture," Proc. IEEE Infocom '86, Miami, Apr. 1986, pp. 364–376.

[60] R. E. Moore, "Problem detection, isolation, and notification in systems network architecture," Proc. IEEE Infocom '86, Miami, Apr. 1986, pp. 377–381.

[61] IBM Syst. J., special issue on Network Management, vol. 27, No. 1, 1988.

[62] B. Don Carlos and J. Winkler, "Token-ring local-area network management," Proc. IEEE Infocom '86, Miami, Apr. 1986, pp. 94–98.

[63] "Writing applications: A design guide," Order No. SC26-4362, available through IBM branch offices.

[64] "Common user access: Panel design and user interaction," Order No. SC26-4351, available through IBM branch offices.

[65] "Systems network architecture—Format and protocol reference manual: Architectural logic," Order No. SC30-3112, available through IBM branch offices.

[66] "Systems network architecture: Technical overview," Order No. GC30-3073, available through IBM branch offices.

[67] IBM Syst. J., special issue on Systems Network Architecture, vol. 22, No. 4, 1983.

[68] V. Ahuja, "Routing and flow control in systems network architecture," IBM Syst. J., vol. 18, No. 2, May 1979, pp. 298–314.

[69] J. D. Atkins, "Path control: The transport network of SNA," IEEE Trans. Commun., vol. COM-28, No. 4, Apr. 1980, pp. 527–538.

[70] G. A. Deaton, "Flow control in packet-switched networks with explicit path routing," Proc. of the Flow Control in Computer Networks Conf., Paris, France, Feb. 1979.

[71] F. D. George, "SNA flow control: Architecture and implementation," IBM Syst. J., vol. 21, No. 2, pp. 179–210, 1982.

[72] J. P. Gray, "Network services in systems network architecture," IEEE Trans. Commun., vol. 25, No. 1, p. 104, Jan. 1977.

[73] J. P. Gray and T. B. McNeill, "SNA multiple system networking," IBM Syst. J., vol. 18, No. 2, pp. 263–297, 1979.

[74] H. Rudin and H. Muller, "On routing and flow control," Proc. of the Flow Control in Computer Networks Conf., Paris, France, February 1979.

[75] T. Schick and R. F. Brockish, "The document interchange architecture: A member of a family of architectures in the SNA environment," IBM Syst. J., vol. 21, No. 2, pp. 220–244, 1982.

[76] N. C. Strole, "A local communications network based on interconnected token-access rings: A tutorial," IBM J. Res. Dev., vol. 27, No. 5, pp. 481–496, Sep. 1983.

[77] T. P. Sullivan, "Communications network management enhancements for SNA networks: An overview," IBM Syst. J., vol. 22, Nos. 1/2, pp. 129–142, 1983.

[78] R. J. Sundstrom and G. D. Schultz, "SNA's first six years: 1974–1980," Fifth International Conference on Computer Communication, Atlanta, GA. Amsterdam: North-Holland, pp. 578–585, September 1980.

[79] R. J. Sundstrom, J. B. Staton, G. D. Schultz, M. L. Hess, G. A. Deaton, L. J. Cole, and R. M. Amy, "SNA directions—A 1985 perspective," Conf. Proc. NCC '85, 1985, pp. 589–603.

Formal Specifications and Their Manipulation

Formal Methods for Protocol Specification and Validation

Gregor V. Bochmann and Carl A. Sunshine

I. Introduction

As evidenced by the earlier chapters in this book, increasingly numerous and complex communication protocols are being employed in distributed systems and computer networks of various types. The informal techniques used to design these protocols have been largely successful, but have also yielded a disturbing number of errors or unexpected and undesirable behavior in most protocols. This chapter describes some of the more formal techniques which are being developed to facilitate design and implementation of protocols.

As they develop, protocols must be described for many purposes. Early descriptions provide a reference for cooperation among designers of different parts of a protocol system. The design must be checked for logical correctness. Then the protocol must be implemented, and if the protocol is in wide use, many different implementations may have to be checked for compliance with a standard. Although narrative descriptions and informal walk-throughs are invaluable elements of this process, painful experience has shown that by themselves they are inadequate.

In the following sections, we shall discuss the use of formal techniques in each of the major design steps of specification, verification, and imple-

GREGOR V. BOCHMANN • Département d'IRO, University of Montreal, Montreal H3C 3J7, Canada   CARL A. SUNSHINE • Unisys, West Coast Research Center, Santa Monica, California 90406.

mentation. Section II clarifies the meaning of specification in the context of a layered protocol architecture, identifies what a protocol specification should include, and describes the major approaches to protocol specification. Section III defines the meaning of verification, discusses what can be verified, and describes the main verification methods. Section IV provides an overview of protocol conformance testing and the main uses of formal methods. Complementary surveys and collections of papers can be found in [12, 22, 45, 54, 64, 75, 76].

II. Protocol Specification

As noted above, protocol descriptions play a key role in all stages of protocol design. This section clarifies the meaning of specification in the domain of communication protocols, identifies the major elements that comprise a specification, and presents the major methods for protocol specification.

A. The Meaning of Specification

We assume that the communication architecture of a distributed system is structured as a hierarchy of different protocol layers, as described in earlier chapters. Each layer provides a particular set of *Services* to its users above. From their viewpoint, the layer may be seen as a "black box" or machine which allows a certain set of interactions with other users (see Fig. 1). A user is concerned with the nature of the service provided, but not with how the protocol manages to provide it.

This description of the input/output behavior of the protocol layer constitutes a *Service Specification* of the protocol. It should be "abstract" in the sense that it describes the types of commands and their effects, but leaves open the exact format and mechanisms for conveying them (e.g., procedure calls, system calls, interrupts). These formats and mechanisms may be different for users in different parts of the system, and are defined by an *Interface Specification*.

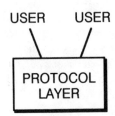

Fig. 1. Services provided by a protocol layer.

1. Service Specifications

Specifying the service to be provided by a layer of a distributed communication system presents problems similar to specifying any software module of a complex computer system. Therefore methods developed for software engineering [43, 49] are useful for the definition of communication services. Usually, a service specification is based on a set of *Service Primitives* which, in an abstract manner, describe the operations at the interface through which the service is provided. In the case of a transport service, for example, some basic service primitives are *Connect*, *Disconnect*, *Send*, and *Receive*. The execution of a service primitive is associated with the exchange of parameter values between the entities involved, i.e., the service providing and using entities of two adjacent layers. The possible parameter values and the direction of transfer must be defined for each parameter.

Clearly, the service primitives should not be executed in an arbitrary order and with arbitrary parameter values (within the range of possible values). At any given moment, the allowed primitives and parameter values depend on the preceding history of operations. The service specification must reflect these constraints by defining the allowed sequences of operations directly, or by making use of a "state" of the service which may be changed as a result of some operations.

In general, the constraints depend on previous operations by the same user ("local" constraints), and by other users ("global" constraints). Considering again the example of a transport service, a local constraint is the fact that *Send* and *Receive* may only be executed after a successful *Connect*. An example of a global constraint is the fact that the "message" parameter value of the first *Receive* on one side is equal to the message parameter value of the first *Send* on the other side.

2. Protocol Specifications

Although it is irrelevant to the user, the protocol designer must be concerned with the internal structure of a protocol layer. In a network environment with physically separated users, a protocol layer must be implemented in a distributed fashion, with *Entities* (processes or modules) local to each user communicating among one another via the services of the lower layer (see Fig. 2). The interaction among entities in providing the layer's service constitutes the actual *Protocol*. Hence a protocol specification must describe the operation of each entity within a layer in response to commands from its users, messages from the other entities [via the lower layer service, also called "protocol data units" (PDU)], and internally initiated actions (e.g., timeouts).

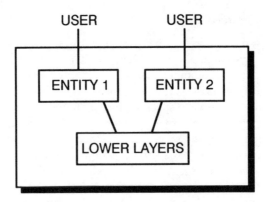

Fig. 2. Internal structure of a protocol layer.

3. Abstraction and Stepwise Refinement

The specifications described above must embody the key concept of *Abstraction* if they are to be successful. To be abstract, a specification must include the essential requirements that an object must satisfy and omit the unessential. A service specification is abstract primarily in the sense that it does not describe how the service is achieved (i.e., the interactions among its constituent entities), and secondarily in the sense that it defines only the general form of the interaction with its users (not the specific interface).

A protocol specification is a refinement or distributed "implementation" of its service specification in the sense that it partly defines how the service is provided (i.e., by a set of cooperating entities). This "implementation" of the service is what is usually meant by the design of a protocol layer. The protocol specification should define each entity to the degree necessary to ensure compatibility with the other entities of the layer, but no further. Each entity remains to be implemented in the more conventional sense of that term, typically by coding in a particular programming language. There may be several steps in this process until the lowest-level implementation of a given protocol layer is achieved [15, 28].

B. What a Protocol Definition Should Include

A protocol cannot be defined without describing its context. This context is given by the architectural layer of the distributed system in which the protocol is used. A description of a layer should include the following items [34]:

1. A general description of the purpose of the layer and the services it provides.
2. An exact specification of the service to be provided by the layer.

3. An exact specification of the service provided by the layer below, and required for the correct and efficient operation of the protocol. (This of course is redundant with the lower layer's definition, but makes the protocol definition self-contained.)
4. The internal structure of the layer in terms of entities and their relations.
5. A description of the protocol(s) used between the entities including:
   a. An overall, informal description of the operation of the entities.
   b. A protocol specification which includes
      i. a list of the types and formats of messages exchanged between the entities;
      ii. rules governing the reaction of each entity to user commands, messages from other entities, and internal events.
   c. Any additional details (not included in point b), such as considerations for improving the efficiency, suggestions for implementation choices, or a detailed description which may come close to an implementation.

Reference [12] presents an example of these items for a simple data transfer protocol.

C. Specification Methods

Descriptions of communication services and protocols must be both easy to understand and precise—goals which often conflict. The use of natural language gives the illusion of being easily understood, but leads to lengthy and informal specifications which often contain ambiguities and are difficult to check for completeness and correctness. The arguments for the use of formal specification methods in the general context of software engineering [49] apply also in our context.

1. Protocol Specifications

Most of the work on formal specification of protocols has focused on the protocol itself and not on the service it provides. A variety of general formalisms such as state diagrams, Petri nets, grammars, and programming languages have been applied to this problem and in many cases adaptations or extensions to facilitate protocol modeling have been made [20]. Most of these techniques may be classified into three main categories: transition models, programming languages, and algebraic methods.

Transition models are motivated by the observation that protocols consist largely of relatively simple processing in response to numerous "events" such as commands (from the user), message arrival (from the lower layer), and internal timeouts. Hence state machine models of one sort or another with such events forming their inputs are a natural model. However, for protocols of any complexity, the number of events and states required in a straightforward finite state transition formalism becomes unworkably large. For example, to model a protocol using sequence numbers, there must be different states and events to handle each possible sequence number [50]. Models falling into this category include state transition diagrams [10, 72], grammars [33a, 68], Petri nets and their derivatives [3, 16, 23, 47, 67], L-systems [27], UCLA graphs [50], and colloquies [42].

Programming language models [11, 29, 32, 63] are motivated by the observation that protocols are simply one type of algorithm, and that high-level programming languages provide a clear and relatively concise means of describing algorithms. Depending on how high level and abstract a language is used, this approach to specification may be quite near to an implementation of the protocol. As noted above, this proximity is a mixed blessing, since unessential features in the program are often combined with the essential properties of the algorithm. A major advantage of this approach is the ease in handling variables and parameters which may take on a large number of values (e.g., sequence numbers, timers), as opposed to pure state machine models.

Hybrid models [9, 14, 21, 35, 57, 66] attempt to combine the advantages of state models and programs. These typically employ a small-state model to capture only the main features of the protocol (e.g., connection establishments, resets, interrupts). This state model is then augmented with additional "context" variables and processing routines for each state. In such hybrid models, the actions to be taken are determined by using parameters from the inputs and values of the context variables according to the processing routine for each major state. For example, the sequence number of an arriving message may be compared with a variable giving the expected next sequence number to determine whether to accept the message, what the next state should be, and how to update the expected sequence number. Bochmann and Gecsei [14] have demonstrated the potential for trading off the complexity of the state model with the amount of context information for a given protocol. (Other techniques for managing the complexity of protocols are discussed in Section III C.)

Other specification methods that do not require definition of explicit states may also be used. I/O history methods define the allowed input and output sequences of the entity and their relation to each other (e.g., that the sequence of messages delivered is a subset of the sequence of messages received [9, 28, 31b]). Algebraic specifications provide another way of defining the allowed sequence of operations [1, 33b, 36, 48, 70], but in contrast to simple state machine models, may allow the incorporation of rules concerning the parameter values of the exchanged interactions [31a, 44]. Some of these methods are related to abstract data type definitions [65].

This discussion cannot be considered complete without mentioning the application of temporal logic specification methods to the protocol area [59, 60], and the use of logic programming techniques as specification language or as a support for building validation tools [7, 61, 69].

2. Service Specifications

The need for comprehensive protocol service specifications has been realized more recently [9, 55, 64, 70, 71]. Initial efforts at formal service specifications were directed toward applying general software engineering methodology. As noted in Section II A, definition of the primitive operations supported by the layer (e.g., Send, Connect) is a basic feature of any specification. In abstract machine type specifications, internal "states" of the layer are also defined. These states are used in defining the effects of each operation, and may be changed as a result of the operation [53].

It is generally believed that the specification methods developed for protocols should also be applicable for the specification of communication services. However, the simple finite state transition formalism, which captures an important part of most protocol specifications, is less useful for the specification of communication services, since the service interactions with two different users (see Fig. 1) are usually not closely synchronized with one another. Instead most specifications of communication services involve some kind of queues, which correspond to the propagation of information from one service user to the other.

3. Standardization

In recent years, the use of formal methods for the description and validation of communication protocols has been discussed in standardization bodies including ISO and CCITT. In the context of the OSI standardization effort, in particular, some work on the standardization of formal description methods, called "formal description techniques" (FDT) has been pursued [24, 70]. Three FDTs are under consideration by ISO and CCITT for application to the specification of OSI protocols and services. These methods are the following (in alphabetical order):

(a) ESTELLE [35] has been developed by Subgroup B of the ISO TC97 SC21 WG1 ad hoc group on FDT. It is a method based on a finite state machine model extended by the use of PASCAL programming language elements to handle data structures and more complex operations. A specified system may consist of a larger number of interconnected state machine modules.

(b) LOTOS [36] has been developed by Subgroup C of the ISO TC97 SC21 WG1 ad hoc group on FDT. It is a method based on the formalism of Milner's CCS [48], which is combined with an abstract data type definition facility, called ACT ONE [26]. It also allows the construction of a specification from several smaller components.

(c) SDL [18] was originally developed by CCITT for the description of switching systems. It has also been found useful for the description of communication protocols [24]. The method is based on an extended finite state machine model, and is largely oriented towards a graphical representation.

Several trial specifications of various communication protocol standards in the OSI area have been developed using these different methods. The possibility has been considered that in the future such a formal protocol specification could become the protocol standard.

III. Protocol Verification

In its broadest interpretation, system validation aims to assure that a system satisfies its design specifications and (hopefully) operates to the satisfaction of its users. Validation activity is important during all design phases, and may include testing of the final system implementation, simulation studies, analytical performance predictions, and verification. Verification is based on the system specification, and involves logical reasoning. Therefore it may be used during the design phase before any system implementation exists, in order to avoid possible design errors. While testing and simulation only validate the system for certain test situations, verification allows, in principle, the consideration of all possible situations the system may encounter during actual operation.

A. The Meaning of Verification

Verification is essentially a demonstration that an object meets its specifications. Recalling from Section II that *Services* and *Protocol Entities* are the two major classes of objects requiring specification for a protocol layer, we see there are two basic verification problems that must be addressed: (1) the protocol's *Design* must be verified by analyzing the possible interactions of the entities of the layer, each functioning according to its (abstract) protocol specification and communicating through the underlying layer's service, to see whether this combined operation satisfies the layer's service specification; and (2) the *Implementation* of each protocol entity must be verified against its abstract protocol specification.

The somewhat ambiguous term "protocol verification" is usually intended to mean this first design verification problem. Because protocols are inherently systems of concurrent independent entities interacting via (possibly unreliable) exchange of messages, verification of protocol designs takes on a characteristic communication oriented flavor. Implementation of each entity, on the other hand, is usually done by "ordinary" programming techniques, and its logical verification represents a more common (but by no means trivial) program verification problem that has received less attention from protocol verifiers. However, the validation of a protocol implementation is usually done by various testing methods. Sometimes called *protocol implementation assessment* or *protocol conformance testing*, this is further discussed in Section IV below.

The service specification itself cannot be verified, but rather forms the standard against which the protocol is verified. However, the service specification can be checked for consistency. It must also properly reflect the users' desires, and provide an adequate basis for the higher levels which use it.

It is important to note that protocol verification also depends on the properties of the lower-layer protocol. In verifying that a protocol meets its service specification, it will be necessary to assume the properties of the lower layer's service. If a protocol fails to meet its service specification, the problem may rest either in the protocol itself, or in the service provided by the lower layer.

Most of the verification work to date has been on design rather than implementation, and we shall focus on design verification in the remainder of this section. While a protocol design need only be verified once, each different implementation must be verified against the design.

B. What Can Be Verified

The overall verification problem may be divided along two axes, each with two categories. On one axis, we distinguish between general and specific properties. On the other we distinguish between safety and liveness.

General properties are those properties common to all protocols that may be considered to form an implicit part of all service specifications. Foremost among these is the absence of deadlock (the arrival in some system state or set of states from which there is no exit). Completeness, or the provision for all possible inputs, is another general property. Progress or termination may also be considered in this category since they require minimal specification of what constitutes "useful" activity or the desired final state.

Specific properties of the protocol, on the other hand, require specification of the particular service to be provided. Examples include reliable data transfer in a transport protocol, copying a file in a file transfer protocol, or clearing a terminal display in a virtual terminal protocol. Definitions of these features make up the bulk of service specifications.

On the other axis, *safety* has the usual meaning that if the protocol performs any action at all, it will be in accord with its service specification.

For example, if a transport protocol delivers any messages, they will be to the correct destination, in the correct order, and without errors. *Liveness* means that the specified services will actually be completed in finite time. In the case of logical verification, which is the subject of this chapter, it is sufficient to ascertain a finite time delay. In the case that the efficiency and responsiveness of the protocol are to be verified, it is clearly necessary to determine numerically the expected time delay, throughput, and other characteristics.

C. Verification Methods

Approaches to protocol verification have followed two main paths: reachability analysis and program proofs. Within the scope of this chapter, we can only outline these two approaches. The references cited in Section IV provide more details on particular techniques.

Reachability analysis is based on exhaustively exploring all the possible interactions of two (or more) entities within a layer. A *composite* or *global state* of the system is defined as a combination of the states of the cooperating protocol entities and the lower layer connecting them. From a given initial state, all possible transitions (user commands, time-outs, message arrivals) are generated, leading to a number of new global states. This process is repeated for each of the newly generated states until no new states are generated (some transitions lead back to already generated states). For a given initial state and set of assumptions about the underlying layer (the type of service it offers), this type of analysis determines all of the possible outcomes that the protocol may achieve.

Reachability analysis is particularly straightforward to apply to transition models of protocols which have explicit states and/or state variables defined. It is also possible to perform a reachability analysis on program models by establishing a number of "break points" in the program that effectively define control states [32]. Symbolic execution (see the following) may also be viewed as a form of reachability analysis.

Reachability analysis is well suited to checking the general correctness properties described above because these properties are a direct consequence of the structure of the reachability graph. Global states with no exits are either deadlocks or desired termination states. Similarly, situations where the processing for a receivable message is not defined, or where the transmission medium capacity is exceeded are easily detected. The generation of the global state space for transition models is easily automated, and several computer aided systems for this purpose have been developed [21, 50, 51, 62, 72, 75]. The major difficulty of this technique is "state space explosion" because the size of the global state space may grow rapidly with the number and complexity of protocol entities involved and the underlying layer's services. Techniques for dealing with this problem are discussed below.

The program proving approach involves the usual formulation of assertions which reflect the desired correctness properties. Ideally, these would be supplied by the service specification, but as noted above, services have not been rigorously defined in most protocol work, so the verifier must formulate appropriate assertions of his own. The basic task is then to show (prove) that the protocol programs for each entity satisfy the high-level assertions (which usually involve both entities). This often requires formulation of additional assertions at appropriate places in the programs [11, 63].

A major strength of this approach is its ability to deal with the full range of protocol properties to be verified, rather than only general properties. Ideally, any property for which an appropriate assertion can be formulated can be verified, but formulation and proof often require a great deal of ingenuity. Only modest progress has been made to date in the automation of this process.

As with specification, a hybrid approach promises to combine the advantages of both techniques. By using a state model for the major states of the protocol, the state space is kept small, and the general properties can be checked by an automated analysis. Other properties, for which a state model would be awkward (e.g., sequenced delivery), can be handled by assertion proofs on the variables and procedures which accompany the state model. Such combined techniques are described in [14, 21].

While a large body of work on general program verification exists, several characteristics of protocols pose special difficulties in proofs. These include concurrency of multiple protocol modules and physical separation of modules so that no shared variables may be used. A further complication is that message exchange between modules may be unreliable requiring methods that can deal with nondeterminism. A few early applications of general program verification methods to protocols are described in [11, 28, 31b, 63, 64, 65].

A particular form of proof that has been useful for protocols with large numbers of interacting entities (e.g., routing protocols) may be called *induction on topology* [45]. The desired properties are first shown to be true for a minimum subset of the entities, and then an induction rule is proved showing that if the properties hold for a system of N entities, they also hold for $N + 1$ entities.

When an error is found by some verification technique, the cause must still be determined. Many transitions or program statements may separate the cause from the place where the error occurs, as for example when the acceptance of a duplicate packet at the receiver is caused by the too rapid reuse of a sequence number at the sender. In some cases the protocol may be modeled incorrectly, or the correctness conditions may be formulated incorrectly. In other cases, undesired behavior may be due to transmission medium properties that were not expected when the protocol was designed (e.g., reordering of messages in transit). Even when an automated verification system is available, considerable human ingenuity is required to understand and repair any errors that are discovered.

Another approach to achieving correct protocols is based on constructive design rules that automatically result in correct protocols. In one approach [72], design rules are formulated which guarantee that the specifications obtained for a set of interacting entities will be complete. For each send transition specified by the designer, the rules determine the corresponding receive transition to be added to the partner entity. In another case [46], the specification of a second entity is determined by a design rule such that it will operate with a specified first entity to provide a given overall service. The derivation of a specification for the protocol from a given service specification has also been considered [38].

A major difficulty for protocol verification by any method is the complexity of the global system of interacting protocol entities, also termed *state space explosion*. The following methods may be used to keep this complexity within manageable limits.

(1) Partial Specification and Verification. Depending on the specification method used, only certain aspects of the protocol are described. This is often the case for transition diagram specifications which usually capture only the rules concerning transitions between major states, ignoring details of parameter values and other state variables.

(2) Choosing Large Units of Actions. State space explosion is due to the interleaving of the actions executed by the different entities. For example, the preparation and sending of a protocol data unit by an entity may usually be considered an indivisible action which proceeds without interaction with the other entities of the system. The execution of such an action may be considered a single "transition" in the global protocol description.

A particular application of this idea is to consider only states where the transmission medium is empty. Such an "empty medium abstraction" [10] is justified when the number of messages in transit is small. In this case, previously separate sending and receiving transitions of different entities can be combined into single joint transitions of both entities. A related approach allows each entity to make "maximal progress" [30].

(3) Decomposition into Sublayers and/or Phases. The decomposition of the protocol of a layer into several sublayers and/or phases of operation simplifies the description and verification, because the protocol of each part may be verified separately [19, 40]. An example of this idea is the decomposition of HDLC into the sublayers of bit stuffing, checksumming, and elements of procedure, and the division of the latter into several components as described in [13].

(4) Classifying States by Assertions. Assertions which are predicates on the set of all possible system states may be formed. Each predicate defines a set (or class) of states which consists of those states for which the predicate

is true. One may then consider classes of states collectively in reachability analysis instead of considering individual states. By making an appropriate choice of predicates (and therefore classes of states) the number of cases to be considered may be reduced considerably. This method is usually applied for proving safety of protocol specifications given in some programming language [11, 63], and forms the basis for *symbolic execution* [17]. Typically, the assertions depend on some variables of the entities and the set of messages in transit (through the layer below).

To illustrate the possible savings in the number of cases to be analyzed, consider the state of an entity receiving numbered information frames. Instead of treating all possible values of a sequence counter variable explicitly as different states, it may be possible to consider only the three cases where the variable is "less than," "equal to," or "greater than" the number in the information frame received.

(5) Focusing Search. Instead of generating all possible states, it is possible to predetermine potential global states with certain properties (e.g., deadlocks), and then check whether they are actually reachable [21].

(6) Automation. Some steps in the analysis process may be performed by automated systems [17, 21, 28, 32, 50, 51, 58, 62, 72]. However, the use of these systems is not trivial, and much work goes into representing the protocol and service in a form suitable for analysis. Human intervention is needed in many cases for distinguishing between useful and undesired loops, or for guiding the proof process.

IV. Uses of Formal Techniques

Formal methods have been used in many cases for designing and implementing data communication and computer network protocols. Some early uses of such methods are reviewed in [31b]. In some cases, the formal specification was made after the system design was essentially finished, and served for an additional analysis of correctness and efficiency, or as an implementation guide. In other cases, the formal specification was used as a reference document during the system design.

If a formal specification of a protocol or communication service is given and can be used as the authoritative reference for all design and implementation efforts, this specification can be used for the following three activities:

(1) For the validation of the protocol design: In this area the verification methods described in Section III may be applied, or simulated executions of the protocol specification may be used for evaluating the performance [25, 41] or detecting any errors or other difficulties in the specification [39, 69].

(2) For the development of an implementation of the protocol: Some specification methods facilitate this goal more than others. Programming language specifications may be quite close to implementations (but often lack the desired degree of abstraction). Direct implementation of transition or hybrid model specifications by some form of interpreter or "compiler" is also relatively straightforward [15]. In many cases these implementation methods have been at least partially automated [2, 4, 5, 8, 38, 57, 68].

(3) For validating an implementation of the protocol and, in particular, for assessing that the implementation conforms to the protocol specification: This activity, also called *conformance testing*, has received much attention recently, in particular in connection with the development of OSI protocol standards [37, 73, 74].

The validation of a protocol implementation is usually performed mainly through testing. Two concerns must be distinguished in this context:

- *Protocol conformance assessment* is concerned with the question whether a given protocol implementation conforms to all rules defined in the corresponding protocol specification, and which of the defined options are supported by the implementation.
- *Implementation assessment* is more general, and is concerned in addition with properties of the implementation that are not part of the protocol specification, such as how the implementation reacts to unexpected (invalid) user commands, how many simultaneous connections can be supported, or performance.

Possible architectures for the testing of higher-level communication protocols have been elaborated to be used for the conformance testing of OSI protocol implementations [37]. The most commonly used architecture [52] employs "remote" or "distributed" testing where the test system acts as the peer protocol entity of the implementation under test (IUT), as shown in Fig. 3. It is important to note that the complete testing of the IUT requires the presence of a test user, also called "upper tester," which verifies that the executed service primitives relate correctly to the protocol data units exchanged between the IUT and the remote test system. Several variations of this basic testing architecture have been proposed [6]. These variations address in particular the question of how the operations of the remote test system and the upper tester may be synchronized.

While in the past most issues in the protocol conformance area have been resolved by pragmatic considerations, formal methods seem to be useful in this area for the following purposes:

(1) Formal specification methods may be used for specifying test sequences.

(2) The formal specification of the protocol standard may be used as the basis for deriving test sequences [56a, 56b].

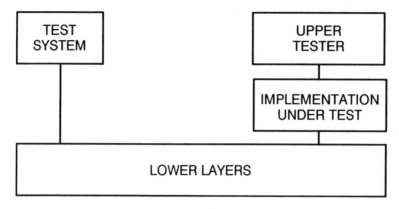

Fig. 3. Protocol testing architecture.

(3) The observed sequence of interactions of an IUT may be checked against the formal protocol specification in order to detect any deviation from the protocol specification [7, 39, 69].

V. Conclusions

The specification of a protocol layer must include definitions of both the services to be provided by the layer, and the protocol executed by the entities within the layer to provide this service. "Design verification" then consists of showing that the interaction of entities is indeed adequate to provide the specified services, while "implementation verification" consists of showing that the implementations of the entities satisfy the more abstract protocol specification. A useful subset of design verification may be described as verification of "general properties" such as deadlock, looping, and completeness. These properties may be checked in many cases without requiring any particular service specification.

Although protocol specifications must serve many purposes, verification and implementation are two critical tasks which require rigorous or formal specification techniques in order to be fully successful. Formal protocol specifications are more precise than descriptions in natural language, and should contain the necessary details for obtaining compatible protocol implementations on different system components. The cases mentioned in Section IV demonstrate that formal methods may be used profitably for the specification, verification, and implementation of communication protocols. However, a great deal of work remains to be done in improving verification techniques and high-level system implementation languages, in integrating performance (efficiency) analysis with analysis for logical correctness, and in automating these analysis techniques.

Many published papers on protocol verification present some particular verification technique, and demonstrate the technique by discussing its application to a simple protocol of more or less academic nature. Some papers treat more complex, realistic protocol examples. Some *a posteriori* verifications of protocol standards or standardization proposals have also been presented pointing out certain difficulties with the adopted or proposed procedures. Most of these verification efforts were based on a state reachability analysis, and in some cases automated systems were used. The results have had some influence on the standardization process, and have probably influenced the implementation of these procedures.

We believe that more effort should be spent on the logical verification of protocols during the design phase. Based on a formal description method, this process may be facilitated by the use of automated systems for protocol analysis. The same formal specification should also serve as the official definition of the protocol. Ideally, the same specification could be transformed, using other automated tools, into a usable protocol implementation, or used to generate conformance tests. Once such tools are available, their use will increase the reliability of protocols, decrease compatibility problems, and lower the cost of implementation. Some recent work on such protocol engineering toolkits is described in [75].

References

[1] S. Aggarwal, R. Kurshan, and K. Sabnani, "A calculus for protocol specification and validation," in [76], 1983.

[2] J. P. Ansart, V. Chari, and D. Simon, "From formal description to automated implementation using PDIL," in [76], 1983.

[3] G. Berthelot and R. Terrat, "Petri net theory for the correctness of protocols," *IEEE Trans. Commun.*, vol COM-30, pp. 2497–2505, 1982.

[4] D. Bjorner, "Finite state automation—Definition of data communication line control procedures," in *Fall Joint Computer Conf.*, *AFIPS Conf. Proc.*, 1970.

[5] T. Blumer and R. Tenney, "A formal specification technique and implementation method for protocols," *Comput. Networks*, vol. 6, pp. 201–217, 1982.

[6] G. Bochmann, E. Cerny, M. Maksud, and B. Sarikaya, "Testing of transport protocol implementations," in *Proc. CIPS Conf.*, Ottawa, 1983, pp. 123–129.

[7] G. Bochmann et al., "Use of PROLOG for building protocol design tools," in [76], 1985.

[8] G. Bochmann, J. M. Serre, and G. Gerber, "Obtaining protocol implementations from formal specifications," in *Proc. CIPS Congress*, Montreal, 1985, pp. 187–193.

[9] G. Bochmann, "A general transition model for protocols and communications services," *IEEE Trans. Commun.*, vol. COM-28, pp. 643–650, April 1980. Also in *Computer Network Architectures and Protocols*, P. Green, Jr. (ed.). New York: Plenum Press, 1982.

[10] G. Bochmann, "Finite state description of communication protocols," *Comput. Networks*, vol. 2, pp. 361–372, Oct. 1978.

[11] G. Bochmann, "Logical verification and implementation of protocols," in *Proc. 4th Data Commun. Symp.*, Quebec, Canada, 1975, pp. 8-5–8-20.

[12] G. Bochmann, "Specification and verification of computer communication protocols," Chap. 5 in *Advances in Distributed Processing Management*, T. A. Rullo (ed.). Philadelphia, PA: Heyden & Son, 1980.

[13] G. Bochmann and R. J. Chung, "A formalized specification of HDLC classes of procedures," in *Proc. Natl. Telecommun. Conf.*, Los Angeles, CA, Dec. 1977, Paper 3A.2.

[14] G. Bochmann and J. Gecsei, "A unified model for the specification and verification of protocols," in *Proc. IFIP Congress*, 1977, pp. 229–234.

[15] G. Bochmann and T. Joachim, "Development and structure of an X.25 implementation," *IEEE Trans. Software Eng.*, vol. SE-5, pp. 429–439, Sep. 1979.

[16] J. Billington, G. Wheeler, and M. Wilbur-Ham, "PROTEAN: A high-level Petri net tool for the specification and verification of communication protocols," pp. 301–316 in [75].

[17] D. Brand and W. Joyner, Jr., "Verification of protocols using symbolic execution," *Comput. Networks*, vol. 2, pp. 351–360, Oct. 1978.

[18] CCITT SG XI, Recommendations Z.100, Z.101, Z.102, Z.103, Z.104, 1984.

[19] C. Chow, M. Gouda, and S. Lam, "An exercise in constructing multi-phase communication protocols," in *Proc. ACM SIGCOMM Symp.*, June 1984, pp. 42–49.

[20] A. Danthine, "Protocol representation with finite state models," *IEEE Trans. Commun.*, vol. COM-28, pp. 632–643, April 1980. Also in *Computer Network Architectures and Protocols*, P. Green, Jr. (ed.). New York: Plenum Press, 1982.

[21] A. Danthine and J. Bremer, "Modeling and verification of end-to-end transport protocols," *Comput. Networks*, vol. 2, pp. 381–395, Oct. 1978.

[22] J. Day and C. Sunshine (eds.), "A bibliography on the formal specification and verification of computer network protocols," *ACM SIGCOMM Comput. Commun. Rev.*, vol. 9, Oct. 1979.

[23] M. Diaz, "Modelling and analysis of communication and cooperation protocols using Petri net based models," *Comput. Networks*, vol. 6, pp. 419–441, 1982.

[24] G. Dickson and P. de Chazal, "Application of the CCITT SDL to protocol specification," in *Proc. IEEE*, Dec. 1983.

[25] M. Didic and B. Wolfinger, "Simulation of local computer network architecture applying a unified modeling system," *Comput. Networks*, vol. 6, pp. 75–91, 1982.

[26] H. Ehrig and B. Mahr, *Fundamentals of Algebraic Specifications 1*, New York: Springer Verlag, 1985.

[27] C. Ellis, "Consistency and correctness of duplicate database systems," in *Proc. 6th Symp. Op. Syst. Principles*, Purdue Univ., West Lafayette, IN, Nov. 1977; *ACM Op. Syst. Rev.*, vol. 11, pp. 67–84, 1977.

[28] D. Good, "Constructing verified and reliable communications processing systems," *ACM Software Eng. Notes*, vol. 2, pp. 8–13, Oct. 1977; also Rep. ICSCA-CPM-6, Univ. Texas at Austin.

[29] D. Good and R. Cohen, "Verifiable communications processing in GYPSY," Univ. Texas at Austin, Rep. ICSCA-CPM-11, June 1978.

[30] M. Gouda and Y. Yu, "Protocol validation by maximal progress state exploration," *IEEE Trans. Commun.*, vol. COM-32, pp. 94–97, 1984.

[31a] J. Guttag, E. Horowitz, and D. Musser, "Abstract data types and software validation," *Commun. ACM*, vol. 21, pp. 1048–1064, Dec. 1978.

[31b] B. Hailpern, "Specifying and verifying protocols represented as abstract programs," in *Computer Network Architectures and Protocols*, P. Green, Jr. (ed.), New York: Plenum Press, 1982.

[32] J. Hajek, "Automatically verified data transfer protocols," in *Proc. 4th Int. Comput. Commun. Conf.*, Kyoto, Japan, Sep. 1978, pp. 749–756; also see progress report in *ACM SIGCOMM Computer Commun. Rev.*, vol. 8, Jan. 1979.

[33a] J. Harangozo, "An approach to describing a link level protocol with a formal language," in *Proc. 5th Data Commun. Symp.*, Utah, 1977, pp. 4-37-4-49.

[33b] G. Holzmann, "A theory for protocol validation," *IEEE Trans. Comp.*, vol. C-31, pp. 730–738, Aug. 1982.

[34] ISO, TC97 SC16, "Guidelines for the specification of services and protocols," Documents N380 and N381, 1981.

[35] ISO, "ESTELLE: A formal description technique based on an extended state transition model, " DP 9074, 1986.

[36] ISO, "LOTOS: A formal description technique," DP 8807, 1986.

[37] ISO TC97 SC21 WG1, ad hoc group on conformance testing, Document N909, 1985.

[38] M. Itoh and H. Ichikawa, "Protocol verification algorithm using reduced reachability analysis," *IECE Trans.*, Japan, pp. 88–93, 1984.

[39] C. Jard and G. Bochmann, "An approach to testing specifications," *J. Syst. Software*, vol. 3, pp. 315–323, 1983.

[40] S. Lam and A. Shankar, "Protocol projections: A method of analysing communication protocols," *Proc. Natl. Telecom. Conf.*, 1981, pp. E3.2.

[41] G. LeLann and H. LeGoff, "Verification and evaluation of communication protocols," *Comput. Networks*, vol. 2, pp. 50–69, Feb. 1978.

[42] G. LeMoli, "A theory of colloquies," *Atla Frequenza*, vol. 42, pp. 493-223E–500-230E, 1973; also in *Proc. 1st European Workshop on Computer Networks*, Arles, France, Apr. 1973, pp. 153–173.

[43] B. Liskov and S. Zilles, "Specification techniques for data abstractions," *IEEE Trans. Software Eng.*, vol. SE-1, pp. 7–18, Mar. 1975.

[44] L. Logrippo, "Specification of transport service using finite-state transducers and abstract data types," CCITT Q39/VII, FDT-77, Geneva, Dec. 1982.

[45] P. Merlin, "Specification and validation of protocols," *IEEE Trans. Commun.*, vol. COM-27, pp. 1671–1680, Nov. 1979.

[46] P. Merlin and G. Bochmann, "On the construction of communication protocols," in *Proc. Intern. Conf. on Comput. Commun.*, Atlanta, October 1980.

[47] P. Merlin and D. Farber, "Recoverability of communication protocols—Implications of a theoretical study," *IEEE Trans. Commun.*, vol. COM-24, pp. 1036–1043, Sep. 1976.

[48] R. Milner, "A calculus of communicating systems," *Lecture Notes in CS, No. 92*, New York: Springer Verlag, 1980.

[49] D. Parnas, "The use of precise specifications in the development of software," in *Proc. IFIP Congress 1977*, pp. 861–867.

[50] J. Postel, "A graph model analysis of computer communications protocols," Ph.D. thesis, Computer Science Dept., Univ. California, Los Angeles, UCLA ENG-7410, 1974.

[51] O. Rafiq and J. P. Ansart, "Vadiloc. A protocol validator and its applications," in [76], 1983.

[52] D. Rayner, "A system for testing protocol implementations," *Comput. Networks*, vol. 6, pp. 383–395, Dec. 1982.

[53] L. Robinson, K. Levitt, and B. Silverberg, *The HDM Handbook*, vol. I-III, SRI Int., Menlo Park, CA, 1979.

[54] H. Rudin, "An informal overview of formal protocol specification," *IEEE Commun. Mag.*, vol. 23, March 1985.

[55] A. Rybczynski and D. Weir, "Datapac X.25 service characteristics," in *Proc. 5th Data Commun. Symp.*, 1977, pp. 4-50–4-57.

[56a] B. Sarikaya, G. Bochmann, and E. Cerny, "A test design methodology for protocol testing," *IEEE Trans. Software Eng.*, vol. 13, pp. 518–531, 1987.

[56b] B. Sarikaya and G. Bochmann, "Synchronization and specification issues in protocol testing," *IEEE Trans. Commun.*, vol. COM-32, pp. 389–395, April 1984.

[57] G. Schultz *et al.*, "Executable description and validation of SNA," *IEEE Trans. Commun.*, vol. COM-28, pp. 661–677, April 1980. Also in *Computer Network Architectures and Protocols*, P. Green, Jr. (ed.). New York: Plenum Press, 1982.

[58] D. Schwabe, "Formal specification and verification of a connection establishment protocol," in *Proc. 7th Data Commun. Symp.*, Mexico City, Oct. 1981, pp. 11–26.

[59] R. Schwartz and P. Melliar-Smith, "From state machines to temporal logic: Specification methods for protocol standards," in [76], 1982.

[60] R. Schwartz, P. Melliar-Smith, and F. Vogt, "An interval logic for higher-level temporal reasoning," in *Proc. ACM Symp. on Princ. of Distr. Computing*, Montreal, Aug. 1983, pp. 173–186.

[61] D. Sidhu, "Protocol verification via executable logic specifications," in [76], 1983.

[62] D. Sidhu and T. Blumer, "Automated verification of connection management of NBS class 4 transport protocol," in *Proc. ACM SIGCOMM Symp.*, Montreal, June 1984, pp. 83–89.

[63] N. Stenning, "A data transfer protocol," *Comput. Networks*, vol. 1, pp. 99–110, Sep. 1976.

[64] C. Sunshine, "Formal techniques for protocol specification and verification," *Computer*, vol. 12, pp. 20–27, Sep. 1979.

[65] C. Sunshine, *et al.*, "Specification and verification of communication protocols in AFFIRM using state transition models," *IEEE Trans. Software Eng.*, vol. 12, Sept. 1982.

[66] C. Sunshine and Y. Dalal, "Connection management in transport protocols," *Comput. Networks*, vol. 2, pp. 454–473, Dec. 1978.

[67] F. Symons, "Modeling and analysis of communications protocols using numerical Petri nets," Dept. Elec. Eng., Univ. Essex, England, Tech. Rep. 152, May 1978.

[68] A. Teng and M. Liu, "A formal model for automatic implementation and logical validation of network communication protocols," in *Proc. Comput. Networking Symp.*, Natl. Bureau Standards, Dec. 1978, pp. 114–123.

[69] H. Ural and R. Probert, "Automated testing of protocol specifications and their implementations," in *Proc. ACM SIGCOMM Symp.*, 1984.

[70] C. Vissers, G. Bochmann, and R. Tenney, "Formal description techniques by ISO/TC97/SC16/WG1 ad hoc group on FDT," *Proc. IEEE*, vol. 71, pp. 1356–1364, Dec. 1983.

[71] C. Vissers and L. Logrippo, "The importance of the concept of service in the design of data communications protocols," in [76], 1985.

[72] P. Zafiropulo *et al.*, "Towards analyzing and synthesizing protocols," *IEEE Trans. Commun.*, vol. COM-28, pp. 651–661, Apr. 1980. Also in *Computer Network Architectures and Protocols*, P. Green, Jr. (ed.). New York: Plenum Press, 1982.

[73] *Proc. Conf. on Introduction of High Level Protocol Standards for OSI*, Dept. of Communications, Ottawa, Canada, May 1984.

[74] *Proc. Int. Conf. on the Introduction of OSI Standards*, Cambridge, UK, Sep. 1985.

[75] "Special Issue on Tools for Computer Communication Systems," *IEEE Trans. Software Eng.*, vol. 14, March 1988.

[76] *Proc. IFIP WG 6.1 Workshop on Protocol Specification, Testing, and Verification*, Amsterdam: North-Holland, annually since 1982.

Index of Acronyms

The page number given is that on which the acronym is defined or first used.

AA, 379
ABM, 110
ACE, 42
ACK, 90
ACSE, 21
ADCCP, 109
ADM, 113
ADMD, 403
AE, 379
ALL, 304
ALOHA, 140
AMPS, 160
ANSI, 40
AP, 378
APDU, 387
APPC, 498
APPN, 457
ARM, 110
ARPANET, 7
ASCII, 89
ASE, 379
ATOMA, 145
AUI, 68

BAC, 120
BBC, 175
BBS, 175
BCC, 98
B-ISDN, 134
BRAM, 172
BSC, 86
BTMA, 155

CASE, 430
CCITT, 4
CCR, 21
CDMA, 143
CEP, 17
CICS, 486
CMIP, 31
CO, 77
COS, 374
CP, 292
CPODA, 182
CQL, 292
CRC, 92
CS, 57
CSMA, 149
CSMA-CD, 151
CUG, 233
CYCLADES, 7

DAF, 477
DAMA, 174
DARPA, 337
DATAPAC, 140
DCA, 409
DCE, 39
DDCMP, 87
DEC, 4
DECNET, 4
DG, 299
DIA, 409
DIS, 8
DISC, 118

DISOSS, 447
DLC, 477
DLE, 90
DM, 57
DNA, 239
DTE, 39
DTFC, 316

EBCDIC, 91
ECMA, 86
EGP, 337
EIA, 42
ELLC, 469
ENQ, 90
EOC, 175
EOT, 90
ER, 305
ESI, 483
ETB, 91
ETE, 387
Ethernet, 140
ETX, 90

FCFS, 158
FCS, 114
FDMA, 143
FDT, 519
FET, 157
FIFO, 169
FOSTAR, 76
FPODA, 182
FRI, 340

533

FRMR, 119
FSK, 71

GAP, 445
GMD, 297
GMDNET, 297
GOSIP, 343
GRA, 183
GSMA, 167

HDLC, 134
HL, 290

IATA, 87
IBL, 314
ICI, 18
ICMP, 340
ICP, 157
IDP, 426
IDU, 18
IEC, 40
IEEE, 68
IFIP, 400
IMP, 293
IP, 4
IPDS, 496
IS, 56
ISDN, 33
ISO, 7
ITU, 39
IUT, 526

LAN, 39
LAPB, 134
LAPD, 237
LFSID, 477
LL, 56
LLC, 136
LME, 30
LRC, 92
LU, 459

MACF, 381
MACS, 182
MAN, 77
MAP, 375
MAU, 69
MD, 403
MDI, 68
MERIT, 266
MHS, 400
MIB, 30
MIC, 72

MLE, 30
MSAP, 171
MSB, 72
MTA, 401

NA, 311
NAK, 90
NAU, 459
NBS, 432
NCE, 245
NCMS, 362
NCP, 3
NDM, 112
NMC, 259
NNI, 251
NRM, 110
NRME, 468
NS, 56
NSDU, 361
NSFNET, 340
NUI, 236

OAF, 477
ODA, 436
ODAI, 477
ODP, 34
OSI, 4

PAD, 333
PARC, 417
PBX, 77
PC, 477
PCI, 18
PDN, 4
PDU, 18
PLU, 490
PN, 250
PODA, 182
PPDU, 390
PRMD, 403
PRNET, 140
PSK, 71
PU, 459
PUCP, 459
PVC, 215

QLLC, 468

RAM, 164
RC, 56
RCP, 157
RD, 57
REJ, 118

REQALL, 304
RES, 444
RFNM, 304
RIM, 119
RL, 56
RM, 164
RNR, 118
RO, 171
ROS, 430
RR, 57
RS, 57
RSET, 118
RT, 57
RU, 489
RUC, 180

SAA, 457
SABM, 118
SABME, 118
SACF, 380
SAO, 380
SAP, 14
SARM, 118
SARME, 118
SATNET, 140
SB, 56
SBP, 295
SC, 56
SD, 57
SDLC, 467
SDMA, 160
SDU, 18
SF, 56
SFD, 407
SIM, 118
SLU, 490
SMA, 292
SMAP, 30
SMXQ, 292
SNA, 239
SNAC, 23
SNADS, 409
SNDC, 23
SNRM, 118
SNIC, 23
SNRME, 118
SOH, 89
SONET, 77
SPADE, 160
SPAG, 375
SPDU, 367
SPP, 426
SREJ, 118

SRUC, 181
SS, 56
SSCP, 459
SSMA, 159
ST, 57
STX, 90
SU, 259
SVC, 215
SXXM, 118
SXXME, 118
SYN, 91

TCP, 4
TCU, 72
TDMA, 143
TELENET, 140
TIA, 42

TM, 56
TOP, 375
TP, 492
TPDU, 360
TPF, 457
TR, 57
TRANSPAC, 258
TSDU, 395
TT, 57
TYMNET, 239

UA, 119
UAC, 120
UBS, 175
UI, 119
UNC, 120
UNIX, 5

UOD, 378
UP, 119

VC, 243
VR, 305
VRC, 92
VSAT, 212
VTAM, 457
VTCP, 445

WAN, 332

XDE, 420
XID, 119
XNS, 341

Subject Index

Abort, 356, 368
Abstract data types, 518
Abstract syntax notation, 22, 387, 428
Acknowledgement, 98, 103, 131, 153–154, 219, 465
ADCCP, 109–134, 214
Addressing
  by attribute, 404
  flat, 334, 341
  hierarchical, 334, 340
  IBM SNA, 460–463
  internet layer, 334–335
  ISO, 27–29, 343–344
  link layer, 114
  mapping, 334
  message handling, 403–404
  X.121, 201, 341
  Xerox, 425–426
Advanced peer-to-peer networking, 457, 502
Advanced program-to-program communication, 488, 498, 502
Algebraic specifications, 518
ALOHA network, 140, 146–148
AMPS system, 160
ANSI, 40, 43, 86, 109
Application layer, 21–22;
  association, 379
  context, 379
  entity, 379, 380–381
    title, 29
  process, 12, 378
  service elements, 379, 381
  See also Association control, Commitment, Directory, File transfer, Message handling, Office document architecture, System management

Architecture: *see* Network architecture
ARPANET
  electronic messaging, 399
  flow control, 293, 304–305, 320
  packet switching, 3, 7, 140
  routing, 242, 252–254, 340
  See also Federal research internet
Association control, 21, 381–385
Asynchronous transmission, 41, 110, 112
Asynchronous response mode, 110
Asynchronous balanced mode, 110, 131
Authentication, 432–434
Autonomous system, 340

Backpressure flow control, 299, 476
Bifurcated routing, 242, 245
Bit oriented link control: *see* Link control layer
Bit stuffing, 114
Block, 89
Blocking, 20, 229, 481
Boundary functions, 457
Brackets, 456, 486
Bridge, 332, 467
Broadcast: *see* Multicast
Buffer class flow control, 291, 296–298, 338
Buffer interference, 278
Burstiness, 145, 146, 224, 300
Busy tone multiple access, 155
Bypass, 72

Capture, 159
Carrier sense multiple access, 149–151
CCITT
  history, 4, 8, 40, 194, 350
  reference model, 7

CCITT (*Cont.*)
  I. Recommendations, 40, 64–66
  S. and T. Recommendations, 350, 407
  V. Recommendations, 40, 44, 46–50
  X. Recommendations, 40, 44, 49–50, 193
  X.21 Recommendation, 59–64, 193–210
  X.25 Recommendation, 4, 24, 50, 121, 134,
    211–238, 258, 301, 467, 468
  X.32 Recommendation, 235–237
  X.75 Recommendation, 339, 341–342
  X.400 Recommendations, 400
Chaining, 456
Channel queue limit flow control, 290,
    292–295
Character codes, 434–435
Character oriented link control: *see* Link
    control layer
Checksum: *see* Cyclic redundancy check
Choke packets, 317, 338, 340
CIGALES network: *see* CYCLADES network
Circuit switching, 4, 193, 235, 273
Class of service: *see* Type of service
Closed user group, 207, 233
Code division multiple access, 145
Collisions, 143, 151
Commitment, concurrency, and recovery, 21,
    385–386
Common channel signaling, 64
Complete packet sequence, 219
Confirmed service, 355
Conformance testing, 33, 46, 526–527
Congestion control, 243, 276; *see also* Multi-
    ple access
Connections
  data transfer phase, 17–20, 355–356
  endpoints, 17, 354
  establishment phase, 17, 18, 354–355
  naming, 17
  release phase, 17, 18, 356
  service primitives, 15, 355
  See also Association control, OSI reference
    model
Connectionless mode, 18, 25, 340, 351,
    356–357; *see also* datagram
Connector, 43–44
Control wire, 175
Corporation for open systems (COS), 374
Credit (for flow control), 287, 320
CYCLADES network
  flow control, 317, 320
  history, 3, 7, 140
  routing, 242
Cyclic redundancy check, 92, 116, 153, 340

Data circuit-terminating equipment, 39, 41,
    285
Data link layer: *see* Link control layer
Data terminal equipment, 39, 41, 285
Datagram, 4, 243–244, 267, 287, 331, 495;
    see also Connectionless mode
DATAPAC
  flow control, 307–308, 316
  routing, 243, 254
Deadlock, 282–283, 295, 302, 303, 494
DECNET
  history, 4
  link control, 87
  routing, 242, 266–269
Decomposition, 524
Defense data network (DDN): *see* Federal
    research internet
Delivery confirmation, 219, 228
Design rules, 524
Directory, 22, 27–29, 334, 340, 341, 403, 409,
    430–432, 451, 471–473
Disconnect: *see* Connections
Document control architecture (IBM DCA),
    409–410
Document description language, 437
Document interchange architecture (IBM
    DIA), 409–411
DoD internet: *see* Federal research internet

Electrical characteristics
  physical layer, 44–48
  ISDN, 66, 134
  RS-449, 57–58
  X.21, 60
Electronic mail: *see* Message handling
End system, 12, 333, 352
Endpoint interconnection, 330–331, 340
Entities, 13, 352, 515
Error control, 20, 361
ESTELLE, 519
Ethernet: *see* Local area networks
Expedited data, 19, 359, 369
Explicit route, 263, 305, 456, 475
Exterior gateway protocol, 340

Fair queuing, 338
Fairness, 281, 314, 322
Fast select, 225, 234
Federal research internet, 5, 337, 339–341;
    see also ARPANET
File transfer, 3, 381, 394, 440–441

Flag, 114
Flooding, 253, 337
Flow control, 20, 218, 228, 233, 273-328
  entry-to-exit level, 302-311
  hop (link) level, 290-302
  network access level, 311-318
  performance measures, 288-290
  transport level, 318-320, 364
Formal description techniques, 11, 33, 63,
    351, 513-531
  conformance testing, 526-527
  protocol and service specification, 514-520
  protocol verification, 520-525
  standards, 519-520
Forms service, 412-413
Fragmentation: see Segmentation
Frame structure, 113-116
Frequency division multiple access, 144
Function, 15
Function oriented protocol: see File transfer,
    Message handling, Terminal support
Functional characteristics
  ISDN, 65
  physical layer, 48-49
  RS-449, 54-56
  X.21, 61

Gateway, 204, 330, 332, 333, 338-339, 445,
    463; see also Internet
Gateway half, 339, 341, 342
Global scheduling multiple access, 167
GMDNET, 297, 301, 306-307, 314

HDLC, 109, 134, 212, 213
History
  ANSI, 40, 43
  ARPANET, 3, 350, 399
  CCITT, 4, 8, 40, 194, 350
  circuit switched network layer, 194-195
  computers, 449-450
  computer networks, 3-5
  EIA, 42, 52
  formal description techniques, 11, 519-520
  IBM SNA, 498-502
  IEEE, 68
  ISO, 4, 7-8, 24, 31, 40
  link layer, 85-86, 107-108, 134
  message handling, 399-400, 403
  packet switched network layer (X.25),
      211-212, 222, 349
  physical layer, 42, 53, 68, 73-77

History (*Cont.*)
  public data network, 4, 211
  transport and session layers, 350
  TYMNET, 254

IBM, 86; see also Systems network architec-
    ture
Identifiers, 16
IFIP, 4, 400, 403
Input buffer limits, 314-317, 338
Input/Output history specifications, 518
Instability, 156
Integrated services, 3, 33, 77, 184, 321, 396
Interface control information, 18
Interface data unit, 18
Internet, 329-346
  addressing, 334-335
  congestion and flow control, 323, 337-338,
      340
  history, 4
  IBM SNA, 463, 475, 490
  ISO, 342-344
  level of interconnection, 331-333
  protocol, 4, 5, 332, 340, 424-426
  public data networks, 341-342
  routing, 335-337, 340
  Xerox, 341, 419, 423-426
Interpersonal messaging service, 405
Interpress, 437-440
Interprocess communication, 9
Interrupt: see expedited data
Interscript, 436
Isarithmic flow control, 312-314
ISDN, 64-67, 75, 135, 212, 235, 237, 322
ISO
  committees, 4, 7, 31, 350
  history, 4, 7-8, 24, 31, 40
  open systems interconnection
    government OSI profile (GOSIP), 343
    internet sublayer, 4, 342, 344
    interpretations, 32
    reference model, 7-36, 40-41, 195, 212,
        352-356, 376-381, 405
    seven layers, 21-24
  standards
    addressing, 343
    application layer, 381-386
    internet protocol, 343
    presentation layer, 387-394
    reference model, 7-8; also see above
    routing, 343
    session layer, 365-373
    transport protocol, 45, 333, 343, 358-364

Layer: *see* Protocol layering
Layer management, 30; *see also* Network
 management
Layered architecture: *see* Network architecture
Least cost routing, 239
Levels of abstraction, 9–11, 516
Link control layer, 24
 bit oriented
  ADCCP, 109–134, 214
  balanced mode, 131–134
  classes of procedure, 120–121
  elements of procedure, 117–120
  error control, 128–130, 133–134
  frame structure, 113–116
  IBM SNA, 465–470
  unbalanced mode, 122–130
 character oriented
  code sets and control characters, 89–91
  deficiencies, 105, 107
  error control, 92, 98, 103–104
  information transfer, 95–99
  link establishment, 93–95
  link termination, 99
  X.21, 197–198
  X3.28, 94–99
 history, 85–86, 107–108, 134
 logical link control, 469
 multiaccess
  classification, 142–143
  fixed assignment techniques, 143–145
  performance, 151–153
  random assignment techniques, 146–159
  demand assignment with central control,
   159–167
  demand assignment with distributed
   control, 168–176
  phases, 92–94
  transparency, 101–103, 108, 114
Load leveling, 262
Local area networks
 characteristics, 142
 distributed queue dual bus, 77
 Ethernet (CSMA), 4, 68–69, 75, 140,
  150–151, 418
 FDDI, 77
 link control protocols, 135
 physical layer, 67–73, 75–77
 routing, 335–336
 STARLAN, 76
 token bus, 70–71, 174
 token ring, 72, 73, 173, 467
 See also Link control layer (multiaccess)

Logical channel, 224
Logical unit, 459, 488
Longitudinal redundancy check, 92
Looping, 242
LOTOS, 519

Mail: *see* Message handling
Management information base, 30
Management domain, 403, 404
MAT-1 system, 160
Mechanical characteristics
 ISDN, 66
 physical layer, 43–44
 RS-449, 59
 X.21, 60–61
Medium attachment unit, 69
Message, 90
Message handling
 addressing, 403–404
 CCITT X.400 series recommendations, 400
 gateway, 333
 history, 3, 399–400
 message structure, 405–406
 message transfer, 407–408
 model, 401–402
 IBM SNA, 409–411
 Xerox, 441–443
Message oriented: *see* Datagram
Message transfer agent, 401–402
Multiaccess link control: *see* Link control
 layer
Multicast, 18, 32–33, 115, 127
Multicommodity flow problem, 244
Multihoming, 329
Multilink operation, 136, 206, 264, 465, 474
Multimedia messaging, 401, 405
Multipeer: *see* multicast
Multiple path routing: *see* bifurcated routing
Multiplexing, 20, 361, 362
Multipoint link, 84, 110, 122

Naming, 16, 27
Negotiation, 364
Network addressable unit, 459
Networks: *see* ARPANET, CYCLADES, Federal
 research internet, Local area network,
 ISDN, Xerox, DECNET, Public data
 network
Network architecture, 3, 7–36, 418, 449
Network interconnection: *see* Internet
Network layer, 23–24
 circuit switched, 193–210
 packet switched, 211–230

Network management, 29–31, 259, 451, 503
Nonconfirmed service, 355
Normal response mode, 110, 122

Offered load, 274, 279, 289
Office applications, 411–413
Office document architecture (Xerox ODA),
    436
Open distributed processing, 34, 396
Open systems interconnection (OSI): see ISO
Out of band signal, 220; see also Expedited
    data

Pacing, 306, 320, 456, 472, 476, 478
Packet radio network, 140, 340
Packet switching, 4, 140, 215, 239, 273
Parallel transmission, 41, 42
Parity, 92
Partitioned network, 253, 336
Path control layer, 456
Path number, 250, 256
Peripheral nodes, 457
Permuter table, 256
Physical layer, 24, 39–80, 196
    electrical characteristics, 44–48
    functional characteristics, 48–49
    history, 42, 53, 68, 73–77
    ISDN, 64–67
    local area networks, 67–73, 75–77
    mechanical characteristics, 43–45
    procedural characteristics, 49–50
    RS-232, 51–53
    RS-449, 53–59
    X.21, 59–64
Physical unit, 459
Physical media, 41
Point of demarcation, 43
Point-to-point link, 84, 110, 131
Polling, 93–95, 111, 115, 125, 160–162
Power, 289
Presentation layer, 22, 387–396
    context management, 390–392
    IBM SNA, 457
    transfer syntax, 388
Primary station, 111, 122
Printing protocol, 440
Priority, 453, 474
Priority oriented demand assignment, 182
Probing, 161, 408
Procedural characteristics
    ISDN, 65
    physical layer, 49–50

Procedural characteristics (Cont.)
    RS-449, 56–57
    X.21, 62–64
Program proof, 524
Protocol, 15, 515
Protocol control information, 18
Protocol data unit, 18, 215
Protocol layering, 3, 7, 12, 349, 420, 454, 497
Protocol specification, 9, 10, 515, 517–519
Provider initiated service, 355
Public data network, 4, 211, 258, 341–342;
    see also DATAPAC, TYMNET, TRANSPAC

Reachability analysis, 522
Reassembly, 242, 302, 338, 481
Redirect route, 340
Relaying, 23
Remote procedure call, 428–430, 487
Repeater, 70, 332
Reservation access methods, 169–171, 174
Reservation upon collision access method,
    180–181
Reset, 20, 221, 229
Resource sharing, 3, 7, 139, 273, 450
Restart, 224, 230
Retransmission, 98, 128, 465
Routing, 23, 239–271
    ARPANET, 252–254, 340
    DECNET, 266–269
    functions, 244–246
    IBM SNA, 262–266, 473–481
    internet layer, 335–337, 340
    ISO, 23, 343–344
    local area network, 335–336
    shortest path algorithms, 246–250
    TRANSPAC, 258–261
    TYMNET, 254–258
Routing table, 240, 248, 260, 335
RS-232 standard, 51–53, 74–75
RS-449 standard, 53–59

Satellite channels, 141, 212, 340
Secondary station, 111, 122
Security, 25–27, 340, 405, 432–434, 492
Segmentation, 338, 340, 341, 343, 363, 481
Selector, 29
Sequencing, 115, 126, 465
Serial transmission, 41, 42
Service access points, 14–17, 354
Service data unit, 18
Service primitives, 15, 357, 515
Service specification, 9, 10, 14–15, 41,
    514–515

Session, 459, 488–495
Session layer, 22, 349–358, 365–376
  activity management, 373
  dialogue control, 365, 367
  functional units, 365
  synchronization, 370–372, 494
  tokens, 367
Shortest path routing, 239, 246–250, 252
Slotting, 147, 150
Slow start flow control, 338
Source routing, 336, 337, 340
Space division multiple access, 160
SPADE system, 160
Split channel reservation multiple access, 163
Splitting, 20, 363
Spread spectrum multiple access, 159
Square root buffer allocation, 294
Standards
  de facto, 5, 86
  goal of, 8
  link layer, 86–87
  message handling, 400
  physical layer, 42–50
  See also ANSI, History, ISO, CCITT
State transition models, 518
Statistical multiplexing, 224; see also Packet
    switching
Stepwise interconnection, 330–331, 342
Structured buffer pool flow control, 295–298
Subarea nodes, 457
Subnetting, 340
Subnetwork access, 23
Subnetwork dependent convergence, 23, 342
Subnetwork independent convergence, 23
Symbolic execution, 525
Synchronous transmission, 41, 59, 107, 193
Synchronization points, 451, 494; see also
    Session layer
System: see End system
System services control point, 459, 490
Systems management, 29–31; see also Net-
    work management
Systems network architecture (IBM SNA),
    449–509
  conversations, 483–488
  design principles, 453–454
  directory service, 471–473
  distribution system (SNADS), 409–411,
    497
  document control architecture (DCA),
    409–410
  document interchange architecture (DIA),
    409–411

Systems network architecture (Cont.)
  flow control (pacing), 305–306, 320, 466,
    489
  history, 498–502
  layers, 455–457
  link layer, 465–470
  network structure, 457–464
  routing, 242, 262–266, 473–481
  session management, 488–495

Temporal logic, 519
Terminal support, 3, 333, 342, 445
Terminal identification, 236
Throughput, 274, 279, 288
Time division multiple access, 144
Time protocol, 434
Timers, 99, 133, 207, 231, 465, 504–505, 517
Token passing access methods, 173–174
Transaction processing, 381, 428, 457, 487,
    495–497
Transfer syntax, 388
Translating gateway, 330, 332, 445
Transmission control protocol (TCP), 4, 5,
    320, 340
Transmission group, 264, 456, 465, 474
TRANSPAC
  flow control, 301
  routing, 258–261, 287
Transport layer, 22–23, 349–364
  classes of protocol, 359
  error control, 361
  expedited data, 359
  flow control, 364
  Xerox, 427–428
  See also Transmission control protocol
Tree retransmission algorithms, 157–158
Two-way alternate transmission, 84, 108
Two-way simultaneous transmission, 84, 108
TYMNET
  flow control, 300–301
  routing, 242, 250, 254–258
Type of service, 262, 478

Unreachable destination, 268, 336, 340
Urn access method, 177–179
User agent, 401
User data, 18
User facilities, 201, 207–208, 233–235

Validation, *see* Formal description techniques
Verification, 520; *see also* Formal description techniques
Vertical redundancy check, 92
Virtual circuit, 4, 215, 243, 258, 286, 331
Virtual circuit flow control, 291, 299–302
Virtual route, 263, 305–306, 456, 475

Window based flow control, 287, 301, 303, 308, 309

Xerox Network System, 4, 417–447
application layer, 430–446
clearinghouse directory, 430–432
design principles, 417–419
future trends, 446–447
internet layer, 341, 423–426
lower layers, 421–423
relation to OSI model, 420–421, 428, 430, 432, 441–443
remote procedure call (Courier), 428–430
X.25: *see* CCITT